Recent Advances In Medicinal Plants Research

Recent Advances In Medicinal Plants Research

Editors :
S.K. Prabhuji,
G.P. Rao, S.K. Patil

SATISH SERIAL PUBLISHING HOUSE

SSPH

403, Express Tower, Commercial Complex, Azadpur, Delhi - 110033 (INDIA)
Phone : 011-27672852 Fax : 91-11-27672046
E-mail : ssp@ndf.vsnl.net.in, hkjain1975@yahoo.com
Website : www.satishserial.com

Published by :

SATISH SERIAL PUBLISHING HOUSE
403, Express Tower, Commercial Complex, Azadpur, Delhi - 110033 (INDIA)
Phone : 011-27672852 Fax : 91-11-27672046
E-mail : ssp@ndf.vsnl.net.in, hkjain1975@yahoo.com

First Published 2005

ISBN 81-902289-9-4

Printed at
SALASAR IMAGING SYSTEMS, DELHI-35

Preface

अयोग्यः पुरुषो नास्ति, नास्ति द्रव्य मनोषम्।
सर्वाणि सुलभाः यत, योजकस्तत् दुलर्भम्।।

"There is no man on this earth who is incompetent and there is no plant which is of no medicinal use. Where every thing is present, in fact, a man to manage them properly is seldom available."

During the past decade, demand for medicinal plants and its products as well as traditional health system attracted the worldwide interest. India, being the botanical garden and treasure-house of Biodiversity with its diverse ecological conditions, rich ethnic diversity and a strong traditional knowledge base, accounts for 45,000 plant species, out of which more than 8,000 species are used in some 10,000 herbal drug formulations. Ninety percent of the herbal industry's requirement is taken out from the forests, resulting into their destruction. There is no reliable assessment of the volume or value of the herb related trade in India, but, the conservatives put the quantity of dry raw material collected at 0.5 million tons each year. The global plant based drug is projected between US$ 30 and 60 billion with a 7% annual growth rate, but, India has only a 2.5% share in it. Even so, there is a growing realization that the demand in fast outstripping supply, putting an unreasonable pressure on our wild phytoresources. This has already placed certain species at risk and others will soon follow, if immediate corrective measures are not taken. Conservation of medicinal plants has to be viewed not as an end in itself, but, as a means to enhance the conservation and cultivation of our precious biowealth for sustainable utilization and in ensuring equitable sharing of the benefits derived from the use of biodiversity.

Our knowledge about the aforesaid is still inadequate and needs further extensive and intensive investigations, so as to have full spectrum at hand. For this more elaborate and research-oriented informations are aptly required.

Being a treasure-house of the phytoresources, India has an insignificant share in the International Herbal Trade. Besides this, the ancient literatures together with the herbal formulations based on folk-medicine and indigenous system of curatives emphasize the fact that our country had been the leader since time immemorial in the knowledge of herbal biowealth. Such contradictions raise a very genuine question before us - how and why the leaders in herbal sciences are lagging behind now? In our view, to find the right answer is to have a broad spectrum introspection in what we have done in the past. Firstly, we have broken the chain by not doing advanced researches using the latest technologies on existing informations on herbal sciences in the past and secondly, we forgot to have a sustainable development of plant resources. Therefore, a special attention is required to revitalize it so that the leader (in the field of herbal sciences) may regain its proper place. It is a matter of immense pleasure that the society is showing awareness for herbal products now and scientists are taking keen interest in working and developing herbal medicare-system. Therefore, the period starting from the year 2001 in 21st century may very well be considered as a period of renaissance in the field of herbal science.

The book, in hand, is just a step ahead which constitutes a part of the giant leap towards a greater goal. It provides a comprehensive and useful informations on various aspects related with medicinal plants, viz., biodiversity, cultivation and agrotechniques, pest and disease management, indigenous knowledge and their importance, pharmacology, pharmacognosy and phytochemistry, marketing patterns, and the biotechnology. Besides these, informations on the newly emerging group - 'Nutraceuticals' has been encorporated along with prospects and problems in the medicinal plant research and some emerging medicinal plants of 21st century which includes phytochemical details of some plants like Safed Musali, Brahmi, Satawar, Stevia, Aonla and Jatamansi.

The two sections, viz., Indigenous Knowledge and their Importance and Pharmacology, Pharmacognosy and Phytochemistry constitute the major portion. The former section includes the account of Indian system of medicines, commercially important plants and the phytomedicines used by the tribals in snake-bite, women folk-health care and for birth-control practice in particular. However, the latter section comprises pharmacological and phytochemical studies on some important plants like *Carica papaya, Bacopa monnieri, Withania somnifera, Cinnamomum verum, Artemisia pallens* and *Cullen corylifolium*. Furthermore, the importance and the techniques of micropropagation of *Coleus forskohlii* and *Stevia rebaudiana* have also been included in the Biotechnology section.

We hope that this updated book will be a valuable source for the botanists, herbalists, drug manufacturers, agriculturists, pharmacologists, biotechnologists, phytochemists as well as anyone interested in medicinal plants' research.

Finally, we wish to thank all the contributors for their help in the preparation of this book.

Gorakhpur
January, 2005

S.K. Patil
S.K. Prabhuji
G.P. Rao

About Editors

Dr. Shakti Kumar Prabhuji is working as *Head*, Department of Botany, Udit Narain Postgraduate College, Padrauna (Dist. Kushinagar), India and has been teaching Ethnobotany, Microbiology and Morphogenesis since 1975. He has earlier served as Lecturer in Department of Botany, St. Andrew's College, Gorakhpur, India. He is a prolific researcher. After doing his Ph.D. in Aquatic Mycology in 1979 from Gorakhpur University his contributions are in various branches of Botany, viz., Aquatic Mycology, Fish-mycopathology, Medical Biotechnology, Biophysics, Biomathematics and Plant Morphogenesis. He has published several research papers in Indian and Foreign Journals of repute and has two text books on Plant Morphogenesis to his credit. He is *Secretary* of Green Blessings and working on endangered medicinal plants of India. He is fellow of various academic societies including Indian Botanical Society. He is currently working on Medicinal Plants, various aspects of Aquatic Mycology and Electrophysiological experiments on plants to develop prior alarm system for natural disasters.

Prof. S.K. Patil was born on 22nd April 1956 at Dharwad, a city known for education in Karnataka. He did his B.Sc. (Agric.) from College of Agriculture, Dharwad in 1977 and M.Sc. (Agric.) in Soil Science in 1979 from the University of Agricultural Sciences, Bangalore. Later in 1980, he joined World Bank aided National Agricultural Research Project as Research Assistant at Agricultural Research Station, Bijapur. In 1983, he was selected as Assistant Professor of Soil Science.

He visited to United States of America in 1987 to pursue his Ph.D. in Wood Science and Technology at the University of California, Berkelay in the prestigious Ford Foundation Fellowship. Then, he did his doctoral research at Dr. Y.S. Parmar University of Horticulture and Forestry, Solan (Himachal Pradesh) and got his Ph.D. in Forest Products and Utilization during 1992. In the same year, he was elevated to the cadre of Associate Professor. Then in 1999, he was selected as Professor of Forest Utilization at College of Forestry, Sirsi (Karnataka). Since 2001, he has been serving as the Director of Instruction (For) and constantly striving to raise the infrastructural facilities and academic standards of Sirsi campus. Dr. S.K. Patil is a member of several Academic Bodies such as Board of Studies, Academic Council, Research Council, Extension Council of UAS, Dharwad and Research Advisory Council of Institute of Wood Science and Technology, Bangalore, etc. Dr. Patil has been instrumental in organizing several national level seminars such as "Production and Utilization of Bamboo and Ratan", "Production, Processing and Marketing of Vanilla", "Cultivation and Marketing of Medicinal and Aromatic Plants in India", etc. in collaboration with other organizations at Sirsi.

Several M.Sc. and Ph.D. students have completed their Post-Graduate degree under his guidance. He has published many research papers and popular articles for the benefit of farmers. He is currently pursuing the research area on "cultivation, processing and marketing of medicinal and aromatic plants", as this can directly benefit the farmers to improve their financial status. He has handled research projects funded by State and Central Government. Dr. Patil has been recognized as a subject matter expert by UPSC and ASRB, New Delhi and State Agricultural Universities.

Dr. Govind P. Rao is *Secretary* of 'Society for Conservation and Resource Development of Medicinal Plants' and member of Editorial Board of 'Journal of Medicinal Plants'. Dr. Rao is working as Senior Scientist at Sugarcane Research Station, Gorakhpur, India. Dr. Rao is a fellow of Indian Phytopathological Society and Sugarcane Technologists Association of India. Dr. Rao has worked on biological properties of many medicinal and aromatic plants and has published more than 100 research publications to his credit. He has also published books on Sugarcane Pathology, Fungal Diversity, Sugarcane Agriculture and Industry and Medicinal Plants. He is also *Editor-in-Chief* of Sugar Tech, an international journal of Sugar Crops and related Industries. He is also regular member of important scientific societies of India and abroad. He has worked and visited many scientific laboratories of USA, France, Yugoslavia, Scotland, UK, China, Sri Lanka, Australia, Singapore, Malaysia and Thailand. He has been involved in the research on sugarcane pathology for the last 17 years. He is currently working on the diagnosis, characterization and management of sugarcane pathogens. He is also currently working on organic cultivation and pests management of medicinal plants especially *Stevia* and Safed musli.

CONTENTS

PHARMACOLOGY, PHARMACOGNOSY AND PHYTOCHEMISTRY

MARKETING PATTERNS

BIOTECHNOLOGY

Recent Advances in Medicinal Plant Research : Vision 21st Century, 2005
Editors : S.K. Prabhuji, G.P. Rao and S.K. Patil
Satish Publication House, New Delhi

Chapter 1

MEDICINAL AND AROMATIC PLANTS — PROBLEMS AND PROSPECTS

Ajay Harlalka
Apex Enterprises, G.P.O. Post Box No. 805, 4th Floor, Room No. 6A, 40, Strand Road, Kolkata - 700 001
e-mail : ajayhar@vsnl.com

Medicinal and aromatic plants have been an integral part of health and healthcare for centuries and will continue to play a vital role in public health for generations to come. Inspite of so many problems being faced by the herbal heritage today, the future is still promising for properly manufactured and researched herbal and ayurvedic preparations. These issues have been discussed pertaining to Indian context in this chapter.

INTRODUCTION

India is proud to be rich in biological diversity. It has approximate 8% of the estimated bio diversity of the world. The basic raw materials of almost two-third of the drugs and perfumery products, used worldwide, are available in India.

There are a number of well established indigenous or traditional systems like Ayurveda, Unani, Siddha, Homoeopathy, Tibetan and Naturopathy that are practiced along with the modern medicine to provide the total healthcare system. In all these systems, a large number of plant drugs are used. In India, medicinal and aromatic plants are an essential part of traditional healthcare systems. They are linked to the region's natural bio diversity. Medicinal plants are among the most important 'Non-Timber Forest Produce' in India.

As per the WHO, 80% population in developing countries relies on traditional natural medicines, and almost 80% of the traditional medicines involve the use of plant extracts.

Medicinal plants are accessible, affordable and appropriate source of primary healthcare for almost two-third population of our country. People who can't afford the high cost of modern medicines are mainly dependent on these culturally familiar, simple, affordable and effective traditional medicines. Because of these favourable factors, there is widespread interest in promoting traditional health system to meet primary healthcare needs. The spiraling cost of modern medicines has made it necessary for the protection and promotion of the cultural and spiritual values of traditional medicines.

Even in developed countries, one in three (34%) of the adult population goes to some form of unconventional therapy.

PROBLEMS

Problem in identification

In the ancient Indian Ayurvedic texts; Charak Samhita, Ashtanga Hridaya Samhita and Susruth Samhita; a large number of medicinal plants have been described that had been used in preparing drugs and curing a number of ailments. Unfortunately, a number of these plant species are known by their vernacular and classical names. A good deal of information had been lost, as it is difficult to correlate these names with acceptable scientific names. One plant species can have many vernacular or classical names and one name may refer to different plant species.

The scientific study of traditional medicines, derivation of drugs through bio-prospecting and systematic conservation of the concerned medicinal plants is very important.

Problems in cultivation

(a) Non availability of Quality Planting materials
(b) Lack of proper guidance

(a) **Non availability of Quality Planting materials :** The first major problem encountered by the growers in the cultivation of medicinal and aromatic plants is the non-availability of good quality planting materials. The high yielding and disease resistant varieties of selected medicinal and aromatic plants introduced by the reputed Research Institutes are not easily available. A fairly large number of growers are not even aware of them and they continue to buy the poor quality planting materials available locally. The yield is lower than what has been described in the books or promised by the experts. Even the quality of the produce is sub-standard that creates problem in marketing. The net return is much less that demoralizes the growers.

(b) **Lack of proper guidance :** Although books have been published on the agronomy of medicinal and aromatic plants, but none of these provide the proper guidance. The cost analysis given in most of these books is unrealistic. There are many medicinal plants like Kalmegh (*Andrographis paniculata*), Muskdana (*Andrographis moschatus*) and Kewanch (*Mucuna pruriens*) to name a few that are easy to cultivate, but difficult to sell. Buyers want to buy these herbs at throwaway prices.

Now-a-days, we find many experts who are ready to provide complete package on medicinal and aromatic plants that include planting materials, cultivation techniques and buy back the produce at attractive prices. With their glib talk they project a rosy picture and assure handsome return to growers. The sub-standard planting materials are sold at exorbitant rates but at the time of harvesting, all these fly-by-night operators would simply disappear in thin air leaving the growers in lurch. The only intention is to mint money by selling the planting materials at exorbitant prices.

The dwindling forest cover, over exploitation and growing demand has led to the cultivation of some of the medicinal and aromatic plants. The trend has gained momentum in the last 6-8 years. There is a growing demand for herbs and herb derived products for variety of other applications like food flavourings, cosmetic and personal hygiene products (in fragrance, creams, lotions), industrial chemicals, food ingredients etc. As a result of such market trends it became rather obvious that herbs deserve serious consideration as potential alternative cash crops.

When the mono cropping of medicinal plants is undertaken, they also suffer from diseases like root wilt, rot, blight etc. and pest attack. The incidence of pest attacks and diseases have been reported from several parts of the country where medicinal plants are being cultivated for the last 3-4 years.

CULTIVATION TIPS

To increase the percentage of germination in herbs that have small seeds like Ashwagandha (*Withania somnifera*), Kalmegh (*Andrographis paniculata*) etc., instead of the common broadcasting method, pellets method and preparation of nursery beds would be more preferable. In nursery method, 6-8 beds of 10'×3'×6' per acre are prepared using FYM.

Seeds are spread on the beds in such a manner that there is no overlapping. A mixture of fine soil and FYM is prepared in the ratio of 1:1. The seeds are covered with a thin layer (1 cm) of this mixture, water is sprinkled and paddy straws are spread over the beds so that seeds should not be washed away in rain or blown away by strong winds. The paddy straw is removed in 3-4 days when the germination begins and plants are transplanted in the main field after 25-30 days. These two techniques can assist in increasing the percentage of germination and yield.

It is interesting to find that the rough environments are most suitable for growing the strongest medicinal herbs, e.g. root crops need little irrigation and colder temperature often produce higher chemical yield.

Precautions to be taken during selection of sites and cultivation :

(a) When selecting a growing site, it's important to avoid areas where chemical residues may be present and where hard to control weeds are established.

(b) In case of aromatic plants used for extracting essential oils, the critical issues are : site selection, adaptability, seed bed preparation, seeding, fertilization, harvesting, weed control, storage and processing.

(c) Most herbs require large amounts of sunlight because the essential oils and alkaloids develop best when the plants receive at least 6-7 hours of sunshine per day.

(d) Growers need to determine how tolerant the various crops are to soil moisture.

(e) Growers must be familiar with the hardiness of different herb varieties.

(f) If seeds are smaller and lighter then the seeding depth should be shallow.

(g) The row spacing for different herbs is different. Some herbs grow well in wide row spacing while other perform well in narrow spacing.

(h) The optimum harvest stage is unique for each crop.

(i) Production cost for individual herb would vary due to location, crops, size, machinery, labour and marketing activities.

(j) Establish a trial plot before investing in any further development.

MARKETING

The marketing of medicinal and aromatic plants is tougher than that of spices, cereals and other agricultural produces. The ayurvedic and pharmaceutial companies are more concerned with Price than Quality.

Although thousands of people are engaged in collection, the unethical trade practices have ensured that only a small margin of profits from medicinal and aromatic plants trickle down to the collectors and growers. Marketing is a very important aspect that is required to be addressed carefully.

We have to participate regularly in the International fairs on herbal medicines, flavours and essential oils etc. to showcase our achievements and potential. Further, tours of overseas trade representatives shall be organized to create awareness and install confidence for outsourcing their requirements of crude herbs and extract of medicinal plants from India.

India has many strategic advantages to emerge as one of the major players in the world herbal trade.

(a) India has centuries old traditions of ayurvedic system of medicines.

(b) There are over 1000 species of aromatic plants available in our country out of which the chemistry and property of 500 species are known but only 50 species are being commercially utilized as source of essential oils.

(c) The nature has been kind enough to provide us with different climatic zones that have added to the rich biodiversity.

REMEDIAL MEASURES

The growers have to focus on two types of medicinal and aromatic plants:

(a) Import substitute
(b) Diminishing availability

The focus should be shifted on St. John's wart, Milk Thistle, *Ginkgo biloba*, Vanilla, Lavender, Ylang ylang, Rosemary, Thyme, Sage to name a few. These medicinal and aromatic plants have worldwide acceptance and the demand is increasing at a healthy rate of 15% per annum.

Another solution to this problem can be the 'Contract Farming'. The ayurvedic and pharmaceutical companies and exporters can enter in 'Contract Farming' with interested farmers for the cultivation of medicinal and aromatic plants that have high value, good export potential, import substitutes or whose availability in wild is rapidly decreasing.

The 'Contract Farming' has the following advantages :

(a) Assured uninterrupted supply.
(b) Insured against fluctuation in prices.
(c) Comparatively less probability of adulteration.
(d) Consistency in Quality of raw materials.
(e) Cheaper or at par with the market rate.
(f) Choice of Organic or conventional crops.

We at Apex Enterprises have 'Contract Farming' for the cultivation of selected medicinal plants.

A number of medicinal plants can be grown in plain and valley; all it takes is the selection of the right variety and the proper combination of soil treatment and growing method.

CHANGING MARKETING TRENDS

From time to time it is valuable to compare International markets to see which success stories have remained restricted to only few countries. One of these success stories is that of 'Black Cohosh', a well-researched herbal product for menopausal symptoms. A leading 'Black Cohosh' product, "Remifemin" has registered fair growth in the last 5 decades in Germany, Italy, France and other EU countries but failed in USA and other countries. Consumers' attitudes and behaviour must be tracked and measured so that companies can produce, position and market the right products to the appropriate target.

Despite there long history of effective use, many herbal medicines have fallen out of favour because of the reported side effects. The recent examples are that of Ephedra and Kava.

Kava has more than 2000 years old history of use in the South pacific as a tranquilizing ritual beverage from the rootstock of the plant. In Europe, Kava's preparations have been used since 1886 to reduce stress and anxiety. Until a short time ago, Kava was one of the top ten best selling herbs worldwide. In 1998, the demand for Kava reached to its peak level, making it one of the fastest growing herbs, posting 473% growth between 1997-98. In 1998, the annual production of Kava was approximately 200 million

USD. There was about 3.2 million users worldwide with an estimated 1.3 million users in Germany alone.

However, in the last months of 2001 the situation changed dramatically and Kava was subjected to actions of the international health authorities. A small number of adverse reactions associated with the use of Kava products were reported in Germany and based on these reports, the German and Swiss health authorities imposed restrictions on the sale of food supplements and herbal medicines containing Kava. Finally, the BfArM in Germany banned Kava on June 14, 2002. Negative publicity from the actions taken by German and Swiss authorities, market recalls and exaggerated press reports caused the sudden downfall of this most promising herb. The backlash of anti-publicity over the last few months was the major cause behind the decision to ban the Kava products much too fast. Although throughout its traditional and medicinal use, no side effects like hepatotoxicity had been reported.

The current restrictions have spelled economic disaster for the growers and traders of Kava producing countries including Fiji, Hawaii, Tonga, Vanuatu and Samoa.

St. John's Wort (*Hypericum perforatum*) containing Hypericin has anti-depressant and anti-viral properties. But, it may have side effects and interactions although much less severe than the synthetic drugs.

Emphasis is always on what makes people lose weight, gain energy, look attractive or have greater sexual capacity rather than on what we need to do to keep our bodies healthy.

There is growing trend of using culinary herbs like rosemary, turmeric, sage, thyme, mint, dill, parsley, cumin, ginger and others that also offer health benefits, by food companies. These culinary herbs have GRAS (generally recognized as safe) status. If an herb is not GRAS then major food companies won't even consider it. Culinary herbs are sold fresh or in dried form.

Flaxseed has suddenly caught the fancy of consumers and it has become one of the top selling speciality foods in the markets of developed countries. The reasons for its success are pleasant nutty taste. The seeds are known source of alpha-linolenic acid (an omega-3 fatty acid), lignan and fiber. With the popularity of flax growing in the developed countries, there is an opportunity for the growers and traders of our country.

The level of consumer knowledge determines the trends. Years ago, the decision to utilize nutraceuticals was often a lifestyle decision. Rapidly this paradigm is changing. In the future, the decision of a consumer to use nutraceuticals either for wellbeing or treatment, will be an economic decision. The market, it seems would gravitate towards greater use of technically researched products that highlight the industry's commitment to verifiable health benefits.

Herbal formulations are leading the way over single herb in terms of growth and products. They are becoming increasingly condition specific. Furthermore, women's herbal formulae in the mainstream channel and women's topical herbal formulae in the natural channel are stealing the spotlight, exhibiting a fair growth rate in the markets of EU and US.

Natural personal care is another category that is growing rapidly. The rise of this gray area category can be largely attributed to the trend towards self-treatment, environment awareness and overall health.

Even then the market has to contend with an uncertain regulatory future, unresolved consumer confidence and the propensity of people to look at herbs as drugs. Another factor that affected the herb industry was the negative campaign launched by pharmaceutical companies towards the industry that view it as their competitors. There is an underlying current working against the herb industry.

UNETHICAL TRADE PRACTICES

In our domestic market, over three fourth of the herbal products are totally ineffective. The products are promoted based on claims that are not substantiated. In number of cases, the overdrive by ayurvedic and herbal companies to adopt aggressive strategies in overselling of products' benefits would raise the consumers' expectations with some of the claims those were made. But, when the consumers didn't experience what was promised in the given timeframe, they were left with unfulfilled expectations. This has ultimately shaken the consumers' confidence and made him suspicious.

There is no standard for research in the present market. As a result the consumers have started loosing faith in the herbal products. Most of the companies do not follow any quality parameters and 'Profit' seems to be the only guiding factor.

Secondly, most of the Indian herbal and ayurvedic companies are ignorant, fearful and inexperienced in research. In majority of cases the people that make the decisions have come to achieve a CEO status through sales and marketing only. The last time they took a science class was in high school.

THREATS

In India, the supply of 70% of medicinal plants is from tropical areas - dry and moist deciduous vegetation, viz. Western and Eastern ghats, Aravalis and Sub-tropical regions of Himalayas whereas, the rest 30% of medicinal plants which include species having high medicinal values are found in the evergreen and temperate habitats.

As the demand of several species of medicinal plants has been increasing and two-third of the collection from forest involves destructive harvesting; this has pushed a number of species in the list of rare and threatened plants. Every year a number of species are becoming extinct. In India, medicinal plants are an essential part of traditional healthcare systems and are linked to the region's natural biodiversity. But, they are increasingly threatened by various environmental, socioeconomic and institutional problems. The expansion of unregulated trade and commerce is posing a threat to biodiversity in the region.

Although some efforts have been made at the national and international levels to address these issues, but it has suffered due to the following reasons:

(a) inadequate funding
(b) lack of government's priority
(c) insufficient information sharing
(d) improper co-ordination among stakeholders

Threats to wild medicinal plants have been classified into 7 categories: Human interference (7.8%), Fragmentation (5%), Loss of habitat (18.7%), Overexploitation (17%), Harvest (19.8%), Trade (24.6%) and others (7.1%).

Along with the destruction of resources, the cultural heritage regarding the use of medicinal and aromatic plants is being eroded. Non-standardized expertise and lack of proper knowledge of traditional systems of medicines, inadequate processing and storage facilities may result in ineffective or unsafe treatments.

CONCLUSION

As per the conservative estimates, the values of medicinal and aromatic plants' related global trade is over 80 billion USD and is likely to double within a decade.

Besides health benefits, medicinal plants also provide crucial livelihood options for millions of rural people in India, tribal and women in general.

Currently more than 95% nutraceutical ingredients have originated from the herb or plant based entity. This fact signifies that great potential exists for herbs.

Herbs have been an integral part of health and healthcare for centuries and will continue to play a vital role in public health for generations to come. In spite of so many turbulence being faced by the herb industry today, the future is still promising for properly manufactured and researched herbal and ayurvedic preparations.

Recent Advances in Medicinal Plant Research : Vision 21st Century, 2005
Editors : S.K. Prabhuji, G.P. Rao and S.K. Patil
Satish Publication House, New Delhi

Chapter 2

EMERGING MEDICINAL PLANTS OF 21ST CENTURY : *SAFED MUSALI, SATAWAR, BRAHMI* AND *JATAMANSI*

S.K. Prabhuji[1], G.P. Rao[2], Shashi P. Singh[1] and S.D. Singh[3]
[1]*Green Blessings, "Ram Kuteer", C-161/263, Tiwaripur, Gorakhpur - 273 001, U.P.*
[2]*Society for Conservation and Resource Development of Medicinal Plant, AKS House, Parha, Mahaddipur, P.O. Railway Colony, Gorakhpur - 273 012, U.P.*
[3]*AKS Herbal Research and Land Development Centre, 'AKS House', Parha, Mohaddipur, Gorakhpur - 273 012, U.P.*

India has been a treasure resource of herbal biowealth since time immemorial. However, it is now lagging behind. Therefore, there is an urgent need to develop proper insight for regaining its appropriate place in herbal science as well as to retain our indigenous knowledge of medicinal wealth. Keeping this in view, we have provided an up-to-date information regarding nomenclature, biological description, medicinal uses, phytochemistry, pharmacology and the cultivation technology about four emerging medicinal plants, viz. Safed Musli (*Chlorophytum borivilianum* Sant.), Satawar (*Asparagus racemosus* Linn.), Brahmi (*Bacopa monnieri* (L.) Pennell) and Jatamansi (*Nardostachys jatamansi* DC.).

INRODUCTION

As per an estimate by the World Health Organization (WHO) about 80% of the population of the developing countries rely on traditional medicines for their healthcare needs which are mostly the plant based drugs. The continuously growing demand of the phytomedicines, food supplements and other healthcare products in the International market has been exerting an outstripping pressure on the Indian phytoresources. As a result, several plant species have, now, arrived at the verge of extinction. This emphasizes the apt need for sustainable development (to have sustainable agriculture) of high demand medicinal plants and active participation of farmers in medicinal plant cultivation. For the purpose advance researches should be done in the fields of Pharmacology and Phytochemistry of medicinal plants and to develop better techniques for their cultivation and mass production.

An upto date information regarding nomenclature, biological description, medicinal uses, phytochemistry and pharmacology and the cultivation technology have been provided here about four emerging medicinal plants, i.e., Safed Musali (*Chlorophytum borivilianum* Sant.), Satawar (*Asparagus racemosus* Linn.), Brahmi (*Bacopa monnieri* (L.) Pennell) and Jatamansi (*Nardostachys jatamansi* DC.).

SAFED MUSALI (*CHLOROPHYTUM BORIVILIANUM* SANT.)

Nomenclature :

Sanskrit - Shweta Musali, Musali; Hindi – Safed Musali; Vishakandri; English – Spider plant; Tamil – Tanirbittang; Telugu – Sallogadda; Marathi – Sufeta, Safed Muswali; Gujarati – Dholi Musali; Malayalam – Rodevalli; Oriya – Bandhau, Bharat Betuli; Botanical name – *Chlorophytum borivilianum* Sant. (Family: Liliaceae/Anthericaceae)

Biological Description :

Santapau and Fernandes (1955) described *C. borivilianum* for the first time from India. Main taxonomic features of this plant, given by Nayar and Shastri (1988) are:

It is a rhizomatous herb with 1-9 roots; tubers upto 5cm long, 1cm in diameter; leaves radially, cariaceous, spirally imbricated at the base, sessile, linear, acute, flat, 8.22 × 0.7 – 1.8 cm; inflorescence is a raceme; flower white, 2.5-3.5cm across, bracteate, pedicellate, in clusters of three, bracts linear, papery, purplish; capsules greenish yellow, obcordate, loculicidal, seeds orbicular, discoid and black in colour. Marais and Reilly (1978), however, placed the genus – *Chlorophytum* close to the genus – *Anthericum* and is distinguished by the enlarged middle part of stamen filament and capsule with three wings.

Flowers emerge after 4-6 days of first rainfall during June in mature plants. Generally, one plant produces only one inflorescence, but, sometimes 3-4 inflorescences may be produced, depending upon the age of the mother plant. Older and thick root-bearing plants may produce more inflorescences. The number of flowers per inflorescence may increase with the increase in size of the underground tuberous roots. *C. borivilianum* produces about 50-110 seeds per plant with 7-10 seeds per fruit; depending on the agro-climatic conditions and age of the plant, however, about 15% seeds remain abortive. The viability of embryos get reduced to 27% on the storage of seeds for one year at 5 °C. Poor viability and germination of the seeds has made this plant an endangered species. Therefore, alternative methods of propagation (the biotechnological approaches) are necessary.

The various features of chromosomes of different species have indicated the primitiveness of the karyotype under 8 basic number. However, Baldwin and Speese (1951) have considered 8 as the basic number on the basis of Liberian material for the genus through duplication of one of the chromosomes. The plants of *Chlorophytum borivilianum*, raised through organogenesis, have been found to contain 2n=24 chromosomes while those derived from somatic embryogenesis contained 2n=21-27 chromosomes. However, previous reports (Arora *et al.*, 1999) have shown that the parent plants are supposed to be triploid with 8 as the basic number.

Medicinal Uses :

Mostly the powdered tuberous roots are widely used in the Indian systems of medicines from the time immemorial for the following properties: (1) as a non-hormonal restorative tonic; (2) in fatigue, general debility, weakness and as a general purpose tonic; (3) used in impotency and sterility and to enhance male potency; (4) as cardiac and brain tonic; (5) as a curative agent in various diseases like piles, diabetes, albuminorrhoea – leucorrhoea, dysmenorrhoea; and (6) as antipyretic, sialogogue, galactogue, diuretic, haemostatic and antiarthritic agent.

Generally, *C. borivilianum* is used along with *Withania somnifera* and *Asparagus racemosus* in various commercial formulations. Due to its rejuvenating and aphrodisiac properties several projects have been launched on *C. borivilianum* for the preparation of root-chips and flakes in European and other developed countries. The sexual stimulant property of the roots has placed it equivalent to 'Viagra', but, without side effects. In general, it has been found to have a very positive effect on human immune system.

The powdered fried roots of *C. borivilianum* are chewed in the ulcerative conditions of mouth and throat. In central Orissa, ladies use the aqueous extract of roots together with boiled rice extract in cases of dysmenorrhoea. Roots are also used as curative in piles, sprue and as blood- purifier. Presently, it is an ingredient in more than hundred herbal formulations.

Moreover, some tribal populations consume the leaves of 'safed musali' in their curry preparations.

Phytochemistry and Pharmacology :

The tuberous roots contain steroidal saponins (neohecogenin, neotigogenin, stigmasterol, tokorogenin), a bibenzyl xyloside (2,4,4'-trihydroxy-2-xylopyranosyl bibenzyl), a disubstituted tetrahydrofuran (4-hydroxy-8, 11-oxidoheneicosanol), nonacosone, tetracosanoic and triacontanoic acid, pentacosyl docosanoate, galactoglucan and sugars like arabinose, glucose and xylose (Fig. 1) (Tandon and Shukla, 1992, 1993; Tandon *et al.*, 1992; Gupta *et al.*, 1979). Saponins and Sapogenins, particularly neohecogenin and neotigogenin, are usually found in genus: *Chlorophytum*. Steroidal saponins, chloromaloside-A-D were isolated from the tubers of *C. malayense* (Li *et al.*, 1990). A steroidal glycoside, chloromaloside-A isolated from *C. malayense,* has been shown to be cytotoxic against several human cancer cell lines (Qiu *et al.*, 2000). Three new spirostanol pentaglycosides were isolated from the tuberous roots of *C. comosum* (Mimaki *et al.*, 1996). A comparison of aglycone structures of the saponins from *C. comosum* with those of *C. malayense,* showed that the former were based on the (25R)-spirostan series, such as in neotigogenin and neohecogenin (Mimaki *et al.*, 1996).

The economic interest, raised in these plants, is from the use of steroidal saponins as a starting material for the industrial production of sex hormones, corticosteroids and steroid derivatives in general. The main source of saponins has been from higher plants, especially those belonging to the family Liliaceae, Solanaceae and Scrophullariaceae. Bibenzyl are commonly present in Orchidaceae, Cannabinaceae and Bryophytes (Sharma and Ramawat, 2000).

The generalized uses of this plant (*C. borivilianum*) have been inferred on the basis of root's use in folk medicine (Jain, 1991), traditional medicine and experiments on men volunteers (Ojha, 1987). Three new saponins, isolated from *C. comosum* (Mimaki *et al.*, 1996), showed an inhibitory activity against ^{32}P incorporation into phospholipids of He La cells as the primary screening test to identify new anti-tumour promotor compounds.

The extracts prepared from the dried root tubers of *C. borivilianum* inhibited dopamine-uptake in striatal synaptosome, and thereby, could lead to enhanced dopamonergic tone in the central nervous system (CNS) to act as psychostimulants. This has a beneficial effect on the brain and as a result the human body shows an increase in alertness, mental ability, intelligence and sexual characters. The active principle, the saponins, are known as stimulants and metabolic enhancers.

Although the extracts of root tubers of *C. borivilianum* have been found to show curative action in impotency and sterility and to enhance male potency, Wang *et al.* (1996) have demonstrated strong spermicidal activity of chloromalosides-A and –B obtained from the roots of *C. malayense*. However, extensive researches are required to be undertaken for the isolation and identification of other unknown active principles from the root tubers of different species of *Chlorophytum* and to assess their effects *in vitro* and finally on human body.

Cultivation :

Attempts have been made to develop methods for cultivation of *C. borivilianum* by seeds as well as the root tubers (Bordia, 1993). Seed-raised plants produce small tuberous roots in their first season and large roots of commercial value are produced at the end of second season. The seeds and the tubers remain in dormant condition for 8-9 months (September-October to March-April) and therefore, germination is not possible during this period. With the advent of monsoon season by the end of June, the environmental conditions become suitable for sowing of seeds. The seed nursery requires sufficient amount of compost manure. The percentage of seed germination in *C. borivilianum* is very low (only 11-24%). Following the one-year growth in a nursery the plants are transplanted in the fields at a distance of 30 × 15 cm in the next season. The plant to plant distance should be adequate enough for the proper development of roots in each plant, otherwise the ultimate yield may be affected. The tubers are sown directly in the fields in 20-30 cm raised beds/bundhs. The tubers without stem-disc do not sprout as they lack apical bud. Yield from the tuber-raised crop is approximately three times more than the seed-raised crop.

The plants require well-drained loamy to sandy soils rich in organic matter. Warm and humid climatic conditions with high soil moisture during the growing period favour luxuriant growth and facilitate the development of fleshy roots.

During presowing soil preparation 20-25 t./ha of compost/dung manure should be mixed. In a recent study, conducted by CIMAP, Lucknow, 60:65:20 kg/ha of N:P:K has been recommended besides the compost manure for better production. The use of green manure for 'safed musali' has been found to be highly beneficial. For good quality luxuriant production of root tubers other recommended organic manures include bone-meal or vermi-compost or micro-nutrients at the rate of 25 kg/ha.

C. borivilianum requires the first irrigation just after sowing and later, after every fortnight. Timely monsoon eliminates the requirement of additional irrigation.

Invasion of foliage-eater insects is very common in 'safed-musali' for which organic pesticides should be sprayed on the aerial parts. Excess of moisture and iron-deficiency in soil may cause chlorosis in leaves which may be overcome with the use of 100 kg/ha of pyrite in soil and foliar spray of 0.05% humic acid 0.5% neutral solution of ferrous sulphate twice a month. For avoiding seed-borne fungal diseases the seeds/tubers should be pre-treated with 4.0 g/kg of theram. Recently in January, 2004 severe root infections of *Mucor* sp. and *Fusarium* sp. have been observed in Bulandshahr (U.P.). For this, soil-treatment should be done with fungicides prior to sowing.

The plant matures in 90-120 days followed by withering of the leaves. For the maturation of root tubers they should remain in the fields upto November. The root tubers should be carefully removed from the soil by the end of November or in early December. Usually 30-40 qtls/ha fresh roots are obtained which on drying comes to 6-8 qtls/ha. Roots are washed thoroughly, peeled off and dried for marketing.

BRAHMI (*BACOPA MONNIERI* (L.) PENNELL)

Nomenclature :

Sanskrit- Brahmi, Neer Brahmi; Hindi-Brahmi, Jal Brahmi; English-Water Hyassop;Bangla-Adha Birmi, Jal Nimba; Tamil-Brimi, Neer Pirmi; Telugu-Sambranichettu;Marathi-Neer Brahmi; Kannada-Neeru Brahmi; Gujarati-Kadavi Luni; Malayalam-NeerBrahmi; Punjabi-Brahmi;Botanical name-*Bacopa monnieri* (Linn.) Pennell (Family: Scrophullariaceae).

Biological Description :

The name- 'Brahmi' is commonly applied to two different plants, viz., *Bacopa monnieri* (Linn.) Pennell of family: Scrophullariaceae and *Centella asiatica* (Linn.) Urban of family: Apiaceae, in different regions of India. However, the true brahmi of Ayurveda is claimed to be *B. monnieri,* which is used extensively in the Indian systems of Medicine for various nervous disorders (Chopra *et al.*, 1956 a).

Bacopa monnieri (Synonyms: *Bramia monnieri* Pennell, *Moniera cuneifolia* Michx., *Herpestis monniera* (Linn.) H.B. & K., *Herpestis spathulata* Blume., *Gratiola monniera* Linn., *Lysimachia monniera* Linn.), is a weak, creeping and somewhat succulent herb which is found in damp and marshy areas in plains throughout India. The plant is bright green in colour and the creeping stem develops roots at nodes touching the soil; leaves are simple, opposite and decussate, sub-sessile, obovate-oblong, obtuse, glabrous, 0.6-2.5 cm long and 0.2-1.0 cm wide and are succulent; flowers actinomorphic, hermaphrodite, white to light violet in colour, solitary, axillary, pedicel thin and 0.6-3.2 cm long; five numbered calyx with unequal lobes; corolla gamopetalous, short funnel-shaped; stamens four in number, didynamous, ovules on swollen, median, axile placenta and fruit is a capsule with persistent calyx.

Medicinal Uses :

The plant is very well known for its use as an effective brain tonic. The dried leaves have a curative effect on the patients of nervous break-down. For the sharpness of brain and memory its leaf extract should be taken daily with milk (Sushruta). In cases of diarrhoea in children the leaf-paste is given together with sugar and cumin (*Cuminum cyminum*) without any side effect. It is used in several ayurvedic formulations for epilepsy, insanity and the brain-fag. Five to seven leaves are chewed daily by those persons who stammer in some tribes of Maharashtra. The scientific studies have shown that the plant (*Bacopa monnieri*) is a good broncho-dilator, mild laxative and is a potent tonic for heart and Central Nervous System. Furthermore, it has anticancerous, anti-inflammatory and antioxidant properties too.

Sushruta (Sushruta Samhitaa) gave the plant as a potherb in fevers, haemoptysis, hiccup, cough, dysponea, urinary diseases, calculi, skin diseases; and externally, for washing ulcers. Several ayurvedic formulations are available of which brahmi is the main ingredient. Saaraswataarishta (Bhaishajya Ratanawali) is available for nervous debility, convulsions and for improving memory; Brahmi Bati (Ras Tantra Saar) for mental weakness, deficient memory; Brahmi Ghrita (Charak Samhitaa) for epilepsy, insanity and for promoting memory.

Phytochemistry and Pharmacology :

The activity of the plant, *B. monnieri*, as a nervine tonic in traditional systems of medicine, is attributed to a complex mixture of dammarane type triterpenoidal saponins it contains (Fig. 2) . Out of these saponins, bacoside-A and bacoside-B have been tested and found to be responsible for the efficacy of the drug. Bacoside-A is laevo-rotatory, while bacoside-B is dextro-rotatory in nature. The haemolytic action of bacoside-B has been found to be double that of bacoside-A, which could be due to the differences in the configuration of carbohydrate part of the molecules (Basu *et al.*, 1967). Bacoside-A, which initially appeared to be a single compound, is a mixture of saponins from which baciside-A_2 and A_3 were separated. Bacoside-A_3 is the main constituent of the mixture (Rastogi *et al.*, 1994; Rastogi and Kulshreshtha, 1998).

The other saponins include bacopasaponins-A,-B,-C,-D (Garai *et al.*, 1996 a, b),-E,-F (Mahato *et al.*, 2000), bacopasaponin-G (Hou *et al.*, 2002), bacopaside I, II (Chakravarty *et al.*, 2001), bacopasides III, IV, V, bacopasides-A,-B,-C (Hou *et al.*, 2002; Chakravarty *et al.*, 2003). Chakravarty *et al.*, (2001) named a saponin as bacopaside III that has jujubogenin as aglycone. Bacopasaponin-A,-E and –F (Fig. 2) are jujubogenin bisdesmosides, with the sugar units attached at two positions of the aglycone (Garai *et al.*, 1996 a, b; Mahato *et al.*, 2000). Bacogenins-

A_1-A_4 are the polar component of the aglycone mixture, obtained following the hydrolysis of bacoside-A and bacoside-B.

Bacogenin-A_1 is an artifact derived from the sapogenin, pseudojujubogenin, during acid hydrolysis (Kulshreshtha and Rastogi, 1973 a, 1974; Kawai *et al.*, 1973; Garai *et al.*, 1996a). Bacogenin-A_2 is an isomer of bacogenin- A_1, differing either in configuration at C-20 or disposition of *vinylic Me* in a side chain (Kulshreshtha and Rastogi, 1974). Bacogenin-A_3 is also an artifact formed during acid hydrolysis (Chandel *et al.*, 1977). Bacogenin-A_4 is identified as ebelin lactone, which has been found to be an artifact, derived from the sapogenin, jujubogenin, during acid hydrolysis (Kulshreshtha and Rastogi, 1973 b). On hydrolysis, bacopasaponins-A, -E and –F yeild ebelin lactone, while bacopasaponins-B, -C and-D yeild bacogenin-A_1 as aglycone (Garai *et al.*, 1996 a, b; Mahato *et al.*, 2000). Of the two sapogenins, viz., jujubogenin and pseudojujubogenin, the structure of pseudojujubogenin has not yet been completely known, as the configurations at C-20 and C-22 have not been determined (Garai *et al.*, 1996 a).

Other chemical constituents of the plant include D-mannitol (Sastry *et al.*, 1959), hersaponin, betulic acid (Chatterji *et al.*, 1963; Rastogi and Dhar, 1960; Jain and Kulshreshtha 1993), alkaloids, viz., brahmine and herpestine (Bose and Bose, 1931; Basu and Walia, 1944; Basu and Pabrai, 1947), flavonoids, viz., luteolin-7-glucoside, glucoronyl-7-apigenin and glucoronyl-7-luteolin (Schulte *et al.*, 1972;Proliac *et al.*, 1991) and phytosterols, viz., stigmasterol, β-sitosterol, stigmastanol (Chatterji *et al.*, 1963).

The ethanolic extract of *B. monnieri* has been found to have improved the performance of rats in motor learning capabilities (Prakash and Sirsi, 1962). It augmented both the cognitive function and mental retention capacity, as revealed by the studies involving acquisition, consolidation and retention of three newly acquired behavioural responses in albino rats, viz., foot-shock motivated brightness discrimination response, active conditioned avoidance response. The two saponins, viz., bacoside-A and bacoside-B have been found to be responsible for these activities. Bacosides have also been found to improve long-term memory (Singh *et al.*, 1990).

Aqueous and ethanolic extracts of the plant did not exhibit any protection from the chemical or electric shock-induced seizures, but, did reduce their severity. They did not produce any sedation or antidepressant effect, but potentiated the barbiturate–induced hypnosis. They also increased the pentylenetetrazole-induced seizure latency, suggesting that a longer duration of treatment with the extracts of the plant may be beneficial in epilepsy (Rao and Karanth, 1992). Moreover, both these extracts (aqueous and ethanolic) have shown a tranquillizing effect in albino rats and dogs. It was found to be a smooth muscle relaxant and an antispasmodic too. The bacosides have exhibited antistress effect and the experiments indicate that they have the potential to modulate the activities of Hsp 70, cytochrome P450 and superoxide dismutase in the brain, thereby preparing the brain to act under adverse conditions such as stress (Chowdhuri *et al.*, 2002).

B. monnieri has been found to have thyroid stimulating activity along with lowering of lipid peroxidation, and increasing the activity of superoxide dismutase and catalase (Das *et al.*,2002). The ethanolic extract of the plant has also found to exhibit anticancer activity against Walker carcinosarcona-256 in rat (Bhakuni *et al.*, 1969) and Sarcoma-180 cell culture (Elangovan *et al.*, 1995). Moreover, the ethanolic extract has the potent antioxidant properties (Tripathi *et al.*, 1996). The extract inhibited $FeSO_4$ and cumene hydroperoxide induced lipid peroxidation in a dose-dependent manner (significant protection obtained at the dose of 500 μg/ml).

Since the saponins are known to have antiprotozoal property, bacopasaponin-C was studied for its anti-leishmanial property both in free form and in different delivery modes, viz., niosomes, microsperes and nanoparticles, and it was found to be very active in all forms. However, free form showed nephrotoxicity while in vesicular form it was found to be safe. Futhermore, the activity was found to be parasite specific (Sinha *et al.*, 2002). Another very important observation was that the alcoholic extract of *B. monnieri* exhibited the hepatoprotective effect (Sumathi *et al.*, 2001).

Cultivation :

Brahmi may be grown in almost every type of soil, however, organic-rich silty-loam is the best. It grows well in moist and swampy areas. A temperature range of 33-40°C and 65-80% relative humidity provide the most suitable conditions for the proper development and luxuriant growth of brahmi. *B. monnieri* is sown using the small cuttings (10-15 cm in length) of its runner plant and it requires watering just after. The best period for sowing is between March to June so that it may get good moist conditions for faster growth during the monsoon season. The plants very quickly cover the complete soil-surface of the field and therefore, once sown it may be harvested for years to come.

For pre-sowing soil preparations 5 t/ha compost/farm manure should be mixed. Besides this, 60 kg/ha phosphorus and 60 kg/ha potash will be required. Furthermore, it may require nitrates at the rate of 30-35 kg/ha following every harvest, i.e. 100 kg/ha of nitrates annually.

The brahmi crop needs high and prevailing moisture conditions. Usually it requires weekly irrigation, so that good amount of foliage may be obtained. To protect the crop from the attack of insects, nematodes and fungi, organic pesticide like neem-cake-extract should be used for spray.

During the harvest in June-July and November-December the foliage should be cut 2-3 inches above ground using sharp sickle. Every time 4.5-6.5 t/ha dry plants may be obtained. For storage the plants are dried under the sun and packed in bags. The best method is to dry the plants for 30 minutes in an oven at 80°C followed by sun/air drying.

SATAWAR (*ASPARAGUS RACEMOSUS* LINN.)

Nomenclature :

Sanskrit-Shatawari, Shweta mool; Hindi-Shatawari, Satawar; English-Asparagus; Bangla-Satmooli; Tamil-Chattawari; Telugu-Chaltagadda, Pilli-gaddalu; Marathi-Asawal, Shatmooli; Kannada-Eharuballi, Sipariberuballi; Gujarati-Satawari, Ekalkanto; Malayalam-Shatawali, Satawari; Oriya-Satmooli, Mohajolo, Sotabori; Punjabi-Bojidan, Satawar; Botanical name-*Asparagus racemosus* Linn. (Family: Liliaceae)

Biological Description :

The genus *Asparagus* of family: Liliaceae comprises several herbs or under-shrubs. A total number of 22 species have been found to be distributed in India. Among these, *A. racemosus*, commonly known as 'shatawar' is a perennial, climbing shrub with fleshy (storage) roots and stem comprising recurved spines/cladodes that are actually a modified stem (Abraham, 1998). The plant blossoms in the last week of November. It develops typical liliaceous flower which is actinomorphic, hermaphrodite, trimerous with two whorls of perianth of three tepals each. The fruits, the berries, contain two-three seeds which are spherical or hemispherical. A ten- year old plant produces about 130g seeds (Kar and Sen, 1984).

The bunch of roots develop into fusiform, fleshy branches. The size is variable-30 cm-2m long and 1-2.5 cm in diameter. The entire plant, particularly the roots contain several inorganic and organic compounds.

Medicinal Uses :

The roots of *Asparagus racemosus* are used for the treatment of several diseases such as diarrhoea, dysentery, rheumatism, nervous breakdown, as an aphrodisiac (Chadha, 1985; Nadkarni, 1976). The studies on the total extracts and the isolated principals have shown a wide range of biological activities, including the antitumour (Shao *et al.*, 1996), antifungal (Shimoyamada *et al.*, 1990), antimutagenic (Edenharder, 1990), ulcer protective (Sairam *et al.*, 2003) and diuretic (Balansand and Rayband, 1987) properties. This plant has also been considered to be lactogogue in lactational inadequacy (Sharma *et al.*, 1996). The crude saponins from this plant (from the edible part) have been found to inhibit the growth of human leukemia HL-60 cells in culture and macromolecular synthesis in a dose-and–time-dependent manner (Shao *et al.*, 1996).

Sushruta (Sushruta Samhitaa) gave the drug in internal abscesses, oedema, respiratory troubles, urethral discharges and seminal discharges and prescribed the decoction boiled with clarified butter internally in urinary calculi. Shataavari Ghrita (Charak Samhitaa) is a reputed rejuvenating and aphrodisiac tonic. Shataavari Ghrita (Sahasrayoga) is given in disorders of the urinary system, urethritis, dysuria and cystitis. Phalaghrita (Bhaishajya Ratanaawali) is another classical compound, prescribed for treating disorders of female genitals, leucorrhoea and infertility. Shataavaryaadi Churna (Yoga Ratanaakar) is given (3-6g) in sexual debility. Shataavari Lehya (Sahasrayoga) is given as a restorative tonic, haematinic, choalagogue and vitalizer in emaciating diseases, menorrhagia, haemoptysis and jaundice, in urinary disorders and as a diuretic (2-5 g once daily).

Phytochemistry and Pharmacology :

The storage roots of *Asparagus racemosus* have been found to exhibit a positive action on female reproductive system and is used as lactogogue and aphrodisiac in traditional systems of medicines. Moreover, it is used as a general tonic. The roots contain several vitamins and free amino acids together with many saponins and sapogenins. Free amino acids, isolated, include arginine, asperagine, aspartic acid, glutamic acid, glycine, histidine, isoleucine, leucine, lysine, methionine, phenylalanine, proline, serine, tryptophane, tyrosine and valine whereas the vitamins are α- and β-carotene, choline, niacin, tochoferol and zeaxanthine.

Besides the vitamins and amino acids, other isolated chemicals are coniferine, α-aminodimethyl-γ-butyrothetin, asparagoside, asparasaponin, asparagusic acid, sinidine-3, 5-diglucoside, sinidine-3-monoglucoside, sinidine-3-rhamnosyl glucoside, rhamnosyl glucosyl glucoside, diosgenin, folacin, guiacol, inositol, inulin, jamogenin, camphirol, lauric acid, lenoleic acid, α-lenoleic acid, officinallicin, felothione, pseudoasparagos peufa, quercetin, rutine, sersapogenin, 2-spirostein-3-β-ol, 4-vinylguiacol, and 4-vinyl phenol.

Cultivation :

A. racemosus has a very flexible cultivation conditions. It may very easily be grown in sandy-loam to silty-loam and slightly acidic to slightly alkaline soils with low or high organic contents. However, the best one is organic matter rich sandy-loam soil having a good water drainage facility.

The soil of the cultivation field should not be hard and for this deep ploughing may be required. Monsoon season is the best for sowing. The cultivation of 'Satawar' may be done by seeds as well as with suckers. Seeds are first sown in nursery-beds during the month of April. Seeds with hard testa should be soaked in warm water for 24 hours prior to sowing. After 2-3 months in nursery the young plants become ready for

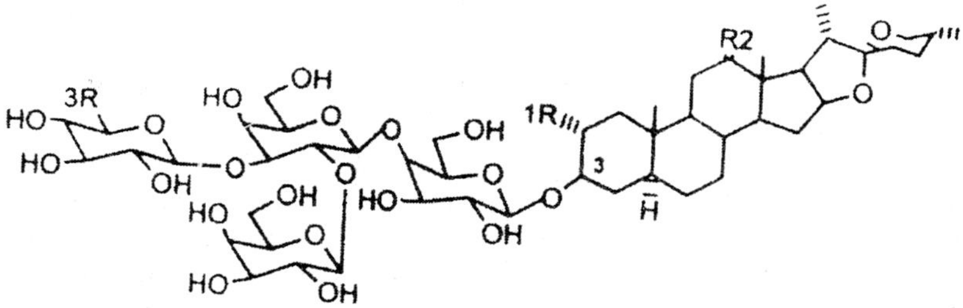

1R= OH, R2=H,H 3R=H

Spirostanol saponins

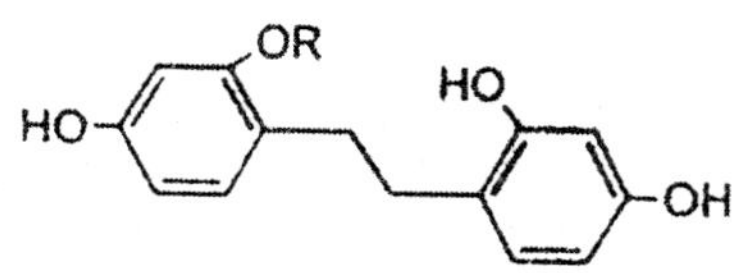

R= Xylose

2,2', 4, 4' - tetrahydroxybibenzyl xyloside

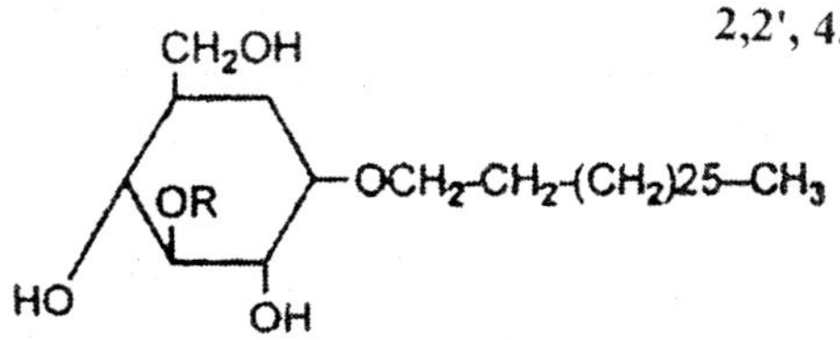

R= PhCO

Arundinoside B

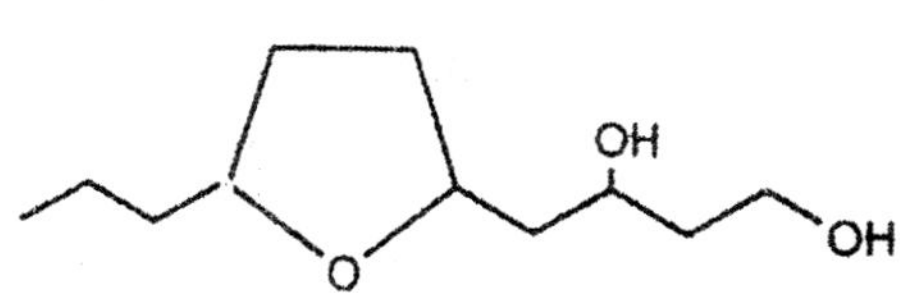

4- hydroxy- 8, 11-oxidoheneicosanol

Fig. 1 : Structure of metabolites from *Chlorophytum* sp.

transplantation to the fields. For sucker plantation long stem suckers should be selected for sowing in the ditches or on soil-bundhs. The aerial climbing portion of the plants require at least 10-12 feet long wooden post for each plant.

During the preparation of the field soil approximately 20-25 t/ha of compost manure should be mixed. As per requirement 120:60:60 NPK may

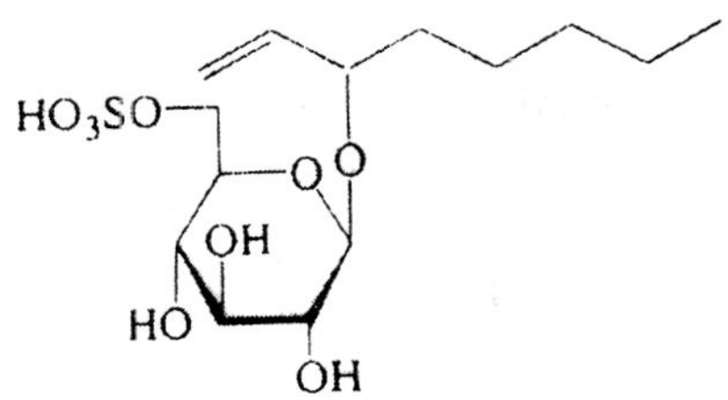

Bacopaside A

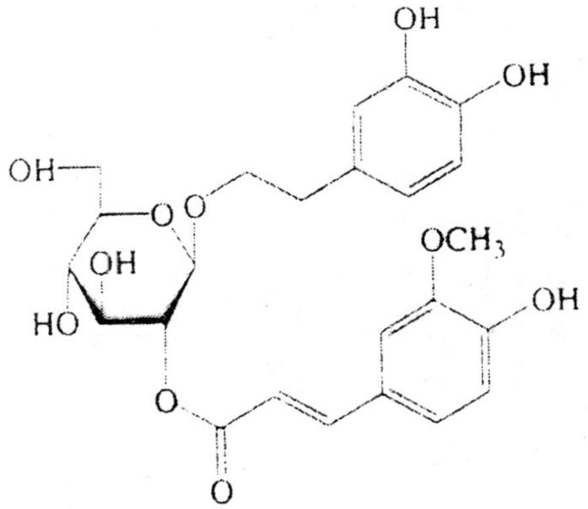

Bacopaside B

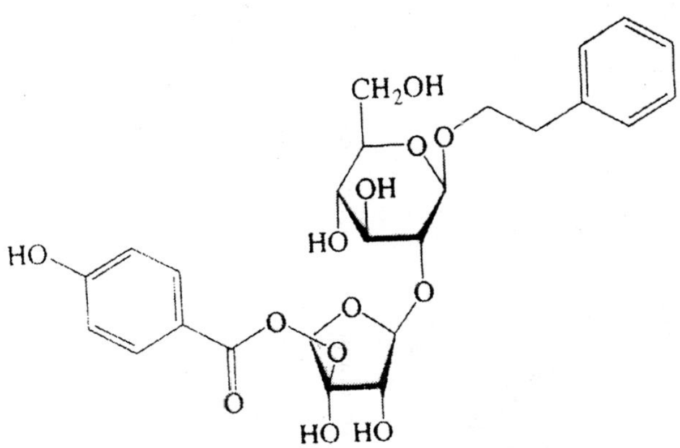

Bacopaside C

Fig. 2 : Some chemical constituents of saponins from *Bacopa monnieri*

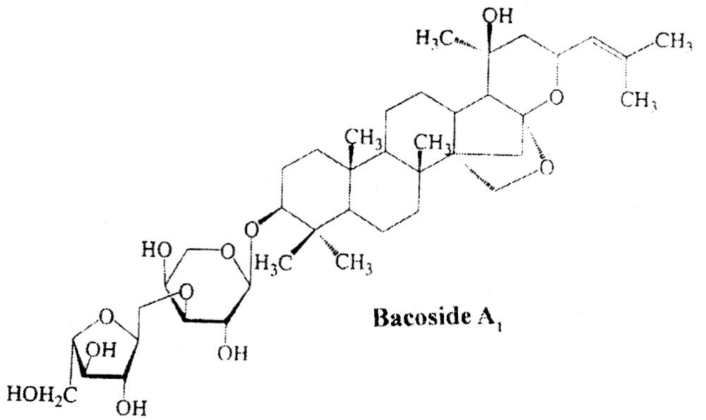

Bacoside A_1

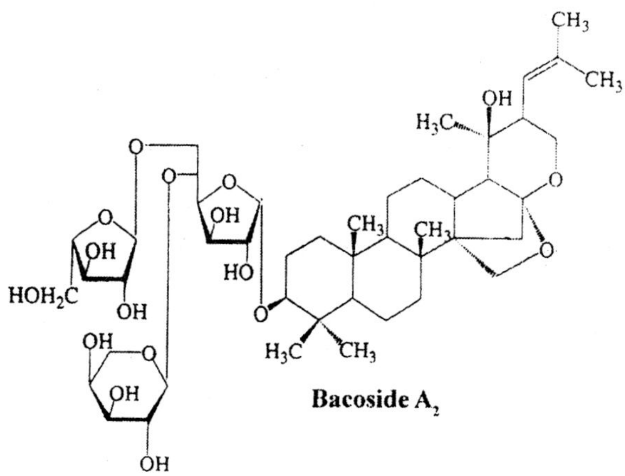

Bacoside A_2

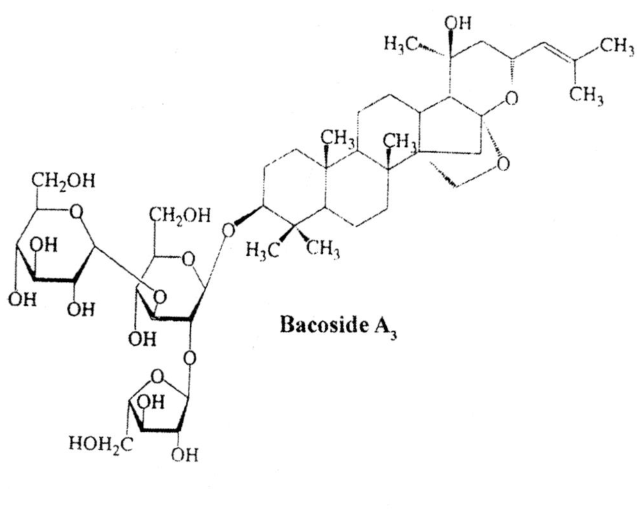

Bacoside A_3

Fig. 2 : Some chemical constituents of saponins from *Bacopa monnieri* (continued .....)

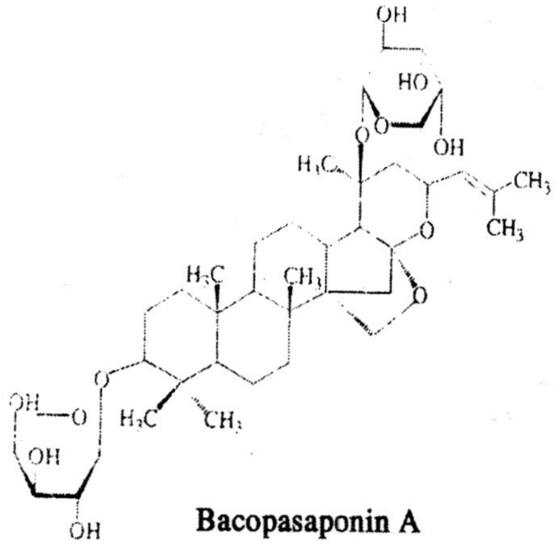

Bacopasaponin A

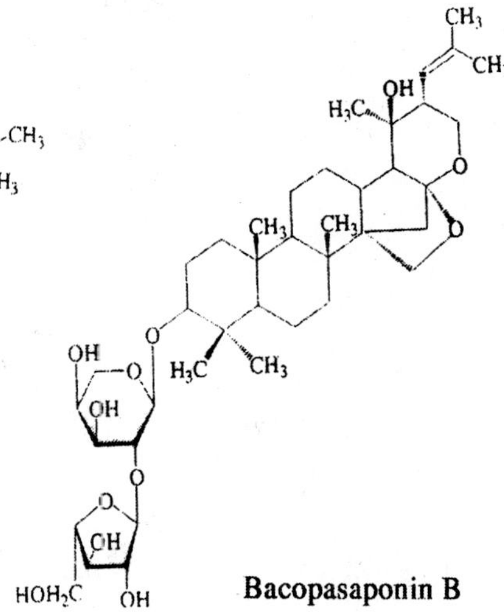

Bacopasaponin B

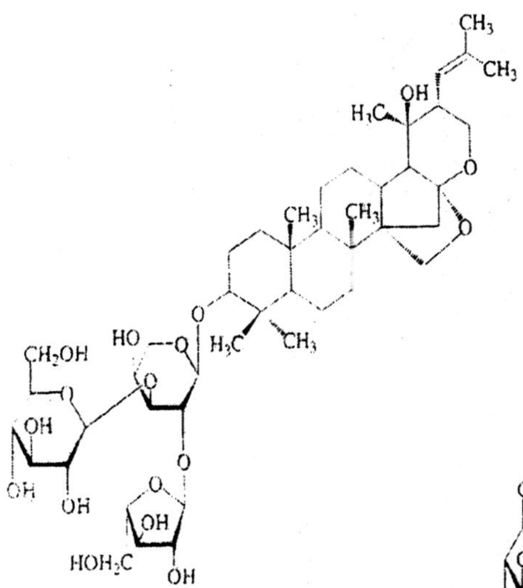

Bacopasaponin C

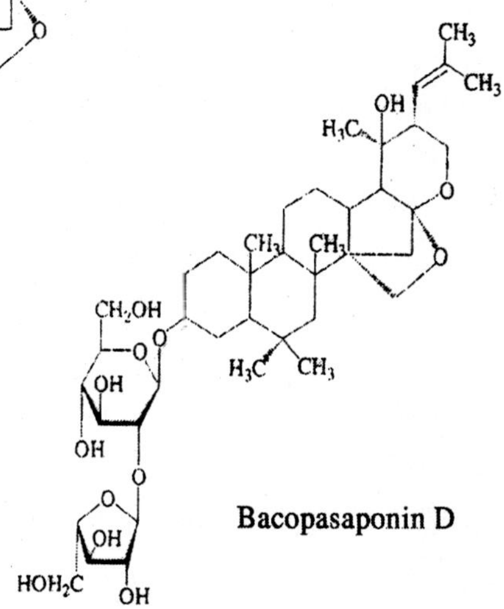

Bacopasaponin D

Fig. 2 : Some chemical constituents of saponins from *Bacopa monnieri* (continued .....)

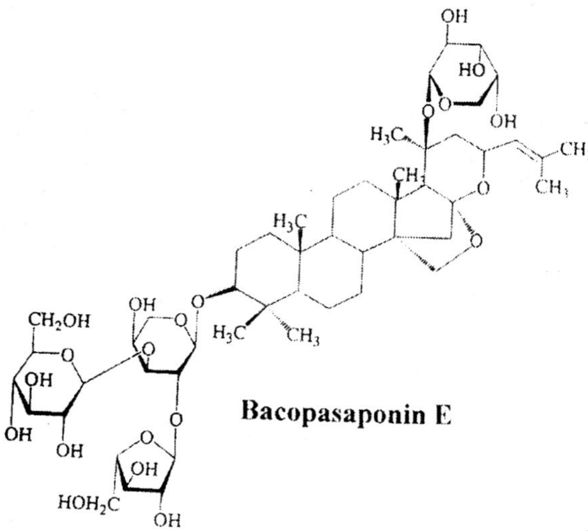

Bacopasaponin E

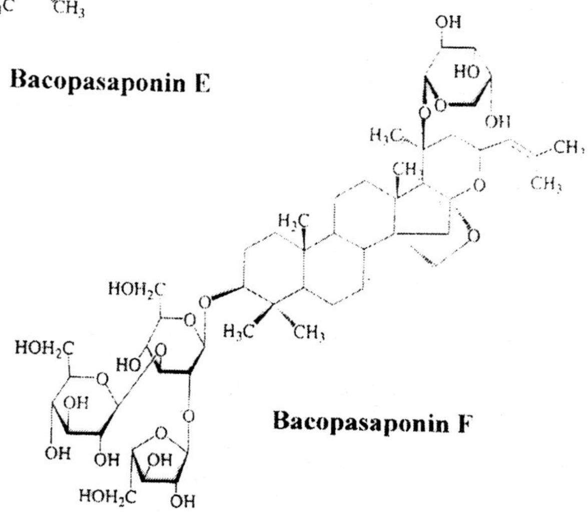

Bacopasaponin F

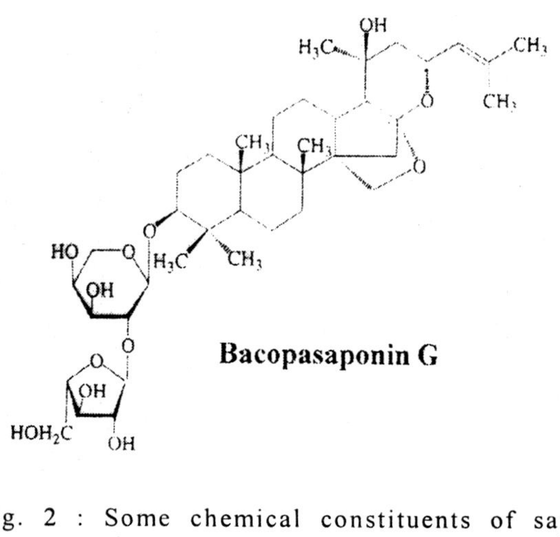

Bacopasaponin G

Fig. 2 : Some chemical constituents of saponins from *Bacopa monnieri* (continued .....)

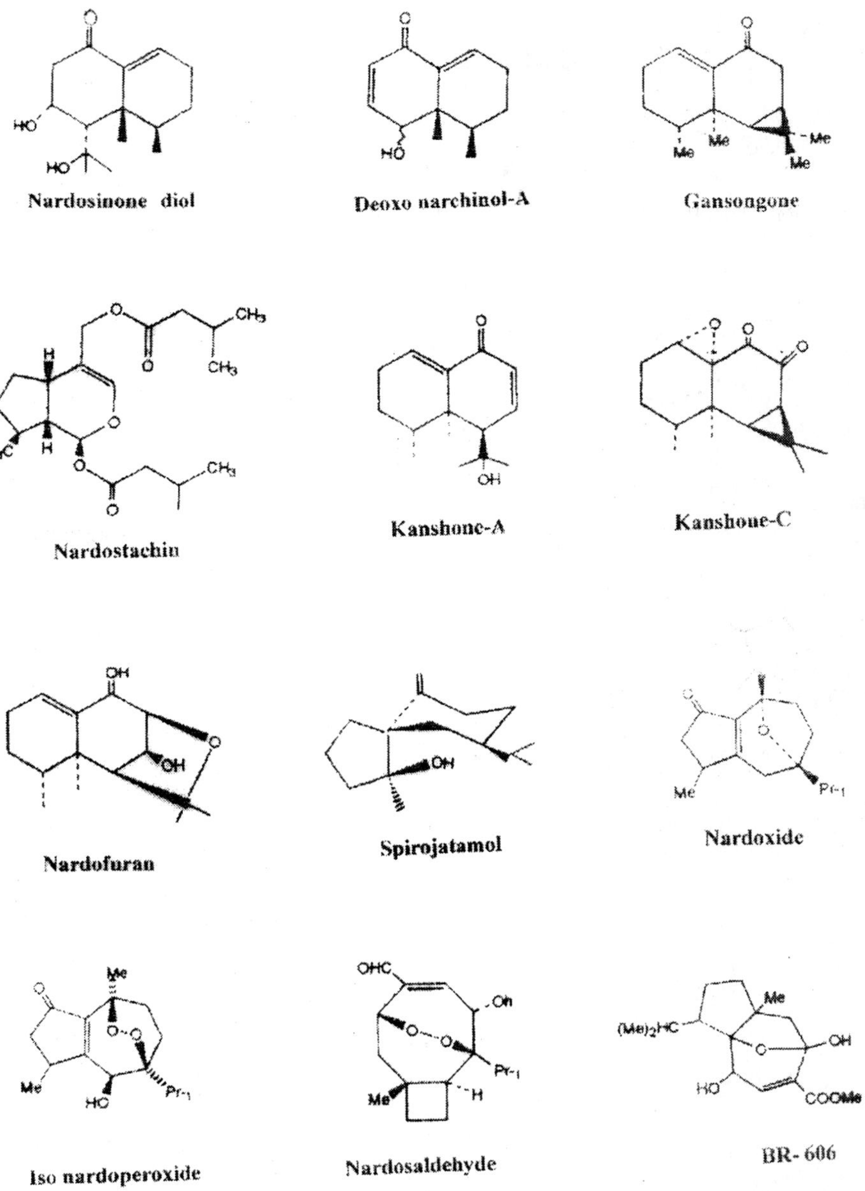

Fig. 3 : Structure of some chemical constituents from *Nardostachys jatamansi*.

also be used. It has been observed that good amount of organic matter in the soil favours the development of good quality root tubers.

After the monsoon season the crop requires irrigation once in a month. Care should be taken during summer season when the soil should be kept

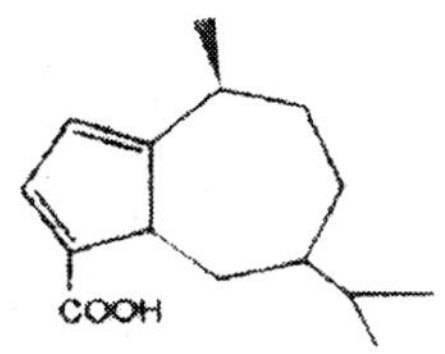

Jatamanshic acid

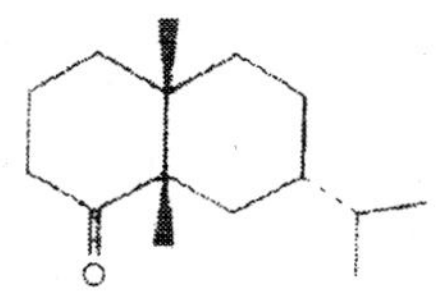

Jatamansone

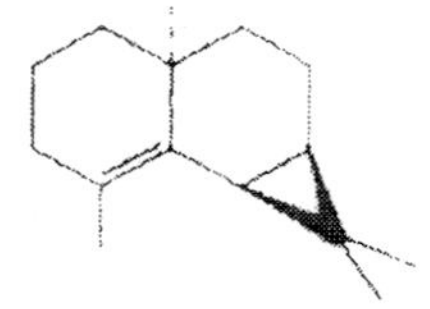

β-Maaliene

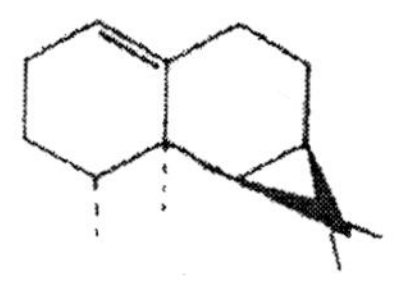

Calarene

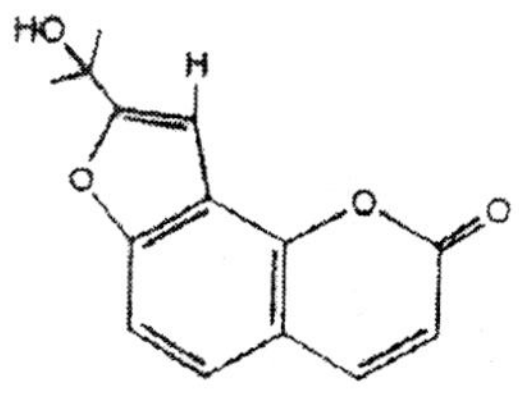

Oroselol

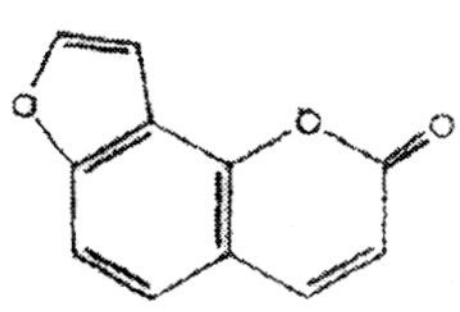

Angelicin

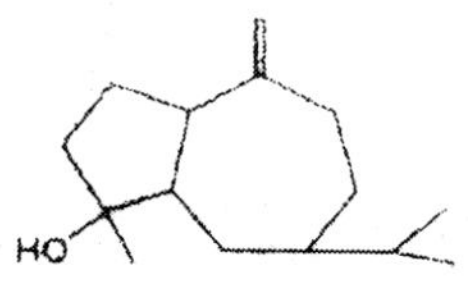

Nardol

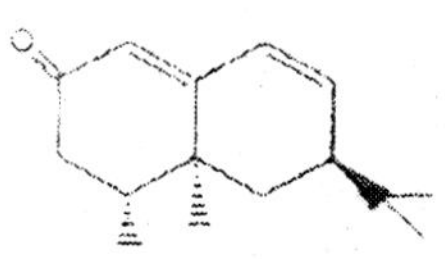

Nardostachone

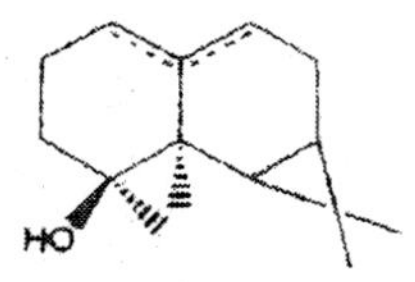

Calarenol

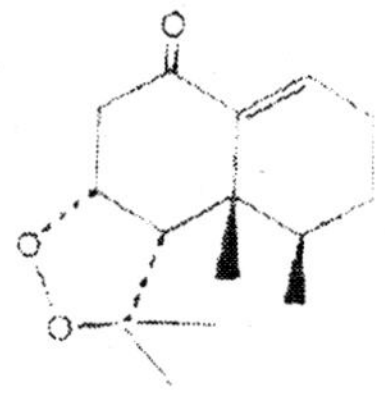

Nardosinone

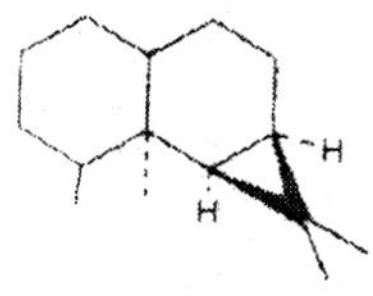

Aristolane

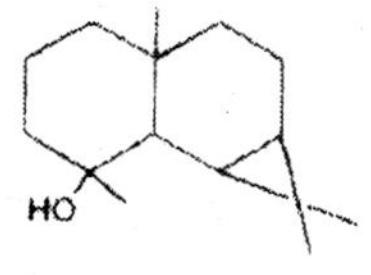

Maaliol

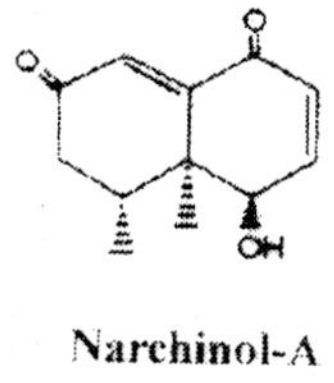

Narchinol-A

Patchouli alchohol

β- Patchoulene

Fig. 3 : Structure of some chemical constituents from *Nardostachys jatamansi* (continued .....)

moist. For the protection against root-rot fungi and nematodes, only organic pesticides like neem-cake extract may be used.

Root tubers are carefully harvested after two years of growth. The harvesting should be done in October-November. The roots are washed thoroughly, cut to small pieces, dried and packed in bags for marketing.

JATAMANSI (*NARDOSTACHYS JATAMANSI* DC.)

Nomenclature :

Sanskrit - Jatamansi; Hindi - Jatamansi, Bal-chir; English - Spikenard, Indian Nard; Bangla - Jatamasi; Marathi - Jatamansi; Telugu - Jatamasi; Gujarati - Jatamansi, Balchhar; Kannad - Bahulgandha; Botanical name - *Nardostachys jatamansi* DC. (Family : Valerianaceae).

Biological Description :

Nardostachys generally grows in alpine Himalayas at an altitude of 2200-5000 m above sea level and in the geographical area from Punjab to Sikkim and Bhutan. It is also found in Tehri Garhwal (Uttaranchal, India) in the Western regions of the Himalayan mountains to Yunnan and South-West Szechuan in the east and Tibet in the north, covering a large area of Nepal, Sikkim and Bhutan (Kihara, 1955; Kitamura, 1954; Nayar, 1964; Tiwari and Joshi, 1974; Weberling, 1975; Shah, 1994; Amatya and Staphit, 1994). *Nardostachys grandiflora* DC. was kept as synonym under *N. jatamansi* for a long time, however, recently both these varieties have been assigned as two distinct species (Rastogi and Mehrotra, 1998).

N. jatamansi is an aromatic perennial herb (Satyawati *et al.*, 1987). The part of the plant stem partly underground (rhizome) has a long, thickened, stout woody root stock generally enclosed with fibres originating from the petiole of the dead and whithered leaves (Fig. 3). The plant may grow upto 60 cm in height. The upper portion of the stem is hairy. Leaves develop from the root stock as well as the stem, differing from each other. Leaves, developing from the root stock, are entire, elongate, spathulate, 15-18 cm long and 1.5-2.5 cm wide, having longitudinal parallel veins and have petiole, whereas those arising from the stem are in one or two pairs, approximately 7-8 cm long and 2-3 cm wide, oblong or ovate in shape and are sessile. Rhizomes are short, thick, and dark grey in colour. Each rhizome is crowned with the reddish-brown coloured tufted fibrous remains of petioles of root-stock leaves. The inflorescence is cymose, having terminal cluster with broad, ovate bracts. The inflorescence axis is red to reddish-brown in colour. Flowers are rosy, pale-pink or blue in dense cymes. The fruits are small and crowned with the persistent calyx teeth (Weberling, 1975). The seeds are obovate and compressed.

Medicinal uses :

The underground parts, i.e., roots and rhizomes of *N. jatamansi* have been used in Indian traditional medicines such as Ayurveda and Unani (Nadkarni, 1954; Satyawati *et al.*, 1987; Ahmad *et al.*, 1993) as a bitter tonic, which acts as a stimulant for the cardiac, respiratory and nervous systems, antispasmodic, carminative, laxative, diuretic and antiseptic. The decoction of the roots of this plant have been used for centuries in Ayurveda for the treatment of epilepsy, hysteria, convulsions, mental disorders and diseases of circulatory system (Nadkarni, 1954; Mukherjee, 1961; Hamied *et al.*, 1962). In addition, the plant has been used as a curative for dysmenorrhoea, flatulence, excessive urination, jaundice and several other disorders of the digestive system. This plant is also well known as a hair tonic (Ashtaanga Hridaya) and is an ingredient of many herbal shampoo and hair-oils.

The roots and rhizomes of jatamansi, on steam distillation, yield a light yellow, peculiar smelling essential oil. This essential oil is known to have antibacterial, fungitoxic (Girgune *et al.*, 1978) and insect-repellent activity (Dixit *et al.*, 1992). Besides this, it is also used in the perfumery industry (Han Dong S.'n *et al.*, 1980) and has a tranquillizing effect when massaged onto the head. It is also applied on leprous wounds.

Charak (Charak Samhitaa) and Sushruta (Sushruta Samhitaa) have prescribed smoking pillets, containing jatamansi, in cough, hiccough and asthama. They have also incorporated jatamansi in many medicinal oils, prescribed in oedema, piles, arthritis, gout, fractures and obstinate skin diseases, mostly for external application. This plant was also used for fumigation for warding off fear-psychosis (Ashtaanga Hridaya).

Phytochemistry and Pharmacology :

Choudhary *et al.* (1951) reported, for the first time, isolation of a water insoluble crystalline acid named jatamanshic acid and essential oil. Govindachari *et al.* (1958), then, isolated a bicyclic sesquiterpene ketone from *N. jatamansi* and named it jatamanscne. Later, Klyne *et al.* (1964) studied the detailed chemistry of jatamansone and valeranone and found both the compounds to be the same. This compound was found to be present in *N. jatamansi* as well as in *Valeriana officinalis* (Horster *et al.*, 1977). Two compounds, viz., β-maaliene and calarene have been isolated from chinese spikenard (*N. chinensis*) by Buchi *et al.* (1962). However, calarene was later on found to be a mixture of two compounds — β-gurjunene and aristolene (Vrkoc *et al.*, 1963). By refrigeration and elaborate column chromatography of the solvent extracted oil from the roots of a greyish brown variety of *N. jatamansi* (collected from Jammu and Kashmir) two new compounds, jatamansin (optically-active coumarin) and oroselol (identical to coumarin) have been isolated (Shanbhag *et al.*, 1965). But, oroselol has been reported earlier by Mitsuhashi and Itoh (1962). Later,

Shanbhag *et al.* (1965) further reported several hydrocarbons, a new oxide, alcohol jatamansinol, a polyoxygenated crystalline compound named angelicin along with other known compounds, i.e., α-pinene, β-pinene, carene, β-eudesmol, elemol, β-sitosterol, angelecin and jatamansin. Jatamansinol (3'-hydroxy-3'-4'dihydroseselin) is also obtained by the hydrolysis of jatamansin with alkali (Fig. 3).

Detailed investigation of the essential oil of roots and rhizomes of *N. jatamansi* has revealed, together with several known compounds, the presence of three new sesquiterpenes - nardol, calarenol and nardostachone (Pai *et al.*, 1971; Rucker *et al.*, 1978; Sastry *et al.*, 1966, 1967a,b,c; Saunders and Pinder, 1977). Calarenol was a new monoethynoid tricyclic tertiary alcohol named thus because of its similarity to calarene. Like calarene, calarenol was also found to be a mixture of two inseparable structural isomers. The presence of a new ketone-nardosinone was reported from the essential oil of *N. chinensis* along with jatamansone (previously reported ketone from the ketonic fraction) and new compounds-aristolene and 9-aristolene in the hydrocarbon fraction. Another important compound, isolated from *N. jatamansi*, was maaliol.

Rucker (1971) and Rucker and Kretzschmar (1971) have identified several compounds from this plant, viz., aristolane, 1(10) aristolene-2-one, 1, 8, 9, 10-tetradehydroaristolan-2-one, 9-hydroxy-1(10) aristolene-2-one (debilon), 9-aristolene-1-α-ol (nardostachnol) and 1, 2, 9, 10-tetrahydro-aristolene along with β-ionone and other previously reported compounds. A new trisnorsesquiterpenic diketo alcohol - narchinol-A was isolated in the form of yellow needles from methanolic extract of roots and rhizomes of *N. chinensis* (Hikino *et al.*, 1972), while the gas chromatographic study led to the isolation of patchouli alcohol and patchoulene (Rucker and Kretzschmar, 1972). Rucker *et al.* (1976) have shown the presence of seychellene and α- and β-patchoulene in *N. jatamansi* and the nardosinone, present in the oil of *N. chinensis* has been shown to be converted into deoxonarchinol, a C12 compound (Rucker and Eberhard, 1974).

An alkaloid - actiridine was isolated from the acetone extract of *N. jatamansi*, while this was reported to be absent in *N. chinensis* (Hirose *et al.*, 1978) and then, a new instable aristolane type sesquiterpene ketone-gansongone (Shide *et al.*, 1987a,b,c) in addition to nardosinone, nardosinondiol, deoxonarchinol-A, β-sitosterol, oleanolic acid and ethyl-β-D glucopyranoside on the basis of 2D NMR-spectroscopy and they have revised the structure of previously known 9-aristolene-1-α-ol to 1(10) aristolene-9-β-ol which, on oxidation, got converted into gansongone. They further reported a new sesquiterpene - nardonoxide from *N. chinensis*. Several new compounds have been isolated, during an exhaustive investigations carried out on the chemistry of rhizomes of *N. chinensis*, by Bagchi *et al.* (1988a,b,c,d,e). The dichloromethane extract yielded nardostachin, two new sesquiterpenoids - Kanshone-A as a colourless oil, Kanshone-B as colourless needles and a highly oxygenated aristolene type sesquiterpenoid - Kanshone-C, whereas the methylene chloride extract has resulted into two minor constituents - Kanshone-D and Kanshone-E in addition to major constituents - nardosinone, isonardosinone, nardosinone-diol and nardofuran. The advancement of structures for Kanshone-D and -E were mainly based on spectral data and chemical transformation and the last two, nardosinone-diol and nardofuran, were earlier reported as reaction products of nardosinone. Bagchi *et al.* (1991) have isolated, from the underground parts of *N. jatamansi*, three neolignans - virolin, erythro-1-(3,4-dimethoxy phenyl)-2-(2 methoxy-4(E) propanylphenoxy) propan-1-ol which was diesteromer of virolin and erythro-1-(4-hydroxy-3-methoxyphenyl)-2-(2-methoxy-4(E) propanylphenoxy)-propan-1-ol. The lignans, isolated were (+) pinoresinol and (+)-1-hydroxypinoresinol. A sesquiterpene having novel spiranic skeleton was spirajatamol and two eudesmane sesquiterpene - jatamol-A (viscous oil-$C_{15}H_{24}O$) and -B (viscous oil-$C_{20}H_{32}O_3$) were also isolated from dichloromethane extract of the same plant species (Bagchi *et al.*, 1988b, 1990, 1991). Some sesquiterpenes, viz., desoxonarchinol-A, nardosinone, debilon, nardosinone-diol and kanshone-A (isolated from *N. chinensis*) and eudesm-11-en-2,4 diol (isolated from *N. jatamansi*) exhibit cytotoxic, anti-tumour activity against P-388 cells. Other three new quaiane type sesquiterpenoids - nardoperoxide, isonardoperoxide and nardoxide have been isolated from *N. chinensis* (Takaya *et al.*, 1998). Out of these three compounds, the first two compounds (nardoperoxide and isonardoperoxide) showed antimalarial activity against *Plasmodium falciparum*. Two more compounds — BR-606 and nardosaldehyde have also been isolated.

Takaya *et al.* (2000a,b) have isolated and elucidated nine new guaiane endoperoxides - nardoguaianone-A, -B, -C, -D, -E, -F, -G, -H and -I from the roots of *N. chinensis* and evaluated their antinociceptive and antimalarial activities. Later, the same group of scientists have isolated two more guaiane-type compounds - nardoguaianone-J and -K and two aristolane-type compounds - Kanshone-F and -G (Tanitsu *et al.*, 2002).

The roots and rhizomes of *N. jatamansi*, on hydrodistillation, yield an essential oil - commonly known as 'spikenard oil' with a peculiar pleasant, sweet and spicy aroma. The type and maturity of rhizomes, site and time of collection greatly affect the yield, aroma character and quality of the oil (Amatya and Staphit, 1994). The oil yield varies greatly from 0.5% to 1.9% of rhizome dry weight and oil resinifies on being exposed to air. As the volatile oil cells are generally located on the fine fibrous hairs of the rhizome, the loss of fine hairs of rhizome during harvest, drying and storage may result into a low yield of oil. Due to non-availability of any commercial or pharmaceutical standards for the spikenard oil only the tradionally-

Table - 1 : Physico-chemical properties of spikenard oil

Colour	Light yellow to greenish blue
Acid value	2.16
Ester value	17.35
Ester value (after acetylation)	32.76
Carbonyl content	22.76
Specific gravity at 17°C	0.9649
Optical rotation at 17°C	20° 20°
Refractive index at 17°C	1.5078
Solubility in alcohol	0.15

accepted physico-chemical properties are available (Table-1). It has been observed that the Indian oil is dextro-rolatory whereas the Japanese oil is laevo-rotatory. The extracts of oil have been found to be the source of many sesquiterpenes. The major constituents are d-nardostachone and valeronone whereas minor constituents are - α-pinene, β-pinene, limonene, 1,4-cineol, linallool, α-gurjunene, β-gurjunene, t-pinocarveol, β-bisabolene, α-curcumene, β-ionone, humulene oxide, iso-valeric ester of alcohol and jatamansi acid. GC-MS (Gas Chromatography-Mass Spectrometry) analysis of oil has shown the occurrence of patchoulene, aromadendrene and δ-cadinene in addition to jatamansone and d-nardostachone. Besides these, aristolene, 8-hydroxypinoresinol, jatamol-A, -B, lomatin, maaliol, nordol, pinoresinol, selinidin and spirojatamol have also been reported.

The spikenard oil possesses fungitoxic (Misra *et al.*, 1995), nematicidal (Saxena *et al.*, 1987) and antiarrhythmic (Arora and Madan, 1956) activity. The oil exerts a hypotensive effect (Arora *et al.*, 1958) and in moderate doses, it has a distinct depressant action on the central nervous system (CNS), however, heavy (excessive) doses cause deep narcosis and ultimately, death within a few hours.

Cultivation :

The plant is propagated by cuttings of underground parts and occasionally also by seeds. The rhizomes are collected from wild plants. Collection of the plant in the wild is highly dependent on snow-fall. Early snow-fall and late spring snow-falls generally affect rhizome collection. Rhizomes may be harvested from wild and cultivated plants from June to October. However, at higher altitudes, the possible collection in the wild frequently lasts for only a few days due to heavy rains and snow-storms.

The domestication from wild sources are being tested and the details of cultivation and actual economics are currently being worked out.

Medicinal plants are used in medicine manufacturing units. Over 70% of the plant collections involve destructive harvesting because of the use of parts like roots, bark, wood, stem and the whole plant in case of herbs. This poses a definite threat to the genetic stocks as also to the diversity of medicinal plants. It is imperative that industries based on plants for obtaining natural product can not sustain only on wild plants, as a regular supply of uniform material can not be available in such circumstances. Therefore, it is necessary to bring the plant under cultivation for the production of biomass which can be achieved through conventional and biotechnological methods.

Thus, the medicinally important plants require more exhaustive studies particularly on proper evaluation of their phytochemical and pharmacological properties based on traditional medicine and ethnobiological information. Furthermore, there is apt need for the development of technology for micro-propagation, reproductive biology and agricultural practices towards domestication of the selected materials.

REFERENCES

Abraham, V. (1998). Biodiversity at Bhabha Atomic Research Centre, Trombay, p. 119.

Ahmad, R., Sheikh, T.U., Ahmad, A. and Ahmad, M. (1993). Medicinal importance of essential oils. *Hamdard Medicus,* **36** : 101-115.

Amatya, G. and Staphit, V.M. (1994). A note on Nardostachys jatamansi. *Journal of Herbs, Spices and Aromatic Plants,* **2** : 39-47.

Arora, R.B. and Madan, B.R. (1956). Antiarrhythmics III. Antiarrhythmic activity of Nardostachys jatamansi (an Indian indigenous drug). *Indian J. Med. Res.,* **44** : 259-269.

Arora, R.B., Singh, K.P., Das, P.K. and Mistry, P.N. (1958). Prolonged hypotensive effect of the essential oil of Nardostachys jatamansi. *Arch. Inter. Pharmacodynamic.,* **113** : 367-376.

Arora, D.K., Suri, S.S., Ramawat, K.G. and Merìllon, J.M. (1999). Factors affecting somatic embryogenesis in long term callus cultures of 'safed musli' (*Chlorophytum borivilianum*), an endangered wonder herb. *Indian J. Exp. Biol.,* **37** : 75-82.

Bagchi, A., Oshima, Y. and Hikino, H. (1988a). Nardostachyn, an iridoid of Nardostachys chinensis. *Planta Medica,* **54** : 87-88.

Bagchi, A., Oshima, Y. and Hikino, H. (1988b). Jatamols A and B : sesquiterpenoids of Nardostachys jatamansi roots. *Planta Medica,* **57** : 282-283.

Bagchi, A., Oshima, Y. and Hikino, H. (1988c). Kanshones A and B, sesquiterpenoids of Nardostachys chinensis. *Phytochemistry,* **27** : 1199-1201.

Bagchi, A., Oshima, Y. and Hikino, H. (1988d). Kanshone C, a sesquiterpenoids of Nardostachys chinensis roots. *Phytochemistry,* **27** : 2877-2879.

Bagchi, A., Oshima, Y. and Hikino, H. (1988e). Kanshones D and E, sesquiterpenoids of Nardostachys chinensis roots. *Phytochemistry,* **27** : 3667-3669.

Bagchi, A., Oshima, Y. and Hikino, H. (1990). Spirojatamol, a new skeletal sesquiterpenoids of Nardostachys chinensis roots. *Tetrahedron,* **46** : 1523-1530.

Bagchi, A., Oshima, Y. and Hikino, H. (1991). Neolignans and lignans of Nardostachys chinensis roots. *Planta Medica,* **57** : 96-97.

Balansand, J. and Rayband, M. (1987). Diuretic action of Asparagus officinalis. *Critical Reviews in Society of Biology,* **126** : 954-956.

Baldwin, J.T. Jr. and Speese, B.M. (1951). Cytogeography of Chlorophytum in Liberia. *Amer. J. Bot.,* **38** : 153.

Basu, N., Rastogi, R.P. and Dhar, M.L. (1967). Chemical examination of Bacopa monniera Wettst. Part III. Bacoside B. *Indian J. Chem.,* **5** : 84-86.

Basu, N.K. and Walia, J.S. (1944). The chemical investigation of the leaves of Herpestis monniera. *Indian J. Pharm.,* **6** : 84-85, 91.

Basu, N.K. and Pabrai, P.R. (1947). Investigations on Indian medicinal plants. Monniera cuneifolia. *Quart. J. Pharm. Pharmacol.,* **20** : 137-139.

Bhakuni, D.S., Dhar, M.L., Dhar, M.M., Dhawan, B.N. and Mehrotra, B.N. (1969). Screening of Indian plants for biological activity : Part-II. *Indian J. Exp. Biol.,* **7** : 250-262.

Bordia, P.C. (1993). Development of suitable culture of safed musli (Chlorophytum species) for root yield and acceptable medicinal quality, 1988-93. Final Technical Report, ICAR Project, RCA, Udaipur.

Bose, K.C. and Bose, N.K. (1931). J. Indian Med. Assoc. 1 : 60 — cited in Sastry *et al.* (1959) and Rastogi and Dhar (1960).

Buchi, G., Greuter, F. and Tokoroyama, T. (1962). Terpenes XVIII. Structure of calarene and stereochemistry of aristolone. *Tetrahedron. Lett.,* **18** : 827-833.

Chadha, Y.R. (1985). The Wealth of India. Publications and Information Directorate, New Delhi, Vol. 1, p. 468.

Chakravarty, A.K., Garai, S., Masuda, K., Nakane, T. and Kawahara, N. (2003). Bacopasides III-V : Three new triterpenoid glycosides from Bacopa monniera. *Chem. Pharm. Bull.,* **51(2)** : 215-217.

Chakravarty, A.K., Sarkar, T., Masuda, K., Shiojima, K., Nakane, T. and Kawahara, N. (2001). Bacopaside I and II : Two pseudojujubogenin glycosides from Bacopa monniera. *Phytochemistry,* **58(4)** : 553-556.

Chandel, R.S., Kulshreshtha, D.K. and Rastogi, R.P. (1977). Bacogenin-A_3 : A new sapogenin from Bacopa monniera. *Phytochemistry,* **16** : 141-143.

Chatterji, N., Rastogi, R.P. and Dhar, M.L. (1963). Chemical examination of Bacopa monniera Wettst. : Part-I — Isolation of chemical constituents. *Indian J. Chem.,* **1** : 212-215.

Chopra, I.C., Handa, K.L. and Sobti, S.N. (1956a). Need for the cultivation of vegetable drugs used in Ayurvedic and Unani medicines. *Indian J. Pharm.,* **18** : 364-377.

Choudhary, G.R., Sharma, V.N. and Salimuzzaman Siddiqui (1951). Chemical constituents of Nardostachys jatamansi I 'isolation of a crystalline acid and an essential oil. *Jour. Sci. Ind. Res.,* **10B** : 48.

Chowdhuri, D.K., Parmar, D., Kakkar, P., Shukla, R., Seth, P.K. and Srimal, R.C. (2002). Antistress effects of bacosides of Bacopa monniera : Modulation of Hsp70 expression, superoxide dismutase and cytochrome P450 activity in rat brain. *Phytother. Res.,* **16(7)** : 639-645.

Das, A., Shanker, G., Nath, C., Pal, R., Singh, S. and Singh, H. (2002). A comparative study in rodents of standardized extracts of Bacopa monniera and Ginkgo biloba : Anticholinesterase and cognitive enhancing activities. *Pharmacol. Biochem. Behav.,* **73(4)** : 893-900.

Dixit, V., Chaturvedi, R.V. and Tripathy, S.C. (1992). Evaluation of some essential oils against pulse beetle (Callosobruchus chinensis). *Nat. Acad. Sci. Lett.,* **15** : 255-257.

Edenharder, R. (1990). Antimutagenic activity of vegetable and fruit extracts against in-vitro benzo(a)pyrene. Zeitschrift fur Gesamte. *Hygiene,* **36** : 144-148.

Elangovan, V., Govindasamy, S., Ramamoorthy, N. and Balasubramanian, K. (1995). In vitro studies on the anticancer activity of Bacopa monniera. *Fitoterapia,* **66(3)** : 211-215.

Garai, S., Mahato, S.B., Ohtani, K. and Yamasaki, K. (1996a). Dammarane-type triterpenoid saponins from Bacopa monniera. *Phytochemistry,* **42(3)** : 815-820.

Garai, S., Mahato, S.B., Ohtani, K. and Yamasaki, K. (1996b). Bacopasaponin D — A pseudojujubogenin glycoside from Bacopa monniera. *Phytochemistry,* **43(2)** : 447-449.

Girgune, J.B., Jain, N.K. and Garg, B.D. (1978). Antifungal activity of some essential oils-II. *Current Science,* **47** : 454-455.

Girgune, J.B., Jain, N.K. and Garg, B.D. (1978). Antifungal activity of some essential oils-II. *Indian Drugs,* **16** : 24.

Govindachari, T.R., Rajadurai, S. and Pai, B.R. (1958). Structure of jatamansone. *Chem. Ber.,* **91** : 56-58.

Gupta, R., Gupta, O.C.D., Gupta, P.C. and Panda, C.S. (1979). A new galactoglucan from the fruits of Chlorophytum arundinaceum. *Planta Med.,* **37(1)** : 94-95.

Hamied, K.A., Bakshi, V.M. and Aghara, L.P. (1962). Pharmacological investigation of Nardostachys jatamansi. *Jour. Sci. Ind. Res.,* **21C** : 100-103.

Han Dong Sun, Ding, Jing-Kai, Zhong-Wen and Che Fang Rong (1980). Study on the chemical constituents of the essential oil of Nardostachys grandiflora DC and N. chinensis Batalin and their uses on the perfume. *Yun nanchih Wu Yen Chin,* **2** : 213-223.

Hikino, H., Hikino, Y., Koakutsu, S. and Takemoto, T. (1972). Structure and absolute configuration of Narchinol A. *Phytochemistry,* **11** : 2097-2099.

Hirose, Y., Yonemitsu, R. and Sonada, T. (1978). Chemical studies on the components of Nardostachys jatamansi de Candolle. *Shoyakugaku Zasshi,* **32** : 121-122.

Horster, H., Rucker, G. and Tautges, J. (1977). Valeranone content in the roots of Nardostachys jatamansi and Valeriana officinalis. *Phytochemistry,* **16** : 1070-1071.

Hou, C.C., Lin, S.J. and Hsu, F.L. (2002). Bacopaside III, bacopasaponin G and bacopasides A, B and C from Bacopa monniera. *J. Nat. Prod.,* **65(12)** : 1759-1763.

Jain, S.K. (1991). The Dictionary of Indian Folk Medicines and Ethnobotany. Deep Pub., New Delhi.

Jain, P. and Kulshreshtha, D.K. (1993). Bacoside A_1, a minor saponin from Bacopa monniera. *Phytochemistry,* **33(2)** : 449-451.

Kar, D.K. and Sen, S. (1984). Organogenesis of diploid Asparagus racemosus through callus culture. Proceedings of the Indian National Science Academy, Part-B, Biological Sciences, **50** : 113-118.

Kawai, K., Iitaka, Y., Shibata, S., Kulshreshtha, D.K. and Rastogi, R.P. (1973). Crystal and molecular structure of bacogenin-A_1 dibromoacetate. Acta Crystallogr. Sect. B. (Pt. 12), **29** : 2947-2953.

Kihara, H. (1955). Fauna and flora of Nepal Himalayas. Scientific result of the Japanese expedition to Nepal Himalayas 1952-53. Fauna and Flora Research Society, Kyoto University, Kyoto, Japan, pp. 235-236.

Kitamura, S. (1954). New species from Nepal collected by Sasuke nakao, a member of the Himalayan expedition III. *Acta Phytotax. et Geobot.,* **15** : 129-134.

Klyne, W., Bhattacharya, S.C., Paknikar, S.K., Narayan, C.S., Kulkarni, K.S., Krepinsky, J., Romanuk, M., Herout, V. and Sorm, F. (1964). The absolute stereo chemistry of the sesquiterpene ketone valeranone. *Tetrahedron Lett.,* **23** : 1443-1448.

Kulshreshtha, D.K. and Rastogi, R.P. (1973a). Chemical examination of Bacopa monniera. IV. Bacogenin-A_1; A novel dammarane triterpene sapogenin from Bacopa monniera. *Phytochemistry,* **12** : 887-892.

Kulshreshtha, D.K. and Rastogi, R.P. (1973b). Identification of ebelin lactone from bacoside A and the nature of its genuine sapogenin. *Phytochemistry,* **12** : 2074-2076.

Kulshreshtha, D.K. and Rastogi, R.P. (1974). Bacogenin A_2 : A new sapogenin from bacosides. *Phytochemistry,* **13** : 1205-1206.

Li, X.C., Wang, D.Z. and Yang, C.R. (1990). Steroidal saponins from Chlorophytum malayense. *Phytochemistry,* **29(12)** : 3893-3898.

Mahato, S.B., Garai, S. and Chakravarty, A.K. (2000). Bacopasaponins E and F : two jujubogenin bisdesmosides from Bacopa monniera. *Phytochemistry,* **53** : 711-714.

Marais, W. and Reilly, J. (1978). Chlorophytum and its related genera (Liliaceae). *Kew Bull.,* **32(3)** : 653-664.

Mimaki, Y., Kanmoto, T., Sashida, Y., Nishino, A., Satomi, Y. and Nishino, H. (1996). Steroidal saponins from the underground parts of Chlorophytum comosum and their inhibitory activity on tumor promoter-induced phospholipids metabolism of HeLa cells. *Phytochem.,* **41** : 1405-1410.

Mishra, D., Chaturvedi, R.V. and Tripathy, S.C. (1995). The fungitoxic effect of the essential oil of the herb Nardostachys jatamansi DC. *Tropical Agriculture,* **72** : 48-52.

Mitsuhashi, H. and Itoh, T. (1962). Constituents of Umbelliferae plants VI. Constituents of Angelica edulis. *Chem. Pharm. Bull. (Japan),* **10** : 514-518.

Mukerjee, B. (1961). Pharmacology and chemistry of Nardostachys jatamansi *Pakistan Jour. Sci. Ind. Res.,* **4** : 277-279.

Nadkarni, A.K. (1954). The Indian Materia Medica, New Delhi. Vol. I, 840 pp.

Nadkarni, A.K. (1976). Indian Materia Medica, Popular Prakashan, Mumbai, p. 151.

Nayar, S.L. (1964). Medicinal plants of commercial importance found wild in Uttar Pradesh and their distribution. *Jour. Bombay Nat. Hist. Soc.,* **61** : 651-661.

Nayar, M.P. and Shastri, A.R.K. (1988). Chlorophytum borivilianum. In : Red Data Book of Indian Plants, Vol. 2, (eds) M.P. Nayar and A.R.K. Shastri, Botanical Survey of India, pp. 142.

Ojha, A. (1987). Botanical Study of safed musli (Aphrodisiac effects). MD Thesis, MMR Ayurvedic College, Udaipur (India).

Pai, P.P., Chakravarti, K.K. and Paknikar, S.K. (1971). Revised structure for nardostachone : identify with 1,8,9,10-tetrahydroaristlane-2-one. *Indian Jour. Chemistry,* **9** : 729.

Prakash, J.C. and Sirsi, M. (1962). Comparative study of the effects of Brahmi (Bacopa monniera) and chlorpromazine in motor learning in rats. *J. Sci. Ind. Res.,* **21C** : 93-96.

Proliac, A., Chabaud, A. and Raynaud, J. (1991). Pharm. Acta Helv. 66 : 153 — cited in Anonymous 1998, Indian Herbal Pharmacopoeia. Vol. I.

Qiu, S.X., Li, X.C., Xiong, Y., Dong, Y., Chai, H., Farnsworth, N.R., Pezzuto, J.M. and Fong, H.H.S. (2000). Isolation and characterization of cytotoxic saponin chloromaloside A from Chlorophytum malayense. *Planta Med.,* **66** : 587-590.

Rao, G.M.A. and Karanth, K.S. (1992). Neuropharmacological activity of Herpestis monniera. *Fitoterapia,* **63** : 399-404.

Rastogi, S., Pal, R. and Kulshreshtha, D.K. (1994). Bacoside A_3 — A triterpenoid saponin from Bacopa monniera. *Phytochemistry,* **36(1)** : 133-137.

Rastogi, S. and Kulshreshtha, D.K. (1998). Bacoside A_2 — A triterpenoid saponin from Bacopa monniera. *Indian J. Chem.,* **38B** : 353-356.

Rastogi, R.P. and Dhar, M.L. (1960). Chemical examination of Bacopa monniera Wettst. *J. Sci. Ind. Res.,* **19B** : 455-456.

Rastogi, R.P. and Mehrotra, B.N. (1998). Compendium of Indian Medicinal Plants. Vols. I to IV, CDRI and NISCOM, New Delhi.

Rucker, G. and Kretzschmar, U. (1971). 9-aristolene-1-α-ol and 1,2,9,10-tetradehydroaristolene — a new aristolane type sesquiterpene. *Liebigs Ann. Chem.,* **748** : 214.

Rucker, G. (1971). 9-hydroxy-1(10) aristolene-2-one (debilon) from Nardostachys chinenesis. *Planta Medica,* **19** : 16-18.

Rucker, G. and Kretzeschamar, U. (1972). Patchouli alcohol and patchoulen aus Nardostachys jatamansi. *Planta Medica,* **21** : 1-4.

Rucker, G. and Dyck Eberhard (1974). Formation of deoxynarchinol A from Nardosinone. *Phytochemistry,* **13** : 1907-1909.

Rucker, G., Tautges, J., Maheshwari, M.L. and Saxena, D.B. (1976). Norseychelanone, α and β patchoulenes and patchouli alcohol from Nardostachys jatamansi. *Phytochemistry,* **15** : 224.

Rucker, G., Tautges, J., Sieck, A., Wenzel, H. and Graf, E. (1978). Studies on isolation and pharmacodynamic activity of the sesquiterpene valeranone from Nardostachys jatamansi DC. *Arzneim Forsch,* **28** : 7-13.

Sairam, K., Priyambada, S., Aryya, N.C. and Goel, R.K. (2003). Gastroduodenal ulcer protective activity of Asparagus racemosus; an experimental, biochemical and biological study. *J. Ethnopharmacol.,* **86** : 1-10.

Santapau, H. and Fernandes, R.R. (1955). A new species of Chlorophytum from Salsette island. *J. Bombay Nat. Hist. Soc.,* **52** : 897.

Sastry, M.S., Dhalla, N.S. and Malhotra, C.L. (1959). Chemical investigation of Herpestis monniera Linn. (Brahmi). *Indian J. Pharm.,* **21** : 303-304.

Sastry, S.D., Maheshwari, M.L. and Bhattacharya, S.C. (1966). Terpenoids LXXXVII. The structure of nardol. Tetrahedron Lett., pp. 1035-1042.

Sastry, S.D., Maheshwari, M.L., Chakravarti, K.K. and Bhattacharya, S.C. (1967a). Terpenoids CXV. Chemical constituents of Nardostachys jatamansi. *Perfum. Essent. Oil. Rec.,* **58** : 154-158.

Sastry, S.D., Maheshwari, M.L., Chakravarti, K.K. and Bhattacharya, S.C. (1967b). Terpenoids. CVI. Structure of calarenol. *Tetrahedron,* **23** : 1997-2000.

Sastry, S.D., Maheshwari, M.L., Chakravarti, K.K. and Bhattacharya, S.C. (1967c). Terpenoid. CVII. The structure of nardostachone. *Tetrahedron,* **23** : 2491-2493.

Satyawati, G.V., Gupta, A.K. and Tandon, N. (1987). Medicinal Plants of India. Vol. II. 312 pp. Indian Council of Medical Research, New Delhi, India.

Saunders, W.D. and Pinder, A.R. (1977). Further studies relating to structure of nardostachone. Tetrahedron Lett., pp. 1687-1690.

Saxena, D.B., Goswami, B.K. and Tomar, S.S. 1987. Nematicidal activity of some essential oils against Meloidogyne incognita. *Indian Perfumer,* **31** : 150-154.

Schulte, K.E., Ruecker, G. and El-Kersch, M. (1972). Components of medicinal plants. XXXVI. Nicotine and 3-formyl-4-hydroxy-2H-pyran from Herpestis monniera. *Phytochemistry,* **11(8)** : 2649-2651.

Shah, N.C. (1994). Cultural and religious aspects of jatamansi, Nardostachys grandiflora DC. in the Kumaon Himalayas, India. In : IV International Congress of Ethnobiology : Ethnobiology in Human Welfare, Lucknow, India, 17-21 Nov.

Shanbhag, S.N., Mehta, C.K., Maheshwari, M.L. and Bhattacharya, S.C. (1965). Terpenoids. LXXXV. Constituents of Nardostachys jatamansi and synthesis of (±) dihydrosemidin and visnadin from jatamansin. *Tetrahedron,* **21** : 3591-3597.

Shao, Y., Chin, C.K., Ho, C.T., Ma, W., Garrison, S.A. and Huang, M.T. (1996). Anti-tumour activity of the crude saponins obtained from asparagus. *Can. Lett.,* **104** : 31-36.

Sharma, S., Ramji, S., Kumari, S. and Bapna, J.S. (1996). Randomized controlled trial of Asparagus racemosus (Shatavari) as a lactogogue in lactational inadequacy. *Indian Pediatrics,* **32** : 675-677.

Sharma, R. and Ramawat, K.G. (2000). Stilbenes : bioactive plant phenolics. In : Role of Biotechnology in Medicinal and Aromatic Plants. III. (eds) Khan and A. Khanum, Ukaz Pub., Hyderabad. pp. 106-115.

Shide, Luo, Mayer, R. and Rucker, G. (1987a). Nardonoxide, a new skeletal sesquiterpenoid of Nardostachys jatamansi. *Tetrahedron,* **46** : 1523-1530.

Shide, Luo, Olbrich, A., Mayer, R. and Rucker, G. (1987b). Gansongone — a new aristolane ketone from Nardostachys chinensis and structure revision of an aristolenol. *Planta Medica,* **53** : 556-558.

Shide, Luo, Mayer, R. and Rucker, G. (1987c). Nardonoxide, a new nardosinane type sesquiterpene ether from Nardostachys chinensis. *Planta Medica,* **53** : 332-334.

Shimoyamada, M., Suzuki, M., Sonta, H., Maruyama, M. and Okubo, K. (1990). Antifungal activity of the saponin fraction obtained from asparagus and its active principle. *Agri. and Biol. Chem.,* **54** : 2553-2557.

Singh, H.K., Srimal, R.C., Srivastava, A.K., Garg, N.K. and Dhawan, B.N. (1990). Neuropsychopharmacological effects of bacosides A and B. Proceedings of the Fourth Conference on the Neurobiology of Learning and Memory, California, 17-20 October, 1990. 79 pp.

Sinha, J., Raay, B., Das, N., Medda, S., Garai, S., Mahato, S.B. and Basu, M.K. (2002). Bacopasaponin C : Critical evaluation of anti-leishmanial properties in various delivery modes. *Drug Delivery,* **9** : 55-62.

Sumathi, T., Subramanian, S., Govindasamy, S., Balakrishna, K. and Veluchamy, G. (2001). Protective role of Bacopa monniera on morphine induced hepatotoxicity in rats. *Phytother Res.,* **15(7)** : 643-645.

Takaya, Y., Kurumada, K., Takeuji, Kim, H.S., Shibata, Y., Ikemoto, N., Watsya, Y. and Oshima, Y. (1998). Novel antmalarial quaiane type sesquiterpenoids from Nardostachys chinensis roots. *Tetrahedron Lett.,* **39** : 1361-1364.

Takaya, Y, Takeuji, Y., Akasaka, M., Nakagawasai, O., Tadeno, T., Kisara, K., Kim, H.S., Wataya, Y., Niwa, M. and Oshima, Y. (2000a). Novel guaiane endoperoxides, Nardoguaianone A-D, from Nardostachys chinensis roots and their antinociceptive and antimalarial activities. *Tetrahedran,* **56** : 7673-7678.

Takaya, Y., Akasaka, M., Takeuji, Y., Tanitsu, M., Niwa, M. and Oshima, X. (2000b). Novel guaianoids, Nardoguaianone E-I, from Nardostachys chinensis roots. *Tetrahedran,* **56** : 7679-7683.

Tandon, M. and Shukla, Y.N. (1992). Sapogenins from Asparagus adscendens and Chlorophytum arundinaceum. *J. Indian Chem. Soc.,* **69** : 893.

Tandon, M. and Shukla, Y.N. (1993). A bibenzyl xyloside from Chlorophytum arundinaceum. *Phytochem.,* **32** : 1624-1625.

Tandon, M., Shukla, Y.N. and Thakur, R.S. (1992). 4-hydroxy-8,11-oxidohemicosanol and other constituents from Chlorophytum arundinaceum roots. *Phytochem.,* **31(7)** : 2525-2526.

Tanitsu, M., Takaya, Y., Akasaka, M., Niwa, M. and Oshima, Y. (2002). Guaine- and aristolane type sesquiterpenoids of Nardostachys chinensis roots. *Phytochemistry,* **59** : 845-849.

Tiwari, K.C. and Joshi, H.C. (1974). Economics of medicinal herbs cultivation in hills of Kumaon and Uttarkhand. *J. Res. Ind. Med.,* **9** : 73-78.

Tripathi, Y.B., Chaurasia, S., Tripathi, E., Upadhyay, A. and Dubey, G.P. (1996). Bacopa monniera Linn. as an antioxidant : mechanism of action. *Indian J. Exp. Biol.,* **34(6)** : 523-526.

Vrkoc, I., Krepinsky, J., Herout, V. and Sorm, F. (1963). The nature of sesquiterpenic hydrocarbon calarene and structure of b-gurjunene. *Tetrahedron Lett.,* pp. 225.

Wang, Y., Yang, H.Y., Wang, J.J. and Yang, C.R. (1996). Inhibitory effect of some steroidal saponins on human spermatozoa *in vitro. Plant Med.,* **62** : 130-132.

Weberling, F. (1975). On the systematics of Nardostachys (Valerianaceae). *Taxon,* **24** : 443-452.

Recent Advances in Medicinal Plant Research : Vision 21st Century, 2005
Editors : S.K. Prabhuji, G.P. Rao and S.K. Patil
Satish Publication House, New Delhi

Chapter 3

NUTRACEUTICALS : AN OVERVIEW

K.S. Patil[1], A.G. Jadhav[1] and S.C. Chaturvedi[2]
[1]*Department of Pharmacognosy and Phytochemistry, K.L.E.S. College of Pharmacy, J.N.M.C. Campus, Nehru Nagar, Belgaum - 590 010*
[2]*Shri G.S.I.T.S., Department of Pharmacy, Indore*

Nutraceuticals are foods, food ingredients or dietary supplements that demonstrate specific health or medical benefits, including the prevention and treatment of diseases beyond basic nutritional functions. Importance and clinical studies on Glucosamine, Octacosanol, Carnitine, Flaxseed and flaxseed oil, Proanthocyanidins and grape-seed extract have been discussed.

INTRODUCTION

The fast-paced technological developments in every domain led to an improvement in the quality of life. The ever-increasing cost of high-tech, intensive medical care and the awareness among the laymen to the negative effects of drugs has resulted in a paradigm shift towards alternative, but, age-old traditional systems of medicine. These systems are growing and mainstream health professionals are reposing faith in the "holistic" approach that these systems provide/take into account not only a person's physiological condition but also psychological, social, environmental and spiritual factors that contribute to illness. The complementary systems include the seemingly mundane acupuncture, naturopathy, and aromatherapy to the more exotic sounding biodynamic massage, orthomolecular therapy, yoga, arid and geomancy.

Another emerging area in this holistic approach is that of Designer foods; also known as Nutraceuticals or Functional foods.

The term Nutraceuticals or functional foods are terms describing foods, food ingredients or dietary supplements that demonstrate specific health or medical benefits, including the prevention and treatment of diseases beyond basic nutritional functions. These products include whole foods, food additives, vitamins and mineral supplements, herbs, phytonutrients and probiotics.

Stephen De Felice, coined the term 'nutraceuticals' in 1979. Dietary supplements are any product (other than tobacco) that contains vitamins, minerals, amino acids of herbs that is intended as supplement to the normal diet' (As per Dietary Supplement and Health Education Act).

NUTRACEUTICAL – FOOD OR MEDICINE?

Food labeling regulations do not allow food labels to carry health claims in many countries. This makes it hard for companies marketing nutraceuticals to advertise the benefits of other products without a medical licence. They may decide either not to do any research or they may research a new product thoroughly and possibly obtain a patent. Unfortunately, most of the companies tend towards the former option due to either expenses, the problems obtaining a patent on a natural product or the fact that they cannot put their claims on the label without a product licence, whether verified or not.

An example was the case of 'cholestin'. A cholesterol-lowering supplement marketed by Pharmanex (Simi Valley, CA) in 1996 (based on a Chinese remedy). E. Merck found the active ingredient to be 'lovastatin' and US FDA subsequently banned its incorporation in Cholestin. As can be seen from this example pharmaceutical and nutraceutical are sometimes difficult to define depending only on regulatory issues.

In Britain

Medical product is the 'substance or combination of substances presented for treating or preventing diseases in human beings or animals.' It further goes on adding ...... 'Any substance or combination administered to human beings or animals with a view to make diagnosis or restoring, correcting, modifying physiological functions in human beings/animals'. From this definition, nutraceuticals are medicines.

In USA

Unlike foods, dietary supplements may carry nutritional support statement provided the label should contain statutory warning that 'this product is not intended to diagnose, treat, cure or prevent any disease'.

In Canada

Nutraceutical is a "Product produced from foods but sold in pills, powders and other medicinal not generally associated with food".

TYPES OF NUTRACEUTICALS

- Dietary supplements and vitamin products.
- Herbal and botanical products.
- Foods for vitality, functional food, medical foods.
- Health food, organic food, dietetic food.
- Sport and energy products.
- Natural cosmetics and body treatment products.
- Natural medicinal products with specific health benefits.
- Nutraceutical antioxidants.
- Vitamins (Vitamin C and Vitamin E)
- Carotenoids (Beta Carotene and Lycopene)
- Thiols (Alpha Lipoic Acid and Glutathione)
- Enzymes (Co-enzyme Q-10 and Superoxide)
- Dismastase (SOD)
- Minerals (Selenium, Chromium) etc.

These cover a gamut of products ranging from engineered grains to simple candy bars or cereals supplemented with vitamins or minerals and even beverages. In Japan, a vitamin C enhanced cola is enjoying wide popularity. Tomato and its products (e.g. ketchup, juices, soups and sauces) are the principal source of lycopene and β-carotene, which have been shown to be beneficial to health, especially in reducing chronic conditions such as coronary heart disease, certain cancers and macular degeneration. An effort was made to enhance the carotenoid content using bacterial carotenoid gene, which was introduced into tomato plants by Agro bacterium-mediated transformation. High protein and low calorie-mushroom have turned out to be a boon to diabetics.

(A) GLUCOSAMINE

Glucosamine is 2-amino-2deoxy-d-Glucose. It can be isolated from its HCl salt by treatment with strong bases like diethlamine or triethylamine (to give free D-glucosamine) [Alternative to this HCl salt is obtained by hydrolysis of chitin which is a polyglucosamine].

It is found naturally in body especially in cartilage, tendons & ligament tissues. It is an essential part of manufacture of glycoaminoglycon (GAG) which forms the most of the cartilage tissue. It is not found in significant amount in diet & must be synthesized in the body. This ability declines with age & predispose the body to arthritis. *Chondroprotection* is the term used when GAGs such as glucosamine are given to protect the joints.

Clinical findings

Clinical studies in Italy & Thailand showed significant pain relief in arthritic knee.

How does it work?

N-acetyl glucosamine inhibits enzyme elastase in a dose dependent manner. Elastase plays an important role in the breakdown of articular cartilage ligaments, tendons and bones in rheumatoid arthritis. It is the building block for the synthesis of GAGs and it offers protection against joint degeneration as well as offering joint protection.

Side effect and contraindications

Endogenous glucosamine is also a product of hexosamine pathway, which is involved in insulin response. Experiments have shown that glucosamine can induce insulin resistance via this pathway. This can cause glucose toxicity, which worsens the diabetic state and makes regulation more difficult and leads to defect in insulin secretion seen in NIDDM in contrast to normal state.

(B) OCTACOSANOL

It is an increasingly popular nutraceutical often used by athletes. It is 28 carbon chain alcohol which is present in the superficial layers of fruit leaves and skin of many plants as well as whole grains. In most of the studies wheat germ oil extract or policosanol [policosanol is natural mixture of primary alcohols purified from sugar cane wax, whose main component is octaconsanol] is used.

Clinical findings

Policosanol, in the doses of 5/50/200 mg/kg in rabbits and 10 mg and 20 mg twice a day in human beings, significantly reduced total cholesterol and LDL – cholesterol in a dose dependent manner. However, HDL-C was unchanged. Both of these studies suggest that it is safe for long duration of treatment of lipid metabolism and cholesterol lowering.

(C) CARNITINE

Carnitine is an essential cellular component, which is sometimes referred to as Vita. Bt although generally not considered as vitamin. It is an amino acid derivative synthesized from lysine and methionine in the liver and kidney from which it is released in the systemic circulation. It is also synthesized in brain. Of the total carnitine found in the human body, 98% is found in cardiac and skeletal muscles. Animal origin food also contains carnitine. It exists in two isomeric forms D and L of which L – form is naturally occurring. The D and L form are often sold as nutritional supplement, however, D form is claimed to be less safe than L form.

Function

The main function of carnitine is in fatty acid metabolism. The biochemical reactions of this nutrient are based on the reaction between carnitine and acyl groups (carnitine + Acyl CoA $\leftrightarrows$ acylcarnitine + coenzyme A) and carnitine is therefore involved in many CoA dependent pathways. It is involved in long chain fatty acid oxidation at the mitochondrial level, providing energy. Carnitine acts as a carrier of acyl and acetyl groups across the mitochondrial membrane before beta-oxidation may provide energy. Indeed, this is the main energy source in the skeletal or cardiac muscles.

It has other metabolic functions too such as branched chain alpha-keto acid oxidation and detoxification of potentially toxic acyl-CoA metabolites from other pathways.

Clinical findings

Carnitine is an increasingly popular supplement, used mainly by athletes to help improve performances. However, other individuals may get benefit, including patients with heart disease and those on haemodialysis. In addition, it shows significant and progressive improvement in cardiac function and quality of life in case of exercise induced stable angina patients (treated with 2g/day orally for 6 months). Other findings also suggest beneficial effect in Congestive Heart Failure (CHF) (2g twice a day) and cardiogenic shock [a state of severe tissue hypo perfusion resulting from underlying pump dysfunction] with marked increase in survival time.

(D) FLAXSEED AND FLAXSEED OIL

Flaxseed or linseed is a good source of alpha-linolenic acid, which is converted to long chain omega-3-fatty acids. Like fish oil it may help to correct deficiency of these fatty acids and could be useful for those wishing to supplement their diets with a plant product.

Flaxseed oil is extracted from dried ripe seeds of *Linum usitatissimum,* family Linaceae.

Chemistry

1. CH_3-CH_2-CH=CH-CH_2-CH=CH-CH_2-CH=CH- $(CH_2)_7$-COOH
   Linolenic acid
2. CH_3- $(CH_2)_4$-CH=CH-CH_2-CH=CH-$(CH_2)_7$-COOH
   Linoleic acid

Flaxseed contains more than 50% alpha-linolenic acids (ALA), one of the two essential fatty acids, which is not naturally found in the human body. In ALA, there are 18 carbon atoms with three unsaturated bonds, starting at third carbon atom which is described by the notation 18:3 W3 (or n-3) Polyunsaturated fatty acid (PUFA). The other essential fatty acid is linoleic acid, which is an omega-6 fatty acid.

The modern western diet contains much greater quantity of omega-6 than omega-3-fatty acids in a ratio of 20-30 to one, whereas the ratio should be almost equal. Several problems related with coronary health have been attributed to this imbalance. Moreover, because of modern industrial and agricultural practice, the omega-3 content of many foods including meat, fish, eggs and vegetable, is much lower than the recommended amount. As a result, many people are deficient in omega-3 ALA that may lead to degenerative diseases.

Both omega-6 and omega-3-fatty acids are precursors of eicosanoides such as prostaglandins, thromboxanes and leukotriens. Those eicosanoides derived from omega-6-fatty acids have opposing properties to those derived from omega-3s and it is therefore, important to have a balance between the two.

Long chain omega-3-fatty acid production from ALA depends on how much omega-6-fatty acid is already present and it proceeds less efficiently when there is an excess of omega-6s. The production of these long chain fatty acids from ALA is, therefore, more significant over a long period. One advantage of ALA over fish oil is that the problem of inadequate vitamin E intake does not occur when plant sources are used.

Clinical findings

1. Whole flaxseed is suggested superior to purified oil as a beneficial supplement in the prevention of cardiovascular problems, including heart attacks and the snack product is good for diabetic population.

2. The flaxseed supplement diet suppressed the immunity from T-cells (Cell mediated immunity) but not from B-cells (humoral immunity). This may have been due to the increase in amount of fat in the diet as well as the ALA supplement. Immunosuppresion could be a health risk if experienced over long periods, but it could also be beneficial for patients with autoimmune disease, such as arthritis.

3. Flaxseed is a good source, not only of ALA but also of lignans. Lignans are one of the main classes of phytoestrogens, the other being isoflavones, which are found in legumes and beans, particularly the soyabean. Enzymes in the gastrointestinal tract convert phytoestrogens to phenolic substances resembling female hormone estrogen. Both flaxseed oil and flaxseed have been observed to reduce the growth of mammary tumours. Lignans have also been shown to offer protection from cancer of colon. Lignans have also been found to reduce the proliferation of tumour cells. It is also suggested that flaxseed supplementation may be helpful to treat chronic renal injury in humans.

(E) PROANTHOCYANIDINS AND GRAPE SEED EXTRACT

Proanthocyanidins (PACs) are a subgroup of bioflavonoids (once called vita. 'P') that has been called "super antioxidants". Diets high in fruits and vegetables containing antioxidants offer protection against many diseases including cancer, autoimmune diseases, psoriasis, stroke and AIDS. PACs are found in the leaves, fruits, bark, seeds, flowers and roots of many plants like Apple (*Malus sp.*), Grape (*Vitis vinifera*), Blackberry, Raspberry (*Rubus sp.*), Rose tips (*Rosa sp.*).

Clinical findings

i) *PACs as antioxidants :* A study showed a good concentration – dependent response for Grape seed proanthocyanidin extract (GSPE) in the inhibitation of superoxide anion and hydroxyl radical production. The free radical scavenging activity of GSPE is superior to Vit. C and Vit. E succinate.

ii) *GSPE and Arteriosclerosis [French paradox] :* In a survey, conducted by WHO in 1989, it has been found that the risk factors such as dietary intakes of saturated fats, blood pressure, obesity and serum cholesterol in the French people were similar to Britain and US countrymen but the death from cardiovascular diseases were much lower in France. This 'French paradox' is attributed to 'Red wine' (20-30g alcohol/day). PACs in the red wine were found to be potent antioxidants preventing the oxidation of LDL.

iii) PACs are known to be potent inhibitors of tumour activity. It is about 90% for epicatechin trimer and 56% for dimmers and 35% for monomer in TPA mice model [i.e. 12-0-tetradecanoyl phorbol-13-acetate carcinogen].

iv) *Hair Growth :* GSPE was found to increase hair growth in isolated mouse hair follicle cell cultures by about 23% compared with minoxidil, a medicine for treating male baldness.

v) *Sweeteners :* A novel sweetener from rhizome of fern *Selligues feei* shows similarity with typical PAC structure and lack of toxicity with mice.

SCOPE OF NUTRACEUTICALS

Recognizing the enormous money-spinning potential of nutraceuticals, many food and pharmaceutical industries are diversifying into this area. In the US alone $ 250 billion is attributed to nutraceuticals; that is 50% of the US food market. An equal market is estimated in Europe. In Japan, nutraceuticals constitute a $ 4 billion market. In India, the pharma-industries are fast catching up the potential of this field. Companies like Shanta Biotech and Parry's Health Line have made impressive development in this segment.

However, lack of proper regulations may prove to be its 'waterloo'. There seems to be a lack of coherent and comprehensive regulatory policy that would serve as guidelines for this sector. The major stumbling block lies in the ambiguous nature of definition of nutraceuticals viz., functional food, dietary supplement, medical food or nutritional supplement. Having a finger on the pulse of the market, companies involved in manufacture of nutraceuticals make exaggerated often-unsubstantiated claims regarding the health benefits accrued to their products. The need of the hour is that the regulatory bodies recognize the potentials of this field and set parameters regarding their production, research, marketing and pricing; not with the intention of clipping the wings, but to standardize the sector. The key players should also maintain ethical obligations and not to focus their products only on the crème-de-le-crème of the society. Developing countries like India, where malnutrition is omnipresent even in the 21st century, these nutritionally enhanced foods will act as a boon for our future generations. In the post-GATT scenario, nutraceuticals portend fascinating implications for the pharmaceutical companies and hence the need of the hour is to develop core competency in this emerging area.

DSHEA regulating nutraceuticals (Dietary Supplement and Health Education Act)

It regulates dietary supplements and nutraceuticals. DSHEA can be regarded as a compromise between patient's autonomy and paternalistic oversight by FDA. DSHEA provides consumer access to herbal product but without the consumer protection provided by FDA drug approval process.

Labeling Conditions For Nutraceuticals

"This statement has not been evaluated by food and drug administration. This product is not intended to diagnose, treat, cure or mitigate any disease."

According to one back ruling on March 1994, FDA now requires an information panel titled "Supplement Fact" to appear on the label of a dietary supplement. [Similar to "Nutrition Facts" panel required on foods] include.....

- Herbs common or usual name or in certain cases its botanical name.
- Part of the plant used.
- Manufacturer's suggested serving size.
- Amount per serving.
- These should appear with % of recommended daily value listed.
- For 'Proprietary blends'-total weight must be listed. Weights of individual ingredients do not have to be listed. Each component of the blend must be listed in order of predominance by weight.

DSHEA currently allows herbal labeling to carry claims about product's ability to affect structure and function of human body or to promote general well being.

e.g. **St. John's Wort** may state "helps enhance general mental well being". But it cannot state "Cures depression".

For ***Ginkgo biloba*** product "Improves memory and concentration" but the caption "Cures Alzheimer's disease" would not be allowed.

However, more study is aptly required in this area. For consumers, some structure/function/claim may be difficult to distinguish from the claims of the disease treatment/prevention. e.g. would most consumers understand that "Supports cardiovascular health" is not synonymous with "Prevents coronary artery disease"?

Some Examples of Marketed Nutraceuticals

Following are the few examples of nutraceuticals from US based Fuji Nut nutraceuticals.

Few Selected Nutraceuticals

Common Name	Botanical Source	Major Use
Ashwagandha	*Withania somnifera* roots	Anti anxiety, health tonic adaptogen
Garlic	*Allium cepa* bulb	Hypolipidemic, antiarteriosclerotic
Shilajit	Asphaltum from rocks of Himalayas and Vindhya mountains	Aphrodisiac, delay ageing, improve vitality
Green tea	*Camellia sinensis* leaves	Anti–oxidant restoring general health
Gingko biloba extract	*Ginkgo biloba*	Vertigo, corrects memory loss

1. *Saw Palmetto Powder :* Sawtab 60: It is a directly compressible without stickiness or oil extrusion powder.
2. *Grape seed extract powder :* Polyphenols and catechin derivatives.
3. *Astaxanthin complex :* Asta REAL: Coined to be a powerful antioxidant available as extract powder as well as emulsion product.

Above example shows that various standardized extracts/powders are available in the market which may be formulated in the form of tablets, syrups, capsules, emulsions etc.

REFERENCES

Woodham, Anne and David, Peters (1997). Encyclopeida of Complementary Medicine, Dorling Kindersley Limited.

Kamboj, V. (2000). *J. Curr. Sci.,* **78(1)** : 35-39.

Baichwal, R.S. (1999). *Pharma Times,* 31, Nov : 1920.

Biotech., May-Jun 2000, 14.

Biotech., Mar-Apr 2000, 16.

Biotech., Nov-Dec 2000, 14.

Romer, S. *et al.* (2000). Nat. biotech. 18, June, 666-669.

Nutraceuticals/Functional Foods-An Exploratory Survey on Canada's Potential, Summary Report, Carol Culhane, International Food Focus Limited., Toronto, June 1995.

Brower, V. (1998). *Nat. Biotech.*, 16, : 728-731.

Upadhyay, K. (2000). The ExternPharm, : 31-35.

Tantry, R. (2001). Express Pharma Pulse, 7 : 10.

Iyer, A. (2001). Express Pharma Pulse, 7 : 18.

Goldberg, I. (1994). ed, functional food, designer food, pharma food, nutraceuticals. New York, capman and Hall, 1994.

www.nutraceutical.com

www.cambridgenutra.com

www.Amerianutra.com

Recent Advances in Medicinal Plant Research : Vision 21st Century, 2005
Editors : S.K. Prabhuji, G.P. Rao and S.K. Patil
Satish Publication House, New Delhi

Chapter 4

MEDICINAL PLANTS DIVERSITY AND THEIR CONSERVATION IN WESTERN GHATS OF KARNATAKA

S.K. Patil[1], V. Suryanarayana[2], G.O. Manjunath[1] and V.M. Dabgar[1]
[1] *Dept. of Forest Utilization, College of Forestry, U.A.S.D, Sirsi - 581 401, Karnataka.* e-mail : diforsrs@sancharnet.in
[2]*Dept. of Forest Protection, College of Forestry, U.A.S.D, Sirsi - 581 401, Karnataka*

Western Ghats forest ecosystem is one of the 25-mega bio-diversity hot spots of the world known for its unique ecosystem . The ecosystem spreads over 1.4 lakh sq. km covering five states of peninsular India. Its distribution in Karnataka is amounted to 0.38 lakh Km^2 and of which 57 per cent area is covered forest. In medicinal plants diversity it is exceeding 27 per cent of the total floristic diversity of Western Ghat part of Karnataka and over 25 per cent of the national MP's diversity level. Field observations and other literature studies have indicated that the Western Ghat region of Karnataka has 1139 MP's comprising 312 trees, 458 herbs, 190 shrubs and 179 climbers distributed over 155 botanical families. Over 177 species were ascertained to fall under one or the other Red listed categories. Among them four species, *Cycas circinalis, Helitropium keralense, Pueraria tuberosa* and *Shorea tumbuggaia* falling under critically endangered category and ten species, *Chonemorpha fragrans, Rauvolfia serpentina, Drosera indica, D. peltata, Michelia champaka, Nothopodytes nimmoniana, Persera macrantha, Saraca asoca, Salacia oblonga* and *S. reticulata* falling under Endangered category need to be conserved immediately. Conservation strategies can be framed mainly with *Ex-situ, In-situ* and In-vitro methods. depending upon the nature of the materials. A perfect integrated structuring of administrative, scientific and local farming communities without a gap in the communication systems act as a potential avenue in dealing conservation issues of medicinal plants from all angles. Formulating a national plan for conservation and sustainable use of medicinal plant resources and integrating them into relevant cross-sectorial programmes need to be framed and executed.

INTRODUCTION

Nature is a treasure of diversified plant species possessing multiple usages to mankind. Globally 12 countries have been recognized as mega diversity countries comprising of 25 hot spots (Myers *et al.*, 2000). It is estimated with a wealth of 2.985 lakh diversified species as given in the Table-1.

Though India is positioned at 11th rank in the global floristic diversity, it occupies over 25% (over 8000 species) of the total global medicinal plants (MP's) diversity (32,000 species) indicating the rich medicinal plants reposition. Out of a floristic diversity of 17,500 in India over 45 percent is of medicinal plants treasure (8000 species) stressing a need for elevating the status of medicinal plants as, "National Resource".

The Western Ghats cover five states of peninsular India with total spread over area of 1.4 lakh sq.km. The Western Ghat, part of Karnataka, has an area of 0.38 lakh km^2 (28 % of the total Western Ghat area). Of which 0.22 lakh km^2 (57% of the total area of Western Ghat part of Karnataka) is covered by the forest. This clearly imply to magnitude of richness of biodiversity as indicated in Table-2.

While cataloguing species richness, Western Ghats alone accounts for 5.8 percent of the national floristic diversity and 43 percent of flowering plant status as depicted in Table-3 indicating its main role in national resource status.

Western Ghats possess 3 per cent and 38 per cent of the global and national level floristic diversity respectively. Out of a floristic diversity of 7500 (angiosperms) in Western Ghats, over 25 per cent is of medicinal plants reserve (2000 species). The richness and diversified MP's make the Western Ghats to crown as one of the two mega bio-diversity centers of the

Table - 1 : Mega diversity countries with Angiosperm status

Sl. No.	Country	No. of flowering species
1.	Brazil	55,000
2.	Columbia	35,000
3.	China	30,000
4.	Mexico	25,000
5.	South Africa	23,000
6.	Soviet union	22,000
7.	Indonesia	20,000
8.	Venezuela	20,000
9.	United states	18,000
10.	Ecuador	18,000
11.	India	17,500
12.	Australia	15,000
	Total	2,98,500

Table - 2 : Geographical area and forest cover in different states falling under Western Ghats (in sq.km.)

State	Geographical area (sq.km.)	Forest cover (%)	Area in Western Ghats (sq.km.)	Forest cover in Western Ghats (%)
Goa	3702	32.60	1724	58.40
Karnataka	1,91791	16.90	38019	57.20
Kerala	38863	26.60	21856	42.20
Maharastra	3,07690	14.30	52643	26.70
Tamil Nadu	1,30058	13.60	22945	29.50

nation. Ecologically this zone is well placed as hilly ranges, which obstruct moisture laden monsoon winds rising from Arabian sea and bring the rains ranging 2000 to 6000mm per year. Even it crosses 8000mm in places like Agumbe in Karnataka. The hot and humid tropical climate coupled with favourable edaphic factors is an another unique feature favouring the

Table - 3 : Status of floristic diversity

Group	Floristic diversity (No.)	
	India	**Western Ghat**
Angiosperms	17,500	7,500
Gymnosperms	64	21
Pteridophytes	1,200	200
Bryophytes	2,850	225
Lichens	2,000	800
Fungi	14,500	600
Rattans	51	25
Orchids	1000	267
Total	39,165	9,638

luxuriant plant growth, which can be hardly seen in many parts of the world.

The problem of "ill effects" due to indiscriminate usage of medicines of allopathic base is deteriorating human health to an aggravating extent day by day. There is an urgent need to focus and patronise "Safe health motive", which warrant the usage of plant-based drugs. In this context, there is still a living culture and tradition in millions of rural folk in using many plants in the form of food or medicine or both with a strong knowledge of herbal home remedies in the Western Ghats area. In addition, there are many well-nourished and flourished local health healers in village communities who are serving the society. But, due to growing trend of unsustainable exploitation, leading to fast depletion of many plants of medicinal importance. So, the medicinal flora has the potentials in protecting the rich heritage of knowledge, value based herbal healing traditions and in meeting growing demands of plant-based drugs. By looking into above problems, the work on detailed documentation of MP's status and in drafting suitable conservation strategies of MP's has been undertaken, which is almost sparse till date.

HISTORICAL REVIEW

India as a rich repository of medicinal plants is well known since time immemorial. Citations in this context have been well described in Vedas. Even in ancient scripts like Charak Samhitha (700 BC) and Sushrutha

Fig. 1 : Map showing Western Ghats part (green patch) in Karnataka.

Samhitha (200 BC) properties and uses of 1100 and 1270 plants had been well documented, respectively.

Phyto-geographically the flora of Western Ghats is paleotropic derived from original Gondwana land mass. In this region four floristically characterized zones have been identified and they are Western Ghats from Tapti to Goa, Kalinadi to Coorg, The Nilagiris, Annamalai, Palani and Cardamom hills. Broadly 11 land scape elements have been recognized in which cultivation is an integral part in some types (Nagendra and Gadgil, 1998).

Luxuriant vegetation with more or less virgin nature is a characteristic feature of this region and is one of the major tropical evergreen forests of India with an exceptional biological diversity and an endemism of 38% in floral diversity (Johansingh, 2001).

The floral habitat constituted of 368 families with a lion share under Fabaceae followed by Poaceae, Euphorbiaceae, Asteraceae and Rubiacae. Nearly 12.5% of the known flowering plants are being threatened with extinction, both at national and global level (Ved *et al.*, 2001; Nair and Daniel, 1986).

At the national level over 95% of medicinal plants, used in plant based industries, are from wild. Hardly very few species are under commercial cultivation in recent years. Presently 15 to 20 per cent of the total vascular flora fall under one or the other category of threatened status. The higher magnitudes of unsustainable exploitation and habitat destruction are responsible for erosion of medicinal plant reserves. Despite all their importance, many MP's are still under exploitation with little or no regard for the future. Indiscriminate harvest in the collections of plant parts like bark, roots, flowers and seeds from wild in particular is making failure of their survival. Increasing and escalating consumers demand because of greater options in adapting to herbal remedies are aggravating the situation further.

The habitat destruction is a prime factor in increasing the threatening status of medicinal plants, as it results in narrowing of genetic variation and collapse of their survival. Some have adoptability to specific traditional ranges only, with a reluctance to survive and establish in new areas. A complete analysis of flora in the literature, herbarium or museum with knowledge of species in gardens and in natural habitat gives clues on status of species.

As per the International Union for Conservation of Natural resources (IUCN), red list of medicinal plants has been categorised on the basis of population status over a period of time standards (IUCN, 2000). The main categories include Extinct, Extinct in wild, Threatened and Low risk. Further the Threatened category has been classified as Critically Endangered, Endangered and Vulnerable. Similarly, Lower risk has been classified as Conservation dependent, Near threatened and Least concerned (Ravikumar and Ved, 2000).

While framing conservation strategies, the importance of *in-situ* conservation of wild relatives of cultivated plants in avoiding genetic erosion has been well narrated by Gadgil *et al.* (1996). Like wise role of sacred groves from the conservation point has been well highlighted by many workers (Ghosh, 2001; Bhakat and Pandit, 2003). In *ex-situ* methods, a potential role of cultivation of MP's in the perfection of conservation has been documented in Madhya Pradesh (Bhattacharya and Regina Hansda, 2003).

METHODOLOGY

The effort was made by a team of experts from the College of Forestry, Sirsi of Karnataka in documenting uniqueness of floristic diversity and medicinal plants in Western Ghat part of Karnataka. The information on the family-wise endemic floristic diversity in different forest types was collected from the literature (Johansingh, 2001). During the study visits were made to forests for understanding the medicinal plants status. The help from local healing practitioners, villagers and foresters was taken for locating and identification purposes. Further, the identifications were backed by the literature prepared by Keshava Murthy (1994). Based on the botanical family and habitat, grouping of documented medicinal plants was done. The information was compressed to habitat-wise number of hosts falling under each of the families fallowed by their summations. From the arranged data, from the point of assessing medicinal plants status and species of immediate concern for conservation the Red list categorization of listed species was made and arranged family-wise based on the literature (Keshava Murthy, 1994; Balakrishne Gowda *et al.*, 1997; Ravikumar and Ved, 2000). Different conservation strategies, involving traditional and improved techniques, have been worked out commonly based on their growth habit and reproduction. Besides this, the information on medicinal plants status was assessed in the Medicinal plant conservation area (MPCA's) and Medicinal plant developing areas (MPDA's). The required additional information was gathered from the available literature for drafting effective conservation strategies (Narong, 1993).

RESULTS AND DISCUSSION

The observations on floristic endemic diversity, medicinal plants flora obtained and the details of suitable conservation strategies frame has been presented and discussed under the following headings :

Endemic floral diversity

A study on assessments of endemic floral diversity was undertaken with the basic knowledge of biodiversity richness. Majority of tree species endemic to this zone fall under different vegetation types (Champion and Seth, 1968) distributed over 11 families is depicted in Table-4.

The pooled information indicates that maximum number of species, which are unique to Western Ghats irrespective of its family, falls under Evergreen type (190 species) followed by semi-evergreen type (175 species). It is evident that the families Myrtaceae, Rubiaceae and Euphorbiaceae are wide spread along both the altitudinal and moisture gradients in the Western Ghats. The family with most restricted distribution is Theaceae, which is largely restricted to the high elevation Sholas. But the families Dipterocarpaceae and Lauraceae are most common in the moist forests. Coming to species uniqueness, it is primed by Euphorbiaceae (120 species) followed by Lauraceae (105 species), Rubiaceae (87 species) and Meliaceae (85 species). However, some species have been found to be sparsely distributed and that too showing uniqueness only to one vegetation type. It is very clear from the above depictions that, biodiversity rate is high in evergreen and semi evergreen due to prevalence of congenial factors.

Table - 4 : Family wise important vegetation types of the Western Ghats.

Family	Shola	Ever green	Semi ever-green	Moist deci-duous	Dry deci-duous	Scrub-Savanna	Mm
Theaceae	5	-	-	-	-	-	-
Dipterocarpaceae	-	15	10	-	1	-	-
Myristicaceae	1	4	2	-	-	-	-
Lauraceae	25	50	25	5	-	-	-
Rutaceae	-	5	10	5	5	5	-
Meliaceae	-	20	25	15	10	5	10
Rubiaceae	5	20	25	15	10	7	5
Anacardiaceae	-	15	10	5	5	5	3
Myrtaceae	10	25	15	3	2	5	3
Combretaceae	-	3	5	5	4	5	2
Leguminosae	-	5	10	20	15	10	15
Euphorbiaceae	5	25	30	10	15	25	10
Arecaceae	-	3	5	1	-	3	3
Rhamnaceae	-	-	3	4	4	5	2

Family and habitat-wise distribution of medicinal plants

In the work done on medicinal plant diversity in Western Ghat part of Karnataka an exhaustive list of habitat-wise and family-wise distribution of medicinal plants are presented in Table-5. From the data an observation was made that medicinal plants are found to distribute across diverse habitats. The results showed that 1139 medicinal plant species, which were observed to spread over in 155 families. In the habitat-wise distribution, herbs ranked first (458 species) followed by trees (312 numbers), shrubs (190 numbers) and climbers (179 numbers). It indicates that out of total medicinal plants flora, more than one third is accounted to herbs, one fourth to trees and approximately equal proportions of shrubs and climbers.

Among the 155 families Fabaceae ranks first by possessing 87 plant species followed by Asteraceae (54), Euphorbiaceae (50), Rubiaceae and Acanthaceae (36), Orchidaceae (32), Poaceae (26), Ceasalpiniaceae (27), Malvaceae (24) and remaining families. In the habitats maximum number

Table - 5 : Red listed medicinal plants, which need immediate conservation

Red list Category	Family	Plant species
Critically Endangered	Cycadaceae	*Cycas circinalis*
	Boraginaceae	*Helitropium keralense*
	Fabaceae	*Pueraria tuberosa*
	Dipterocarpaceae	*Shorea tumbuggaia*
Endangered	Apocyanaceae	*Chonemorpha fragrans, Rauvolfia serpentina*
	Droseraceae	*Drosera indica, D. peltata*
	Magnoliaceae	*Michelia nilgirica*
	Icacinaceae	*Nothapodytes nimmoniana,*
	Lauraceae	*Persera macrantha*
	Hippocrateaceae	*Salacia oblonga, S. reticulata*
	Caesalpinaceae	*Saraca asoca*
Vulnerable	Passifloraceae	*Adenia hondala*
	Rutaceae	*Aegle marmelos*
	Meliaceae	*Aphanomixis polystachya*
	Aristolochiaceae	*Aristolochia tagala*
	Euphorbiaceae	*Baliospermum montanum*
	Myrsinaceae	*Embelia ribes*
	Clusiceae	*Garcinia morella*
	Colchicaceae	*Gloriosa superba*
	Asclepiadaceae	*Holostemma adakodien*
	Sapotaceae	*Madhuca longifolia, M. neriifolia*
	Myristicaceae	*Myristica dactoides*
	Convolvumlaceae	*Operculia turpenthum*
	Bignoniaceae	*Oroxylum indicum*
	Piperaceae	*Piper mullesua*
	Fabaceae	*Pseudarthria viscida*
	Araceae	*Raphidophora pertusa,*
	Santalaceae	*Santalum album,*
	Oleaeae	*Schrebera switeniodes*
	Symplocaceae	*Symplocos recemosa*
	Menispermaceae	*Tinospora sinensis*

of trees of medicinal value fall in Euphorbiaceae (21) followed by 17 each under Rubiaceae and Moraceae. Like wise Asteraceae is accounted to possess maximum number of herbs (48) followed by Fabaceae (36) and Orchidaceae (32). But, maximum number of shrubs was observed to fall under Fabaceae (20) followed by Acanthaceae (15). However Convolvulaceae and Asclepiadaceae was observed to possess maximum number of climbers, i.e., 17 and 16 respectively.

Red list categorization

Emerging consciousness of socio-cultural and economic importance of medicinal plant resources is becoming a serious threat in the survival of vital flora of medicinal importance. That is why, medicinal plant resource is fast depleting in recent years by falling under one or the other level of threatening status. Lack of regular updating in herbaria and museums, absence of regular field surveys, natural decline in the populations, failure to locate a particular species in field for research are few signals of rarity.

In Western Ghats part of Karnataka a total of 177 species are found to fall under one or the other Red listed category distributed over 71 families (Table-5). Among all the families, Burseraceae and Clusiaceae are subjected to maximum commercial exploitation. This is because of their high commercial value like resins and vegetable butter respectively. It has also revealed that 12 families possessing a single species and all of them have been identified to fall under one or the other categories of Red list. Further, among them *Gloriosa superba, Cycas circinalis* and *Santalum album* are highly exploited for pharmaceutical industries. From the pooled data the species which need immediate attention of conservation has been tabulated in Table-5 based on the Red listed data (Ravikumar and Ved, 2000).

The above observations clearly spell how much it is crucial now to develop immediate strategies for conservation of these plants otherwise they will be dragged into extinction phase.

Conservation strategies

Despite knowing that large proportion of local bio-diversity in our entire ecosystem is used for medicinal purpose, a very little attention has been given in knowing about their conservation status in the wild. Basically the meaning of conservation goes as, "The management of human use of the bio-diversity, so that it may yield the greatest sustainable benefit to present generation while maintaining its potentials to meet the needs and aspirations of future generations."

The most widely accepted scientific technologies for medicinal plant conservation are *in-situ* and *ex-situ* methods, nursery network, cultivation and control of over exploitation through legislation.

Table - 5 : Families and habit wise distribution of the medicinal plants in Western Ghats of Karnataka.

Family	Tree	Herb	Shrub	Climber	Total	Red listed species
Acanthaceae	-	18	15	3	36	*Adathoda beddomi, Andrographis echinoids, Niligirianthus ciliatus, Blepharis ginifolia*
Agavaceae	-	-	1	-	01	-
Aizoaceae	-	4	-	-	04	-
Amaranthaceae	-	9	-	-	09	-
Amaryllidaceae	-	3	-	-	03	-
Anacardiaceae	7	-	-	10	17	*Buchanania lanzan, Semicarpus anacardium, S. travencorica*
Ancistrocladiaceae	-	-	-	1	01	-
Annonaceae	2	1	-	2	05	-
Apiaceae	1	5	-	-	06	*Heracleum ringens, H. candoueanum*
Apocynaceae	8	2	3	5	18	*Alstonia scholaris, Chonemorpha fragrans, Rauvolafia serpentina*
Araceae	-	3	3	1	07	*Acorus calamus, Remusatia vivipara, Amorphopholus commutatus, A. paeoniifoeius, Rhapidophora pertusa*
Araliaceae	-	-	1	-	01	-
Areceae	6	5	-	11	22	-

Table - 5 : Continued .....

Family	Tree	Herb	Shrub	Climber	Total	Red listed species
Aristolochiaceae	-	1	-	3	04	*Aristolochia bracteata, A. tagala*
Asclepiadacae	-	2	4	16	22	*Decalepis hamiltonii, Holostemma adakodien, H. annularae, Sarcostemma acida, Gymnema khandalense,G. montanum*
Asteraceae	-	48	3	3	54	*Artemisia nilagirica, Lamprachaenium microcephalum*
Balanaphoraceae	-	1	-	-	01	*Balanophora fungosa*
Balsamniaceae	-	1	-	-	01	-
Bignoniaceae	7	-	2	-	09	*Oroxylum indicum*
Bombacaceae	4	-	-	-	04	-
Boraginaceae	1	4	-	-	5	*Heliotropium keralense*
Brassicaceae	-	2	-	-	02	-
Buddlejaceae	1	-	-	-	01	-
Burseraceae	5	-	-	-	05	*Boswellia serrata, Canarium strictum, Commiphorà mukul, C. Wightii*
Cactaceae	-	-	1	-	01	-

Table - 5 : Continued .....

Family	Tree	Herb	Shrub	Climber	Total	Red listed species
Caesalpiniaceae	12	6	7	2	27	*Saraca asoca, Kingiodendron pinnatum, Caesalpinia bonducella, Bauhinia racemosa,Humboldtia vahliana, H. brounonis*
Campanulaceae	-	1	-	-	01	-
Capparaceae	2	-	3	2	07	-
Capparidaceae	1	-	-	-	01	*Crataeva magna*
Caprifoliaceae	2	-	-	-	02	-
Caryophyllaceae	1	2	-	-	03	-
Celastraceae	1	-	2	3	06	*Celastrus paniculatus*
Chenopodiaceae	-	1	-	-	01	-
Cleomaceae	-	3	-	-	03	-
Clusiaceae	10	-	-	-	10	*Mesua ferrea, Mammea suriga, Garcinia indica, G. morella, G. travencorica, G. xanthochymus, G. gummigutta, Calophyllum inophyllum, C. capetatum*
Cochlospermaceae	1	-	-	-	01	-
Combretaceae	5	-	-	1	06	*Terminalia chebula, T. arjuna*
Commelimaceae	-	8	-	-	08	-

Table - 5 : Continued .....

Family	Tree	Herb	Shrub	Climber	Total	Red listed species
Connaraceae	-	-	1	1	02	-
Convolvulaceae	1	5	3	17	26	*Operculina turpenthum*
Cordiaceae	3	-	-	-	03	-
Cucurbitaceae	-	3	-	15	18	*Citrullus colocynthis, Corollocarpus epigaeus*
Cuscutaceae	-	-	-	1	01	-
Cycadaceae	1	-	-	-	01	*Cycas circinalis*
Cyperaceae	-	16	-	-	16	-
Dillenaceae	2	-	-	-	02	-
Dioscoraceae	-	1	-	7	08	*Dioscorea bulbifera, D. Glabra,*
Dipteracarpaceae	6	-	-	-	06	*Dipterocarpus indicus, Vateria indica, V. macrocarpa, Shorea tumbuggaia.*
Droseraceae	-	4	-	-	04	*Drosera burmmani, D. indica, D. peltata*
Ebenaceae	8	-	-	-	08	*Diospyros ebenum*
Elaeocarpaceae	3	-	-	-	03	*Elaeocarpus serratus*
Ehretiaceae	1	-	-	-	01	-
Elaeagnaceae	3	-	2	-	05	*Elaeagnus conferta*

Table - 5 : Continued .....

Family	Tree	Herb	Shrub	Climber	Total	Red listed species
Eriocaulaceae	-	2	-	-	02	-
Erythroxylaceae	-	-	1	-	01	-
Euphorbiaceae	21	12	14	3	50	*Baliospermum montanum, Tragia bicolor*
Fabaceae	16	36	20	15	87	*Abrus precatorius, Desmodium gangeticum, Indigofera tinctoria, Ougenia dalbergioides, Pterocarpus santalinus, Pseudarthria viscida, Pueraria tuberosa*
Flacourtiaceae	6	-	1	-	07	*Flacourtia indica, Hydnocarpus alpina, H. macrocarpa, H. pentandra.*
Flagellariaceae	-	-	1	-	01	-
Gentianaceae	-	13	-	-	13	*Gentiana kurro, Canscora diffusa, Swertia corymbosa, S. lawii*
Geraniaceae	-	1	-	-	01	-
Gesneriaceae	-	1	-	-	01	-
Heliortropiaceae	-	2	-	-	02	-
Hippocrateaceae	-	-	2	-	02	*Salicia oblonga, S. reticulata*
Hydrocharitaceae	-	2	-	-	02	-
Hydroleaceae	-	1	-	-	01	-

Table - 5 : Continued .....

Family	Tree	Herb	Shrub	Climber	Total	Red listed species
Hypericaceae	-	1	-	-	01	-
Hypoxidaceae	-	1	-	-	01	*Curculigo orchioides*
Icacinaceae	2	-	-	1	03	*Nothopodytes nimaoniana*
Lamiaceae	-	27	5	1	33	*Coleus vetiverioides, Plectranthus nilighaerricus, P. vettiveroides*
Lauraceae	10	-	-	1	11	*Cinnamomum macrocarpum, C. sulphuratum, C. wightii, Persea macrantha,*
Lecythidaceae	1	4	-	-	05	*Baringtonia racemosa*
Leeaceae	-	1	2	-	03	-
Lemnaceae	-	2	-	-	02	-
Lentibulariaceae	-	1	-	-	01	-
Liliaceae	-	2	1	2	05	*Aloe barbadensis, Asparagus gonocladus, A. rofilari, Gloriosa superba*
Linaceae	-	-	2	-	02	-
Lobeliaceae	-	-	1	-	01	-
Loranthaceae	-	2	3	-	05	-
Lythraceae	3	4	1	-	08	-

Table - 5 : Continued .....

Family	Tree	Herb	Shrub	Climber	Total	Red listed species
Magnoliaceae	3	1	-	-	04	*Michelia nilagirica, M. champaka*
Malpighiaceae	-	-	-	1	01	-
Malvaceae	3	15	6	-	24	*Sida cordifolia*
Marantaceae	-	1	-	-	01	-
Melastomataceae	-	1	3	-	04	-
Meliaceae	12	2	2	-	16	*Dysoxylum malabaricum, Naregemia alata, Aphanamixis polystachya*
Menispermaceae	1	-	-	6	07	*Coscinium fenestratum, Cyclea peltata, Tinospora malabarica, T. sinensis*
Menyanthaceae	-	2	-	2	04	-
Mimosaceae	9	2	2	1	14	*Acacia catechu, Albizzia lebbek,*
Molluginaceae	-	1	-	-	01	-
Moraceae	17	-	4	-	21	*Artocarpus gomezianus, A. hirsutus, Ficus tomentosa*
Moringaceae	1	-	-	-	01	*Moringa concanensis*
Myrsinaceae	4	-	3	1	08	*Embelia basal, E. ribes, E. tsjeriam-cottam,*
Myrtaceae	5	-	-	-	05	*Syzygium travencorecum*

Table - 5 : Continued .....

Family	Tree	Herb	Shrub	Climber	Total	Red listed species
Myristicaceae	4	-	-	-	04	*Myristica dactyloides, Myristica malabarica, Knema attenata*
Nelumbonaceae	-	1	-	-	01	*Nelumbo nucifera*
Nyctaginaceae	-	2	-	1	03	*Boerhaavia chiensis*
Nymphaeaceae	-	2	-	-	02	-
Ochnaceae	1	-	-	-	01	-
Olacaceae	-	-	1	-	01	-
Oleaceae	4	-	-	8	12	*Schrebra sweitenioides*
Onagraceae	-	2	-	-	02	-
Ophioglossaceae	-	1	-	-	01	*Helminthostachys zeylanica*
Orchidaceae	-	32	-	-	32	*Eulophia cullenii, E. ramentacea, Nervillaara goana, Paphiopedilum drumyii.*
Oxalidaceae	-	1	-	-	01	-
Pandanaceae	1	-	-	-	01	-
Papararaceae	-	1	-	-	01	-
Passifloraceae	-	1	1	3	05	*Adenia hondala*

Table - 5 : Continued .....

Family	Tree	Herb	Shrub	Climber	Total	Red listed species
Peperomiaceae	-	2	-	5	07	-
Periplocacea	-	-	2	1	03	*Decalapis hemitolnii, Janakia arayalpathra, Utleria saliceria*
Piperaceae	-	-	1	-	01	*Piper chaba, P. longum, P. barberi, P mullesua, P. nigrum*
Piprtosporaceae	2	-	-	-	02	-
Plantaginaceae	-	1	-	-	01	-
Plumbaginaceae	-	-	2	-	02	*Plumbago rosea*
Poaceae	1	25	-	-	26	-
Polygalaceae	-	3	1	-	04	-
Polygonaceae	-	7	1	-	08	-
Pontederiaceae	-	2	-	-	02	-
Portulacaceae	-	2	-	-	02	-
Ranunculaceae	-	1	1	6	08	-
Rhamnaceae	1	-	6	2	09	-
Rhizophoraceae	5	-	-	-	05	*Carallia brachiata*

Table - 5 : Continued .....

Family	Tree	Herb	Shrub	Climber	Total	Red listed species
Rosaceae	2	-	2	-	04	-
Rubiaceae	17	10	9	1	37	*Hymenodyctyon excelsa, Gardenia gummifera, Ochreinauclea missionis, Anthocephalus indicus*
Rutaceae	7	-	3	2	12	*Glycosimis macrocarpa, Zanthoxylum rhetsa, Aegle marmelos*
Salicaceae	1	-	-	-	01	-
Salvadoraceae	-	1	-	-	01	-
Santalaceae	1	-	-	-	01	*Santalum album*
Sapindaceae	4	-	2	1	07	-
Sapotaceae	8	-	-	-	08	*Madhuca diplostemon, M. insignis, M. longifolia, M. neriifolia*
Scrophulariaceae	-	18	-	-	18	*Bacopa monnieri*
Seiophulariaceae	-	1	-	-	01	-
Simaroubaceae	3	-	-	-	03	*Belanites roxburghii*
Smilacaceae	-	-	-	2	02	-
Solanaceae	-	8	8	-	16	-
Spihenocleaceae	-	1	-	-	01	-

Table - 5 : Continued .....

Family	Tree	Herb	Shrub	Climber	Total	Red listed species
Sterculiaceae	5	2	1	-	08	*Sterculia urenus*
Strichnaceae	5	-	-	1	06	*Strychnos potatorum, S. nux-vomica, S. aenea, S.dalzelli*
Symplocaceae	3	-	-	-	03	*Symplocos recemosa, S. cochin-chinensis*
Taccaceae	-	1	-	-	01	-
Theaceae	-	-	2	-	02	-
Thymelaeaceae	1	-	-	-	01	-
Tiliaceae	-	5	4	-	09	-
Trapaceae	1	-	-	-	01	-
Trichopodacea	-	1	-	-	01	*Trichopus zeylanicus*
Turneraceae	-	1	-	-	01	-
Tythaceae	-	1	-	-	01	-
Ulmaceae	6	-	-	-	06	*Holoptelea integrifolia*
Urticaceae	-	5	-	-	05	*Mappia foetida*
Valerianaceae	-	1	-	-	01	*Valerina leschenultii*
Verbenaceae	10	4	9	-	23	*Clerodendrum serratum*

Table - 5 : Continued .....

Family	Tree	Herb	Shrub	Climber	Total	Red listed species
Violaceae	-	3	-	-	03	-
Vitaceae	-	-	5	7	12	*Ampelocissus arneosa, A. indica, Cayratia pedata*
Xyridaceae	-	3	-	-	03	-
Zigophyllaceae	-	1	2	-	03	-
Zingiberaceae	-	5	2	-	07	*Kaempherea galanga, Alpinia galanga, Curcuma caesia, C. pseudomontana, Hedichium coronarium*
155 families	312	458	190	179	1139	177

IN-SITU CONSERVATION

It conveys the promotion of medicinal plants to grow in their natural habitat amidst prevailing ecosystem. This serves in conserving the diversity at gene, species as well as at ecosystem level. This can be channelized through establishment of Biosphere reserves, National parks, Wild-life sanctuaries, Sacred groves and network of other protected areas.

Biosphere reserves, National parks, Wild-life sanctuaries and Gene sanctuaries

In Western Ghats part of Karnataka, there is one biosphere reserve and it has 5 National parks viz., Rajiv Gandhi National park, Anshi National park, Bannerghatta National park, Bandipur National park and Kudremukh National park. Besides this, 21 Wild-Life Sanctuaries are distributed in this region. The channel of national parks and wild-life sanctuaries may check the genetic erosion of MP's growing in their natural habitats by restricting human interference through protected area concept. Sanctuaries for an individual species or group of threatened species, wherever it is possible to establish them, is another best means of conservation. Such sanctuaries have been established for *Nephenthes* in Khasi hills of Meghalaya and for *Primulus* and *Rhododendron* in Sikkim.

Sacred groves

Commonly called in Uttara meaning thereby Kannada as *Kan* and Devara *kadu* in Kodagu regions, "Patches of forests secured as by a religious feeling or sense of justice against any defamation, violation, or intrusion by means of ancient shrines within forests comprising solitary stones, terracotta icons and miniature tridents (trishul) placed under the trees endemic to a particular region". In Uttara Kannada, *Kans* have been referred to as "the historical sacred forests as people continue to have faith in the deity". In some areas the Government has declared them as reserved or protected forests. Totally 1424 sacred groves across 5947 ha have been reported (Anonymous, 2003). So, it helps in reducing the loss of MP's diversity with the concept of Socio-cultural-religious bindings on the people.

Medicinal Plant Conservation Areas (MPCA's)

MPCA's imply to the protection of marked areas possessing maximum diversity of endemic medicinal flora. In the concept of protected area net work, MPCA's involving local people is suggested by Goraya and Tandon (1996). With this idea Karnataka Forest Department established 13 MPCA models with the help of Foundation for Revitalization of Local Health Traditions (FRLHT, a Bangalore based NGO) as indicated in Table-6. These help in avoiding shrinkage of endemic medicinal floral diversity, which enable to fulfill the long run conservation strategy.

Table - 6 : Medicinal Plant Conservation Area (MPCA) net work in Karnataka

Category	Details
No. of MPCA	13
MPCA net work area (ha.)	2,800
Species listed from MPCAs	1,400
Medicinal plant species in MPCAs	950
MPCA hosted Percentage of state medicinal plants	46
Threatened species in MPCAs	42
Globally threatened species in MPCAs	21

Medicinal Plant Development Areas (MPDA's)

It has emerged recently with the aim of developing degraded forest areas with medicinal plants through Joint Forest Planning and Management (JFPM) involving local people. This helps in eco-restoration besides fulfilling conservation strategy. In Karnataka 4 such sites have been identified with an area of 5-20ha. each. In these areas the principle of "sustainable harvesting" is being strictly followed to restore the "Concept of conservation" lively.

EX-SITU CONSERVATION

It implies to conservation of medicinal plants outside its natural habitat. This can be channalized through a network of Botanical gardens, Ethno medicinal plant gardens, Gene bank (seed gene bank), Nursery network and cultivation.

Botanical garden

Specially, establishment of botanical gardens have a potential role in conserving medicinal plants, which are at the verge of extinction. It is the best concept in creating not only aesthetic value but also in creating awareness among the visitors about the importance of plants of medicinal value and need for their conservation. Such a motive of creating self-awareness is the best option in conserving our richest treasure of MP's in an efficient manner.

Ethno-Medicinal Plant Gardens (EMPG)

Establishment of EMPG is again a valuable means of ex-situ conservation. It is a network of regional and sub-regional EMPG set to have a broad base of all the known medicinal plants in various ethnic communities. Such sites help in extending the glorifying regional traditions and cultural rhythm by serving as rich repositories at regional level. A network of 15 such sites are operating in Western Ghats comprising 3 southern states.

Institutional gardens

It implies to establishment and maintenance of gardens in the premises of educational institutions with locally grown important species of high medicinal value by the students themselves with the theme of conservation. Implementation of this concept will have a strong bearing in seeding the theme of MP's conservation in the minds of young generation. This can be the potential way in bringing a mass revolution on the conservation of MP's for the use of future generation without giving any scope for further loss. This concept checks the loss of locally available endemic species of medicinal importance and serve as the miniature base in perfecting the ideology of Ethno Medicinal Plant Gardens.

Nursery network

Nursery network is an organized means of raising and supply of quality planting stock for the species, which have been introduced under cultivation to various end users. To start with, a nursery network of region-wise endemic species of medicinal value has to be geared up with an intension of bringing more area under their cultivation. This helps in meeting primary health care demands on one side and to develop packages for economic plants of industrial importance on the other side.

Cultivation

Bringing the species from its wild habitat to commercialization can be done through the concept of cultivation. This is the best method in not only improving the status of species in the areas surrounding its habitat, but also makes a provision for further spread to new areas on commercial scale. This concept helps in improving the economic status of farming community, in meeting industrial raw material demands and in elevating national MP's export potentials. The efforts of National Medicinal Plant Board (NMPB) in encouraging commercial cultivation of medicinal plants which are under threatened status is a good move in achieving conservation strategies. The farming community are being motivated from the point of economic gains, which aids in conserving MP's indirectly. However, for maximising the concept of commercial cultivation, development of agro-techniques and marketing infrastructure needs to be well perfected.

Gene banks

Gene bank concept of *ex-situ* conservation is confined to 'Seed gene banks' which has high applicability in conserving Orthodox seeds. It can be done by drying the seeds so as to maintain the optimum seed moisture level, which helps in protecting the seeds from pathogenic infections under storage. So, dried seeds should be packed in airtight containers and can be kept in cold storage. This helps in extending the seed viability period for longer time. So, it is the simple and most viable technique for the long run conservation of orthodox seeds. National Bureau of Plant Genetic Resources (NBPGR) located at New Delhi has actively taken part in maintaining seed gene banks of medicinal plants.

IN-VITRO CONSERVATION

In-vitro conservation is helping in the conservation of medicinal plants under varied degrees of threatening out side its parental habitat. *In-vitro* conservation is a novel technique to conserve germplasm of medicinal plants, which do not produce any seeds or produce seeds, which are recalcitrant. It also offers in storing large number of accessions in a relatively small space in a disease free state. The concept of gene bank is confined to pollen banks, DNA libraries, plant tissue or protoplast culture repositories and cryopreservations. This came into existence with a motive of long term-protected preservation of genetic variability. The facility of depositing seeds of wild medicinal plants are created at NBPGR- New Delhi, CIMAP-Lucknow and TBGIR-Tiruvananthapuram.

REFERENCES

Anonymous (2003). Sacred groves for fragments' sake. *A Down to Earth suppliment,* **3** : 72-75.

Balakrishna Gowda, Nissar, M.M. Seetharam, Y.N. and Ramesh, S.R. (1997). The threatened plants of peninsular India. *My Forest,* **33(1)** : 327-334.

Bhakat, R.K. and Pandit, P.K. (2003). Role of a scred grove in conservation of medicinal plants. *Indian forester,* **129(2)** : 224-232.

Bhattacharya, A.K. and Regina Hansda (2003). Ex-situ conservation of medicinal and aromatic plants in India with special reference to Madhya Pradesh. *Indian forester,* **129(1)** : 93-101.

Champion, H.G. and Seth, S.K. (1968). A revised survey of forest type of India. Govt. of India press, New Delhi.

Gadgil, M., Singh, S.N., Nagendra, H. and Chandran, M.D.S. (1996). *In-situ* conservation of wild relatives of cultivated plants : guiding principles and a case study, Bangalore: FAO, Rome and Indian Institute of Science. P21.

Ghosh, S. (2001). Sacred groves and their role in plant conservation. *Sci. & Cult.,* **67** : 347-349

Goraya, G.S. and Tandon, V. (1996). *In-situ* conservation of Medicinal plant diversity in tropical forests of southern India. *My Forest,* **32(4)** : 247-257.

IUCN (2000). *Red List of Threatened Species.* Published by The world conservation Union, Gland, Switzeland.

Johansingh, A.J.T. (2001). The Kalakad-Mundanthurai Tiger reserve: a global herotage of biological diversity. *Current Science,* **80(3)** : 378-388.

Keshava Murthy, K.R. (1994). *Medicinal plants of Karnataka.* Publ. by Karnataka Forest Department, Bangalore. P 399.

Myers, N., Mittermeier, R.A. and Jennifer (2000). Bio-diversity hot spots for conservation priorities. *Nature,* **403(24)** : 853-857.

Nagendra, H. and Gadgil, P. (1998). Linking regional landscape scales for assessing biodiversity : Acase study from the Western Ghats. *Current Science,* **75(3)** : 264-271.

Nair, N.C. and Daniel, P. (1986). The floristic diversity of the Western Ghats and its conservation. *Proc. Indian Acad. Sci.* New Delhi, 127-163.

Narong, Chomchalow (1993). Medicinal and aromatic plants conservation and Utilization in Asia. *In : Medicinal and aromatic plants in Asia* (Ed.) Narong Chomchalow and H.V. Henle. Oxford and IBH publishing Co. Pvt. Ltd., New Delhi, pp. 1-5.

Ravikumar, K. and Ved, D.K., (2000). Illustrated field guide ti 100 red listed medicinal plant of conservation concerne in South India. FRLHT, Bangalore. P 467.

Srinivasa Murthy, T.S., Mohan Karant, Prabhakaran, V., Jadhav, S.N., Satish Elkunchwar, Ravikumar, K. and Utkarsh Ghate (2003). Medicinal plants conservation and sustainable use through forest gene banks. *Indian Forester,* **129(2)** : 179-186.

Ved, D.K., Prathima, C.L., Nancy Morton and Darshan Shankar (2001). Conservation of India's medicinal plants diversity through novel approach of establishing *in-situ* gene banks. Forest Genetic Resources; Status, Threat and Conservation strategies [R. Umashankar, R. Ganeshaiah and Bawa, K.S. (Eds.)]. Oxford and IBH, New Delhi. Pp 183-190.

Recent Advances in Medicinal Plant Research : Vision 21st Century, 2005
Editors : S.K. Prabhuji, G.P. Rao and S.K. Patil
Satish Publication House, New Delhi

Chapter 5

WEEDS OF MEDICINAL IMPORTANCE

Pratibha Katiyar, S.S. Kolhe, S.P. Tiwari and M.I. Khan
Department of Plant Physiology, College of Agriculture,
Indira Gandhi Agricultural University, Raipur - 492 006, Chhattisgarh

Chhattisgarh is rich for the wide availability of natural occurrence of medicinal flora that have been in existence, identified and utilized since hundreds of years by tribals, begas and local people for the health care. The study was conducted in crop fields of wet (kharif) and dry (rabi) season to find out the quantitative evaluation and dominance of that medicinal weed flora and its occurrence in crop fields in natural conditions. The study revealed that there are great possibilities to grow them on commercial level to uplift the socio-economic status in rural areas. There is an urgent need for the development of infra-structure in the form of a center and it's proper commercialisation at different levels and establishment of cell of separate unit to promote the cultivation and protection of medicinally important flora of the state, it's proper marketing and export in the state, national and global market which may improve the socio-economic status in rural areas and would be possible way to change the economic scenario of the state.

INTRODUCTION

Chhattisgarh is well known in Asia for it's second highest collection of about 23,000 accessions of precious rice germplasm and also famous for natural occurrence of medicinal flora in the state. However, rice is major crop of this state and other crops viz., wheat, lathyrus, sunflower, soybean, linseed, moong, chickpea, berseem, cotton are also grown in the state. The weed survey of 36 countries represented approximately 90% of upland area in the world, and weeds are reported to be major biological constraints to rice production in 25 countries of them (Arraudeau, 1986). The socio-economic condition of the average farmer is very poor in this state because they are totally dependent on rice cultivation. They have neither enough potential to bear the cost of herbicide in weed management practices nor aware for it's proper use. Therefore, the study has been aimed to cover the villages to find out the dominance of various weeds flora in the field which are the major constraints in the crop production. To meet out these chalanges the study has been made to highlight the identification of weed flora available in the farmers' and experimental fields and its dominance along with the measures for its management.

METHODOLOGY

The study was undertaken in weed survey programme of AICRP-WC during the years 1997-2001 in all types of cultivation practices of rice and other crops in wet and dry seasons in the experimental and farmers, fields in the region. To evaluate the weed flora, present in different types of soil in different locations in majority and minority, the frequency, density, abundance, relative frequency, relative density, and relative abundance have been studied (Katiyar and Kolhe, 2001). The Importance Value Index (IVI) was estimated for deciding the actual dominance of the particular medicinal weed species in the crop field in majority or in minority. The medicinal importance of weed flora was given on the basis of it's local use and its description available in literature.

RESULTS AND DISCUSSION

Plant has been a major source of therapeutic agent for alleviation or cure of the human diseases and widely utilized throughout the world in modern and traditional systems of medicine. The traditional system of medicine still continue to cater in a significant manner to the medicinal needs of about 80% of the world's population. As per reports the traditional and modern systems of medicine are all dependent upon the medicinal plants (Singh *et al.*, 2001).

Chhattisgarh is very rich state for its herbal flora for more than a century period even before the development and spread of modern medicines. More than hundred plant species are available in the state having high medicinal value and they are being cultivated and may be commercialised in large quantity to fulfil the global demand.

The study revealed that the medicinal weeds like *Anagallis arvensis* was found in wheat (30.02), berseem (44.87) in majority while in other crops viz., rice, maize cotton and onion it was available as minor weed of the field. Pei Shengii (1984) also reported the medicinal importance of *Anagallis arvensis, Chorchorus spp, Eclipta alba* and *Spilanthes spp.* The *Achyranthus aspera, Alternanthera sessilis, Ageratum conyzoides* were found growing in the field of wheat crop in dry season. The survey in wheat field was done in three different phases of crop growth and and it was observed that the weeds viz., *Achyranthus aspera, Alternanthera* sessilis, *Ageratum conyzoides, Borreria spp., Chenopodium album, Commelina benghalensis, Echinochloa colona, Euphorbia hirta* were available in majority and as well as in minority and also possessed the high medicinal value and use (Marita *et al.*, 1999). However, *Amaranthus viridus* and *Ageratum conyzoides* were also observed in banana and berseem fields respectively in mid December to late February months. *Alternanthera sessilis* was also observed in minority in the soybean field. *Borreria spp.* and *Chenopodium album* both medicinal weed species were invariably present in dry season crops of this region. Direct seeded paddy has also found to have *Borreira spp.* and *Commelina diffusa* in the field. However, in transplanted conditions *Commelina benghalensis* was available. *Corchorus capsularis* was a common medicinal weed of rice crop. However, it was present as a minor weed in direct seeded paddy, banana and soybean crop. *Cynodon dactylon* was available in minority in cotton, rice (direct seeded) and banana. But, it was available as major weed in maize. *Cyperus iria* and *Cyperus difformis* was found growing in direct seeded rice and in soybean crop. Most of the dry and wet season crops have *Cyperus rotundus* as major weed, viz, maize, rice (direct seeded) and berseem while it was available in minority in rice (transplanted), chickpea and onion. *Dactyloctenium aegyptium* was observed as major weed of chickpea, banana and berseem. *Echinochloa colona* was the most common weed of all the crops of this region whereas *Eclipta alba* and *Euphorbia hirta* were the minor weeds of cotton and chickpea. These were also available in soybean and banana fields, respectively. *Fimbristylis miliacea* was the most common and major weed of direct seeded rice while it was present in minority in transplanted situation. *Sida aquata* was not found as a common weed of all crops. However, it was present in chickpea field in majority. The *Spilanthes spp.* was the minor weed of many crops, i.e., rice, berseem, cotton, maize and sunflower while it was major weed in onion. *Sonchus spp.* and *Solanum nigrum* were available in chickpea as major and minor weeds, respectively. *Tridax procumbens* was available in cotton field as minor weed. The medicinal uses of the weeds as well as it's control measures are listed in Table-1 (Katiyar and Kolhe, 1997-2001).

Table - 1 : Weed flora, it's importance and control measures (survey 1997-2001).

Botanical name	Crop	IVI	Family	Control measures	Medicinal Use	LU	PW
Achyranthus aspera	Wheat	34.66 M	Amaranthaceae	-	Plant powder used as diuretic, piles, skin erupton warts Leaves-cuts, wounds, roots-loose motion and fever	+	-
Amaranthus viridus	Banana	14.62 Mi	Amaranthaceae	2,4-D @ 0.5-0.8 kg/ha applied within 20-30 d after emergence or MCPA @0.4 kg/ha applied within 20-30 d after emergence (Marita *et al.*, 1999)	Whole plant used as medicine to cure wounds young shoots are poisonous to livestock due to high nitrate content (Marita *et al.*, 1999) Leaves-increases eye-sight, shoot-used as restorative	+	-
Alternanthera sessilis	Wheat Soya-bean	45.90 M 14.62 Mi	Amaranthaceae	-	Tender shoot and leaves-digestion, control stomachic dysentry	+	-
Anagallis arvensis	Berseem Wheat Rice Onion Cotton Maize	44.87 M 30.02Mi 20.26Mi 20.15Mi 16.98Mi 14.15Mi	-	-	Skin trouble	+	-

Table - 1 : Continued .....

Botanical name	Crop	IVI	Family	Control measures	Medicinal Use	LU	PW
Ageratum conyzoides	Wheat Berseem	28.29Mi 26.29Mi	Asteraceae	Apply butachlor and oxadiazon at planting stage or spray 20% NaCl mixed with detergent completely eliminate (Marita *et al.*, 1999)	Used as medicine (Marita *et al.*, 1999)	+	-
Bidens pilosa	Berseem	8.04Mi	Asteraceae	2,4-D @ 0.5 to 0.8 kg/ha applied within 20-30 d after emergence gives effective control (Marita *et al.*, 1999)	Leaves are used to treat inflammation and rheumatism. Two acylated glucosides may be associated with therapeutic effects (Marita *et al.*, 1999)	+	-
Blumea lacera	-	-	Asteraceae	-	Tender leaves and shoots-bodyache	+	-
Borreria spp. Annual broad leaf	Wheat Berseem Rice (DSP) Sun-flower Cotton Maize	90.90 M 58.90 M 12.90Mi 39.05 M 7.90 Mi 8.95 Mi	Brubiaceae	MCPA @ 0.4 kg.ha at 25-30 d after emergence	-	-	-

Table - 1 : Continued .....

Botanical name	Crop	IVI	Family	Control measures	Medicinal Use	LU	PW
Cassia tora	-	-	Fabaceae	-	Seeds-rheumatism	+	+
Caesulia axillaris	Soya-bean	7.98 Mi	Asteraceae	-	Styptic	+	
Celosia argentia C_3	Rice (DSP)	6.12Mi	Amaranthaceae	-	Root-cough, cold fever asthma, relapsing fever	+	
Chenopodium album	Onion Moong Maize Sun-flower Wheat	33.83M 40.90M 51.78M 21.90M 10.90Mi	Chenopodiaceae	-	Tender shoot-increase eye sight, vermicide for children, febrifuge in constipution	+	+
Commelina benghalensis Monocot	Rice (TP) Banana Wheat	49.30M 15.38Mi 10.90Mi	Commelinaceae	Butachlor–good pre emergence control. Propanil @ 2kg/ha at 10 d after emergence. 2,4-D @ 0.5-0.8 kg/ha applied within 20-30 d after emergence or MCPA @0.4 kg/ha applied within 20-30 d after emergence.	Leaves used as human food as medicine for treating burns, sore eyes, and sore throat (Marita *et al.*, 1999)	-	-

Table - 1 : Continued .....

Botanical name	Crop	IVI	Family	Control measures	Medicinal Use	LU	PW
				Mixture with thiobencarb 1-3 kg/ha or 2,4-D with propanil can increase effectiveness (Marita *et al.*, 1999)			
Commelina diffusa Monocot	Rice (DSP)	8.56Mi	Commelinaceae	Butachlor–good pre emergence control. Propanil @ 2kg/ha at 10 d after emergence. 2,4-D @ 0.5-0.8 kg/ha applied within 20-30 d after emergence or MCPA @0.4 kg/ha applied within 20-30 d after emergence. Mixture with thiobencarb 1-3 kg/ha or 2,4-D with propanil can increase effectiveness (Marita *et al.*, 1999)	Used as medicine	-	-
Corchorus capsularis	Rice (TP) Rice (DSP) Banana Soybean	44.90M 5.60 Mi 11.50Mi 3.9 Mi	Tiliaceae	-	Dried leaves and shoots- for curing livercirrhosis, blood purifier hyper tension also prevent pox and measeles	+	-

Table - 1 : Continued .....

Botanical name	Crop	IVI	Family	Control measures	Medicinal Use	LU	PW
Corchorus olitorius	+	-	Tiliaceae	-	Dried leaves and shoots- blood purifier	+	-
Cuscuta reflexa	Berseem	14.96Mi	Convolvulaceae	Pendimethalin 1.0 kg/ha as pre emergence or 0.50 kg/ha as early post emergence 10 d after sowing	-	-	-
Cynodon dactylon Monocot C_4	Maize Cotton Rice (DSP) Banana	35.95M 23.36Mi 15.84Mi 16.15Mi	Poaceae	Difficult to eradicate Use Glyphosate	Good for fodder due to have high crude protein 8-15%. (Marita *et al.*, 1999) Skin trouble, styptic	+	-
Cyperus difformis Monocot C_4	Rice (DSP)	15.84Mi	Cyperaceae	-	-	-	-
Cyperus iria Monocot C_4	Rice (DSP) Soya-bean	6.92 Mi 12.12Mi	Cyperaceae	Early post emergence of chemical mixture of butachlor and propanil, or 2,4-D 20.5 to 0.8 kg/ha, MCPA @ 0.4	Juice is used as a tonic and treat stomachaches. (Marita *et al.*, 1999)	-	-

Table - 1 : Continued .....

Botanical name	Crop	IVI	Family	Control measures	Medicinal Use	LU	PW
				kg/ha, propanil 2kg/ha applied at 20-30 d after emergence (Marita *et al.*, 1999)			
Cyperus rotundus Monocot C_4	Maize Rice (DSP) Rice (TP) Chickpea Bersem Onion	47.26M 54.08M 27.18M 9.25M 25.85M 12.4M	Cyperaceae	Repeated use of 2,4-D at 30d intervals followed by soil disturbance reduced tuber population by 86%. This can only be done in successfully in fallows (Marita *et al.*, 1999)	Diuretic, cooling cough, bronchial asthma (Marita *et al.*, 1999)	-	+
Dactyloctenium aegyptium Monocot	Chickpea Banana Berseem	26.10M 17.42M 25.85M	Poaceae	Pre emergence application of oxadiazon @ 0.75-1.0 kg/ha or pendimethylin @ 1.5–2.0 kg/ha or early post emergence (3-5 leaves or 15-25 d after emergence). Application of cyphalofop butyl @ 2 kg/ha or fenoxaprop-P-ethyl 55g/ha (Marita *et al.*, 1999)	The herbaceous part of the plant are applied externally to treat ulcers. A decoction of seed used to releive pain in the kidney region. Purched grain are administered to women who suffer from stomach-ache after childbirth (Marita *et al.*, 1999)	+	-

Table - 1 : Continued .....

Botanical name	Crop	IVI	Family	Control measures	Medicinal Use	LU	PW
Echinochloa colona Monocot C_4	Moong Sun-flower Chickpea Onion Berseem Banana Cotton Maize Wheat Rice (DSP) Rice (TP)	152.8M 38.49M 12.00M 38.31M 12.50M 95.81M 23.94M 31.42M 44.90M 30.46M 20.18M	Poaceae	Pre emergence application of oxadiazon @ 0.75-1.0 kg/ha and pendimethalin @ 1.5-2 kg/ha or post emergence (3-5 leaves or 15-25 d after emergence) application of cyphalofop butyl @ 0.1-0.15 kg/ha.	Used as good fodder grass for stimulating lactation in animals. It is used as faminine food.	-	-
Eclipta spp.	Cotton Onion Chickpea Soybean	4.67M 32.88M 8.99M 8.90M	Asteracea	2,4-D @ 0.5-0.8 kg/ha applied within 20-30 d after emergence or MCPA @ 0.4 kg/ha applied within 20-30 d after emergence. Oxadiazon at @ 0.75-1.0 kg/ha (Marita *et al.*, 1999)	Shoot juice for dysentery Oil used in hair fall (Katiyar and Kolhe, 2001). It is source of coumestan – type compounds used in phytopharmaceautical formulations of medicine prescribed for treatment of Cirrhosis of the liver and infection hepatitis.	+	-

Table - 1 : Continued .....

Botanical name	Crop	IVI	Family	Control measures	Medicinal Use	LU	PW
					The leaves are used for dyeing the hair black and as a hair tonic (Marita *et al.*, 1999)		
Eleocharis spp. Monocot	Rice (DSP)	51.59 M	Cyperaceae	-	Vapours of flower-nontoxic and have antimicrobial action		
Euphorbia heterophylla C_4 Dicot	+	-	Euphorbiaceae	2,4-D @ 0.5-0.8 kg/ha applied within 20-30 d after emergence or MCPA @ 0.4 kg/ha applied within 20-30 d after emergence. Oxadiazon at @ 0.75-1.0 kg/ha (Marita *et al.*, 1999)	Whole plant for respiratory trouble	+	-
Euphorbia hirta C_4 Dicot	Cotton Banana Chickpea	4.67M 3.62 M 5.21M	Euphorbiaceae	Oxadiazon @0.75-1.0 kg/ha applied after harrowing and sowing of rice. It can also be controlled by 2,4-D @ 0.5-0.8 kg/ha applied	Latex are used as a traditional medicine against intestinal disease, ulcers, bronchitis and conjunctivitis (Marita *et al.*, 1999) Whole	+	-

Table - 1 : Continued .....

Botanical name	Crop	IVI	Family	Control measures	Medicinal Use	LU	PW
				within 20-30 d after emergence or MCPA @ 0.4 kg/ha applied within 20-30 d after emergence (Marita *et al.*, 1999)	plant used for lactation, used in curing convulsion and cancerous disease		
Fimbristylis miliacea C_4 Monocot	Rice (DSP) Rice (TP)	102.5 M 12.50 M	Cyperaceae	Early post sowing application of commercial butachlor and propnil mixture, or MCPA at 0.4 kg/ha at 20-30 d after emergence (Marita *et al.*, 1999)	Leaves are used for poulticing during fever (Marita *et al.*, 1999)	+	-
Leucas indica	Chickpea	9.2 M	Lamiaceae	-	Leaves helps in digestion	+	-
Phyllanthus niruri	Cotton	4.67 M	Euphorbiaceae	-	Plant powder used in Jaundice, diuretic and for urinary tract disease	+	-
Portulaca oleracea C_4	+	-	Portulacaceae	Early post emergence of chemical mixture of butachlor and propanil,	Leaves- in diabetes, laprosy, to kill intestinal worms	+	-

Table - 1 : Continued .....

Botanical name	Crop	IVI	Family	Control measures	Medicinal Use	LU	PW
				or 2,4-D 20.5 to 0.8 kg/ha, MCPA @ 0.4 kg/ha, propanil 2 kg/ha applied at 20-30 d after emergence (Marita *et al.*, 1999)			
Sida aquata	Chickpea	29.17 M	Malvaceae	It can also be controlled by 2,4-D @ 0.5-0.8 kg/ha applied within 20-30 d after emergence or MCPA @ 0.4 kg/ha applied within 20-30 d after emergence (Marita *et al.*, 1999)	Root for rheumatism	+	
Spilanthes paniculata	Onion Berseem Cotton Maize Sun-flower	30.63 M 17.78 M 4.67 M 8.53 M 8.3 M	Astraceae		Young shoot-bodyache, toothache and colic	+	-
Sonchus spp.	Chickpea	17.61 M	Asteraceae	It can also be controlled by 2,4-D @ 0.5-0.8	Young leaves are used in dysentry and whole	+	-

Table - 1 : Continued .....

Botanical name	Crop	IVI	Family	Control measures	Medicinal Use	LU	PW
				kg/ha applied within 20-30 d after emergence or MCPA @ 0.4 kg/ha applied within 20-30 d after emergence (Marita *et al.*, 1999)	plant for abscesses, boils, colds, influenza, burns and snakebite (Marita *et al.*, 1999)		
Solanum nigrum	Chickpea	9.12 M	Solanaceae	-	Tender shoot for dysentery, helps in bowel movement	+	+
Trianthema spp.	+	-	Aizoaceae	Propanil @ 2 kg /ha applied at 2 wk after rice emergence	Eaten as vegetable and to cure wounds, diuretic.	-	-
Tridax procumbens	Cotton	10.35 M	Asteraceae	It can also be controlled by 2,4-D @ 0.5-0.8 kg/ha applied within 20-30 d after emergence or MCPA @ 0.4 kg/ha applied within 20-30 d after emergence (Marita *et al.*, 1999)	Used to stop bleeding (Marita *et al.*, 1999)	-	-

DSP = Direct seeded paddy, TP = transplanted paddy, M = Medicinal value, LU = Local use, PW = Potential weed.

CONCLUSIONS

The studies during 1997-2001 highlighted most common medicinally important weeds of different crops of this region viz., *Anagallis arvensis, Borreria spp, Chenopodium album, Commelina benghalensis, Corchorus capsularis, Cynodon dactylon, Cyperus rotundus, Dactyloctenium aegyptium, Echinochloa colona, Eclipta spp. Euphorbia hirta, Spilanthes paniculata.* They were invariably present in the different crop fields. These weeds can neither be fully utilized for medicinal purposes nor they have been controlled and properly eliminated from the fields because of high cost inputs and lack of knowledge regarding the sophisticated and appropriate dose and use of herbicides. Therefore, the present study is useful for weed management along with knowledge of the medicinal uses of that weed flora. Thus, it's the choice of farmers to control them by proper management practices or they may utilize the medicinal weeds as an important medicinal commodity for sustainable economic development of the state. Hence, there is need of an organized marketing system of medicinal plants, particularly weeds, and its various products in this region.

REFERENCES

Arraudeau, M. (1986). Final report of International upland rice genetic survey (IURGS) Los Banos Philippines, International Rice Research Institute.

Katiyar, P. and Kolhe, S.S. (2001). Medicinal weeds of Raipur region of Chhattisgarh plain. *Journal of Medicinal and Aromatic Plants,* **22/4 A** : 455-460.

Katiyar, P. and Kolhe, S.S. (2001). Five years compiled report (1997-2001) of AICRP-WC of Raipur (Chhattisgarh).

Kirtikar, K.R. and Basu, B.D. (1993). Indian Medicinal Plants. Vol. 1, 2, 3 and 4, Periodical Experts Book Agency, New Delhi.

Lindley, J. (1931). Flora Medica, Scientific Publishers, Jodhpur.

Marita, I., Galinato, Keith, Moody and Colin M. Piggin (1999). In: Upland Rice Weeds of South and East Asia, IRRI, Phillinines.

Pei, Shengji (1984). Plant Product and ethnicity in the market of Xishuangbanna, Yummam Province, China paper presented in conference of Ethnic Diversity and the control of natural resources in South East Asia. The University of Michigan, Ann Arbor, MI. USA. Aug. 22-24, 1984.

Ramnath (1992). Vanoshadhi Shatak, Ed. Ramnath.Ayur Pharmacy, New Delhi.

Singh, J., Srivastava, R.K., Singh, A.K. and Kumar, S. (2001). Futuristic scenario in production and trade of major medicinal plants in India. *JMAPS*, **22/4A and 23/ 1A** : 564-571.

Tiwari, D.N. (2000). Medicinal Plants for providing health and improving ecology and economy. In: Abstracts and background papers. National exposition of herbal products (Regional event of Swadeshi Vigyan Mela). Central Drug Research Institute, Lucknow. pp BP 5-11.

Recent Advances in Medicinal Plant Research : Vision 21st Century, 2005
Editors : S.K. Prabhuji, G.P. Rao and S.K. Patil
Satish Publication House, New Delhi

Chapter 6

MEDICINAL PTERIDOPHYTES – AN OVERVIEW

Shweta Singh[1], R.D. Dixit[1] and T.R. Sahu[2]
[1]*Botanical Survey of India, 10-Chathum lines, Central Circle, Allahabad (U.P)*
[2]*Department of Botany, Dr. Hari Singh Gaur Vishwavidyalaya, Sagar (M.P.)*
e-mail : shweta_2oct@rediffmail.com

Pteridophytes are primitive vascular plants. These fascinating groups of plant are always attracting the botanists and naturalists all over the world not only because of its beautiful and unique foliage but also for their useful and medicinal aspects. In comparison to higher plants they have been found to be of very little applications in medicine and are still neglected. However, the tribal communities, ethic groups and folklore throughout the world are utilizing their plant parts like rhizome, stem, fronds, sinnae and spores in various ways for the treatment of various ailments since ancient times. We have compiled the information on distribution and medicinal uses of 137 medicinal plants in the present chapters, which shows the promising scope of utilizing these plants as herbal medicinal in our day-to-day life.

INTRODUCTION

Plants have always been the source of food, medicine, shelter, and other necessities of man's life since ages. The primitive human societies of the World even today depended for all their medicinal and other necessities of life on surrounding plant wealth including pteridophytes. In the 20th century, the field studies on edible, economic and medicinal plants have been advanced enormously. The Arurvedic system of medicine has been in use for over 3000 years. Even today, it is very much popular among the youth and new generation because the treatment aims not only curing the disease but also enhancing the body vitality to combat with the disease and minimize the chance of relapse.

Pteridophytes are primitive vascular plants. These fascinating groups of plant are always attracting the botanists and naturalists all over the globe not only because of its beautiful and unique foliage but also for their useful aspects. They are also known as botanical snake or plant reptiles. They grow luxuriantly in moist tropical and temperate forests and their occurrence in different eco-geographically threatened regions from sea level to the higher mountains are of much interest. They are very specific in distribution and generally occur in moist and shady places near water channels throughout the World. In comparison to higher plants they have been found to be of very little applications in medicine and are still neglected. The tribal communities, ethic groups and folklore throughout the World are utilizing their plant parts like rhizome, stem, fronds, pinnae and spores in various ways for the treatment of various ailments since ancient times. According to a recent survey in Member States of the European Union, the herbal medicinal preparations from the plants including Pteridophytes are widely used by local population in Europe. Herbal preparations have been found to be highly popular in primary healthcare in China, Malayasia, Nepal, Myanmar, Belgium, France, Germany and Netherlands. Although late, the developed countries too are turning towards the use of traditional medicinal systems that involve the use of herbal drugs and remedies.

Unfortunately, such studies have been mainly confined to flowering/ higher plants and no serious efforts have been made so far to explore medicinal properties of Pteridophytes. The number of contributions about the taxonomy, ecology and distribution of pteridophytes have been published from time to time in various ways, but, enough attention has not been paid towards their useful aspects. Our main aim is to explore medicinally useful important pteridophytes and properly document their useful aspects for the benefit of human welfare.

India harbours about 48,000 plant species including about 17,000 angiosperms, 23000 fungi, 2700 bryophytes, 5000 Alage, 1,000 lichens and 1200 Pteridophyes (Dixit, 1984; Nayar, 1996; Chandra, 2000; Sarat Babu

Table - 1 : Medicinal values of Indian Pteridophytes

	Botanical name	Family	Vernacular/ English name	Uses	Distribution	References
1.	*Acrosticum aureum* L.	Acrostichaceae	Leather fern	Fertile fronds used for syphilitic ulcers and emollient topically. Rhizome is vulnerary used in healing inveterate ulcers and in wounds and boils.	W. India Peninsular India Andaman and Nicobar	Quisumbing, 1955. Usher, 1971
2.	*Actiniopteris radiata* (Sw.) Link	Actinopterid-aceae	Morpankhi	Fronds chewed for sore throat. Rhizome styptic anthelminthic and its decoction used in cured of dandruff.	N.W Himayaya Gangetic plain Peninsular India	Dymock, 1890. Lloyd, 1964. Dixit, 1975 Singh *et al.*, 2003
3.	*Adiantum aethiopicum* L.	Adiantaceae	——	Leaves and rhizome smoked for cold and chest disease in South India. Fronds chewed for sore throat. Rhizome boiled cured dandruff.	Peninsular India N. E India	Watt, 1962. Vasudeva, 1999. Manickam, 1999.
4.	*A. capillus-veneris* L.	Adiantaceae	Southern maiden hair fern	Whole plants used in cure of Ricketts in children and steamed for smallpox cure.	Throughout India	Scully, 1970. Remero, 1954; Singh *et al*, 2003

Continued .....

	Botanical name	Family	Vernacular/ English name	Uses	Distribution	References
				It used as diuretic, tonic, febrifuge expectorant, demulcent and astringent.		Dixit, 1975. Vasudeva, 1999.
5.	*A. caudatum* L.	Adiantaceae	Mayur Shikha	Leaves externally remedy for skin disease, internally for diabetes. Also used in cure of chest complains, cough and fever.	Throughout India	Quisumbing, 1955. Vasudeva, 1999. Manickam, 1999.
6.	*A. edgeworthii* Hook.	Adiantaceae	Adhe-sir-di-jari	Paste of fronds used in cure of headache.	N.W. Himalayas	Kirn *et* Kapahi, 2001
7.	*A. ensiforme* Wall. *ex* Hook. *et* Grev.	Adiantaceae	———	Fronds and rhizome used as food by tribals	N.E. Himalayas N.W. Himalayas	Das, 2003
8.	*A. flabellulatum* Hook.	Adiantaceae	——	Herbs is used for cure of cough and Rhizome are anthelminthic.	E. India	Vasudeva, 1999.
9.	*A. incisum* Forssk.	Adiantaceae	——	Fronds used in cure of skins disease, cough	Throughout India	Vasudeva, 1999.

Continued .....

Botanical name	Family	Vernacular/ English name	Uses	Distribution	References
			and diabetes. Fronds are chewed to cure mouth blisters and juice in bronchitis.		Singh *et al*, 2003 Pande *et al*, 1994.
10. *A. pedatum* L.	Adiantaceae	Northern maiden hair fern	Fronds used to stop bleeding, in chest and stomach troubles. Also used as tonic, demulcent, Expectorant, astringent, catarrhal and pectoral. Rhizome used in respiratory ailments	N.W. Himalayas N.E. Himalayas	Vasudeva, 1999. Smith, 1924.
11. *A. philippense* L.	Adiantaceae	Kali-Jhant	Fronds used in fever, asthama, bronchitis, dysentery, epileptic fits, leprosy and erysipelas. Rhizomes gives against dog-bite and Snake-bite as an antidote.	Throughout India	Dixit, 1975. Vasudeva, 1999. Singh *et al*, 2003 Manickam, 1999
12. *A. tenerum* Sw.	Adiantaceae	Brittle maiden hair fern	Plants used as expect-orant and demulcent.	Throughout India	Vasudeva, 1999.

Continued .....

Botanical name	Family	Vernacular/ English name	Uses	Distribution	References
13. *A. venustum* D. Don	Adiantaceae	Hansraj	Fronds used in biliousness, disease of chest cold, headache, ophthalmia, hydrophobia, inflammation and tumors Vapour bath of leaves is done in fever. Used as tonic, diuretic, emetic, resolvent, expectorant and against wounds, piles, and scorpion sting.	N.E. Himalayas	Dixit, 1975. Asolkar *et al*, 1992 Vasudeva, 1999
14. *Ampelopters prolifera* (Retz.) Copel.	Thelypterid-aceae	——	Plants have anti-bacterial properties.	Throughout India	Singh *et al*, 2003
15. *Angiopteris evecta* (Frost.) Hoffm.	Angiopterid-aceae	——	Pith and rhizomes boiled with mustard oil is used as massage for rib and bone pain. Lamina bounded to fractured bone.	Throughout India	Singh *et al*, 2003

Continued .....

	Botanical name	Family	Vernacular/ English name	Uses	Distribution	References
16.	*Asplenium adiantum-nigrum* L.	Aspleniaceae	Black spleen wort	Plant bitter, laxative, diuretic, useful in jaundice, opthalmia, cold, chest pain and diseases of spleen.	Throughout India	Vasudeva, 1999; Kirn *et* Kapahi, 2001
17.	*A. aethiopicum* (Burm f.) Bech	Aspleniaceae	——	Whole plant is used in the treatment of enlarged spleen, calculus, Jaundice and malaria.	N.W. Himalayas Peninsular region	Das, 2003
18.	*A. falcatum* Lam.	Aspleniaceae	Spleen wort	Fronds used in malaria, jaundice, enlargement of spleen, urine trouble and calculus.	S. India, Sikkim, Andaman and Nicobar	Dixit *et* Sinha, 2001 Chopara, *et al*, 1956 Singh *et al*, 2003
19.	*A. falginosum* Hook.	Aspleniaceae	——	Rhizome used as anthelminthic.	S. India, Sikkim, Andaman and Nicobar	Vasudeva, 1999
20.	*A. macrophyllum* Sw.	Aspleniaceae	——	Fronds decoction is powerful diuretic, used in cure of defective urine secretion in beriberi	N.W. Himalayas	Das, 2003

Continued .....

Botanical name	Family	Vernacular/ English name	Uses	Distribution	References
21. *A. ruta-muraria* L.	Aspleniaceae	Wall Rue	Fronds expectorant and used as cure for Rickets. Rhizome anthelmintic and astringent	Kashmir	Vasudeva, 1999 Chopara *et al,* 1956 Jacobs, 1964.
22. *A. nidus* L.	Aspleniaceae	Ruai-mangma	Leaves juice given to cure labour pain. Extraction of leaves are useful in removal ice.	N.W. Himalayas	Dixit, 1975. Singh *et al,* 2001.
23. *A. trichomanes* L.	Aspleniaceae	Common spleen wort	Plant bitter, laxative, expectorant used in pulmonary disease and in abscess of uterus. Rhizome used as anthelminthic and for hypochondriac infections. Whole plant extract is used against enlargement of spleen.	N.E. India Peninsular, India	Kirn *et* Kapahi, 2001. Vasudeva, 1999. Asolkar *et al,* 1992. Singh *et al,* 2003 Bir *et al,* 1983.

Continued .....

	Botanical name	Family	Vernacular/ English name	Uses	Distribution	References
24.	*Athyrium filix-faemina* (L.) Roth	Athyriaceae	Lady fern	Rhizome vermifugal and used as anthelminthic.	Throughout India	Vasudeva, 1999. Chopara *et al*, 1956; Singh *et al*, 2003
25.	*Alsophila albosetacea* (Bedd.) Copel.	Cyatheaceae	—	Leaves are locally applied on wounds.	Andaman and Nicobar	Manickam, 1999.
26.	*A. gigantea* Wall. *ex* Hook.	Cyatheaceae	—	Fronds are anti-inflammatory. Rhizomes are antidote.	N.E. Himalayas	Singh *et al*, 2003
27.	*A. spinulosa* Wall. *ex* Hook.	Cyatheaceae	—	Leaves are locally applied on wounds.	Andaman and Nicobar	Manickam, 1999.
28.	*Azolla pinnata* R. Br.	Azollaceae	—	The plant is diuretic and rich in antibacterial properties.	Throughout India	Singh *et al*. 2003
29.	*Blechnum orientale* L.	Blechnaceae	Hastajori	Whole plant used as medicine for diarrhoea and stomach disorders. Plants used as poultice	N.W. Himalayas E. Himalayas Peninsular India Central India	Dixit, 1975. Vasudeva, 1999. Quisumbing,

Continued .....

Botanical name	Family	Vernacular/ English name	Uses	Distribution	References
			for boils. It is used in impotency in men, urinary disorders, cure of sanipat (Delirium)		1951. Kaushik *et* Dhiman, 1995. Singh *et al.*, 2003.
30. *Botrychium lunaria* (L.) Sw.	Botrychiaceae	Moon wort	Fronds are good vulnerary, used in dysentery ruptures, for heading cuts and wounds. Rhizome juice used in treatment of breast cancer.	N.W. Himalayas Peninsular India	Vasudeva, 1999. Dixit, 1975. Kaushik *et.* Dhiman, 1995 Asolkar *et al,* 1992.
31. *B. ternatum* (Thumb.) Sw.	Botrychiaceae	China-yin-tichuch	Fleshy rhizome applied on cuts and wounds. Also used in cure of dysentery.	N.W. Himalayas N.E. India	Vasudeva, 1999; Cauis, 1935; Dixit, 1975
32. *B. virginianum* (L.) Sw.	Botrychiaceae	Rattle snake	Flashy rhizome applied on cuts and bruises. Also used in dysentery.	N.W. Himalayas N.E. India	Vasudeva, 1999. Dixit, 1975.

Continued .....

	Botanical name	Family	Vernacular/ English name	Uses	Distribution	References
33.	*Cephalomanes javanicum* (Bl.) Vd. Bosch.	Hymeno-phyllaceae	——-	Whole plant with garlic and onion smoked to cure headache.	Eastern India	Dixit *et* Sinha, 2001.
34.	*Ceterach officinalum* Willd.	Aspleniaceae	Finger fern	Whole plant used as astringent and diuretic. It is used in infirmities of spleen, help in strangely, anal wastes, stones in bladder and hiccough. Decoction of plant used in Melancholia.	N.W. Himalayas	Vasudeva, 1999. Caius, 1935. Kaushik *et* Dhiman, 1995.
35.	*Ceratopteris thalictroides* (L..) Ad. Brongn.	Parkeriaceae	Water fern	Whole plant used as tonic and styptic. Fronds are used as poultice in skin complains.	N.E. India Central and Peninsular India	Vasudeva, 1999; Dixit, 1975; Singh *et al,* 2003
36.	*Cheilanthes albomarginata* Clarke	Cheilanthaceae	Nanha	Whole plants used against tuberculosis and. It is a strength tonic.	Throughout India	Vasudeva, 1999. Vyas *et* Sharma, 1988; Singh *et al,* 2003

Continued .....

	Botanical name	Family	Vernacular/ English name	Uses	Distribution	References
37.	*C. bicolor* (Roxb.) Griff. *ex* Fras.- Jenk.	Cheilanthaceae	Silver fern	Fronds are useful in seasonal and cold fever.	Throughout India	Asolkar *et al,* 1992. Vasudeva, 1999; Singh *et al,* 2003
38.	*C. farinosa* (Forsk.) Kaulf.	Cheilanthaceae	——	Paste of root applied in cure of eczema and wounds. Decoction of root used in stomachache	Throughout India	Das, 2003 Singh *et al,* 2003
39.	*C. fragrans* (L. f.) Sw.	Cheilanthaceae	——	The whole plant is emmenagogue, used in cold and cough	N.W. Himalayas	Das, 2003
40.	*C. swartzii* Webb.	Cheilanthaceae	——	The fern yields valuable alkaloids cheilanthone, used in the manufacture of various medicines	Central and Peninsular India	Das, 2003
41.	*C. tenuifolia* (Burm. f.) Sw.	Cheilanthaceae	Dodhari	There is a superstition among the tribal that preparation made from roots is given for sickness attributed to	Throughout India	Vasudeva, 1999; Caius, 1935. Manickam, 1999. Singh

Continued .....

	Botanical name	Family	Vernacular/ English name	Uses	Distribution	References
				"evil eye" or "witch craft". Also used as tonic.		*et al,* 2003
42.	*Christella arida* (D. Don) Holtt.	Thelypterid-aceae	——	Rhizome decoction is useful in veterinary larval infections.	Peninsulaa and Central India, N.W. Himalayas	Manickam, 1999; Singh *et al,* 2003
43.	*C. parasitica* (L.) Lev.	Thelypterid-aceae	——	Fronds are useful against rheumatism.	Throughout India	Manickam, 1999. Singh *et al,* 2003
44.	*Cibotium assamicum* Hook.	Dicksoniaceae	Kow chiak	Young stem sore starchy, pith are eaten in times of famine.	N.E. India	Dixit, 1995.
45.	*C. barometz* (L.) J. Sm.	Dicksoniaceae	Mini tree fern	Stem used as styptic and tonic. Rhizome vermifuge, treats tropical wounds, ulcers, stimulating to liver, kidneys and "Old man's remedy". It is said to exercise a special action on genitourinary organs.	N.E. India	Vasudeva, 1999. Usher, 1971. Quisumbing, 1951 Kaushik *et* Dhiman, 1995.

Continued .....

	Botanical name	**Family**	**Vernacular/ English name**	**Uses**	**Distribution**	**References**
46.	*Cyathea brunoniana* (Wall. *ex* Hook.) Clarke *et* Bak.	Cyatheaceae	——	The soft pith used as food and tonic for strength.	N.E. Himalayas Peninsular India	Das, 2003
47.	*Cystopteris fragilis* (L.) Bernh.	Athyriaceae	Bladder fern	The decoction of rhizome is used as anthelminthic enema. It is pectoral, mucilaginous, expectorant and refrigeration tonic.	Himalayas region	Kaushik *et* Dhiman, 1995 Vasudeva, 1999
48.	*Cyclosorus dentatus*	Thelypterid-aceae	——	Fronds are rich in having antibacterial properties.	Throughout India	Das, 2003
49.	*Crytomium falcatum* (L. f.) Presl	Aspleniaceae	Holy fern	Rhizome have anthelminthic properties and chiefly used for expulsion of tapeworm.	Throughout India	Vasudeva, 1999.
50.	*Davallia tricho-manoides* Bl.	Davalliaceae	——	Whole plant is antidote for food poisoning	Andaman and Nicobar N.E. India	Manickam, 1999.

Continued .....

	Botanical name	Family	Vernacular/ English name	Uses	Distribution	References
51.	*Dicranopteris liearis* (Burm. f.) Underw.	Gleicheniaceae	Thicket fern	Fronds are useful in cure of asthama also having antibacterial properties. Young fronds with cow's milk given in woman's sterility. Rhizome anthelminthic Spores used as favorite remedy for diarrhea in children.	Eastern, Central and Peninsular India	Vasudeva, 1999. Asolkar *et al*, 1992. Dixit, 1975. Kaushik *et* Dhiman, 1995 Singh *et al*, 2003
52.	*Diplazium esculentum* (Retz.) Sw	Athyriaceae	Lingra or Kheri	Young tips of fronds are used as tonic for health. Decoction of rhizome and young leaves are useful haemoptysis and constipation.	Throughout India	Kaushik *et* Dhiman, 1995 Quisumbing, 1951; Singh *et al*, 2003
53.	*D. dilatatum* Bl.	Athyriaceae	——	The extraction of fronds is diuretic.	Peninsular and N.E. India	Vasudeva, 1999.
54.	*Dennstaedtia puntilobula*	Dennstaed-tiaceae	——	Nodules at crown of rhizome used for blood hemorrhages of lungs.	Throughout India	Romero, 1954. May, 1987.

Continued .....

	Botanical name	Family	Vernacular/ English name	Uses	Distribution	References
55.	*Drymoglossum piloselloides* (L.) Presl	Polypodiaceae	Polypody	Whole plant is useful against itching and relieving pain. Leaves are used as antiseptic. Used in cure of healing eczema, cough, constipation and gonorrhea.	Peninsular India	Dixit *et* Sinha, 2001.
56.	*Drynaria quericifolia* (L.) J. Sm.	Drynariaceae	Ashvakatri	Whole plant used in hectic fever, dyspepsia, cough, dyspepsia, poulticing swelling and typhoid fever. Rhizome anthelminthic, expectorant, pectoral and astringent	Throughout India	Vasudeva, 1999. Kirtikar *et* Basu, 1918 Dixit, 1975. Chopara *et al*, 1956. Singh *et al*, 2003
57.	*Dryopteris barbigera* (Moore *ex* Hook.) O. Ktz.	Dryopterid-aceae	——	Rhizome have anthelminthic properties. Substitute for the European male fern. Rhizome vermifuge	Himalayas regions	Vasudeva .1999. Kaushik *et* Dhiman, 1995; Chopara *et al*, 1956.

Continued .....

	Botanical name	Family	Vernacular/ English name	Uses	Distribution	References
58.	*D. blanfordii* (Hope) C. Chr.	Dryopterid-aceae	——	Rhizome have anthelminthic properties chiefly used for expulsion of tapeworm.	Himalayas regions	Vasudeva, 1999. Kaushik *et* Dhiman, 1995; Usher, 1971.
59.	*D. chrysocoma* (Christ.) C. Chr.	Dryopterid-aceae	——	Rhizome have anthelminthic properties. Substitute for the Europian male fern. Fronds are good source of "Filcin".	Himalayas regions	Vasudeva 1999; Bir *et at*, 1983 Kaushik *et* Dhiman, 1995
60.	*D. cochleata* (Buch. Ham. *ex* D. Don.) C. Chr.	Dryopterid-aceae	Jatashankari	Juice extract is given in epilepsy and leprosy. Whole plant paste used against snake bite. Rhizome are used in swellings and pain and have antifungal properties.	Himalayas regions	Kaushik *et* Dhiman, 1995. Verma *et al*, 1995. Asolkar *et al*, 1992; Singh *et al*, 2003
61.	*D. filix-mas* (L.) Schott	Dryopterid-aceae	Male fern	Rhizome and stipes yield the drug know	Cosmopolitan	Vasudeva, 1999.

Continued .....

	Botanical name	Family	Vernacular/ English name	Uses	Distribution	References
				as "Male fern". This is an oleo resinous substance that has been used since long for expelling tapeworms.		Verma *et al*, 1995
62.	*D. juxtaposita* Christ.	Dryopterid-aceae	——	Rhizome used as anthelminthic	N.W. Himalayas, N.E. India E. Himalayas, Peninsular India	Vasudeva, 1999.
63.	*D. marginata* (Wall. ex Christ) Christ.	Dryopterid-aceae	——	Rhizome have anthelminthic properties.	E. Himalayas Eastern India N.W. Himalayas	Vasudeva 1999; Kaushik *et* Dhiman, 1995
64.	*D. odontoloma* (Bedd.) C. Chr.	Dryopterid-aceae	——	Rhizome used as anthelminthic.	E. Himalayas Peninsular India	Kaushik *et* Dhiman, 1995
65.	*D. sparsa* (Ham. *ex* D. Don) O. Kunt.	Dryopterid-aceae	——	Rhizome have anthelminthic properties.	N.W. Himalayas Eastern India, N.E. India, Central India	Manickam, 1999. Singh *et al,* 2003

Continued .....

	Botanical name	Family	Vernacular/ English name	Uses	Distribution	References
66.	*Equisetum arvense* L.	Equisetaceae	Gold rush	Dried plant useful in dropsy, gravel and kidney infections. Decoction used in nasal polypus and breast cancers. Ashes are useful in acidity of stomach and dyspepsia. Also used for washing tumors and cancerous lesions of bones. Stems used for treatment of bone fracture.	Himalayas regions	Vasudeva, 1999. Asolkar *et al,* 1992; Lloyd, 1964; Kirn *et* Kapahi, 2001; Burkill, 1966.
67.	*E. diffusum* D. Don	Equisetaceae	——	Whole plants are useful in acidity and dropsy.	Throughout India	Manickam, 1999; Singh *et al,* 2003
68.	*E. ramossissimum* Desf. ssp. *debile* (Roxb. *ex* Vauch.) Hauke	Equisetaceae	Horsetail	The plant is useful in cure of gonorrhoea and treatment of fracture.	Throughout India	Kaushik et Dhiman, 1995; Kirn *et* Kapahi, 2001; Bir *et at*, 1983; Vasudeva 1999; Singh *et al,* 2003

Continued .....

Botanical name	Family	Vernacular/ English name	Uses	Distribution	References
69. *Helminthostachys zeylanica* (L.) Hook.	Helmintho-stachyaceae	——	Fronds juice is used for relieving from blisters Plants is intoxicant, aperients and anodyne used in sciatica. Rhizome 1995; used as a brain tonic for vitality and as a cure for waist pain. Rhizome paste with cow's urine is used in cure of skin disease. It is also useful is spermatorrhea and for improving memory. Also useful in cure of Malaria. It given to promote strength and vitality in male.	N.W. Himalayas N.E. India Peninsular India. Gangetic plain	Vasudeva 1999; Kaushik *et* Dhiman, 1995; Dixit, 1975; Singh *et al.* 1989. Dixit *et* Sinha, 2001. Singh *et* Maheshwari, 1992; Quisumbing, 1951; Upnof, 1968.
70. *Hemionitis arifolia* (Burm.) Moore	Hemionitid-aceae	——	Crushed juice from the fronds used in skin burns	Central and W. India Peninsular India	Das, 2003 Singh *et al,* 2003

Continued .....

	Botanical name	Family	Vernacular/ English name	Uses	Distribution	References
71.	*Hypodematium crenatum* (Forsk.) Kuhn	Hypodemat-aceae	——	Whole plants is used for gynecological disorders and against insect bite. Scales of this fern are used against the effect of "Witch craft or evil eye". Leaves used to facilitate conception in women.	N.W. Himalayas E. Himalayas Central and Peninsular India Gangetic plain	Vyas *et* Sharma, Kaushik *et* Dhiman, 1995. Vasudeva 1999; Singh *et al,* 2003
72.	*Isoetes coromendeliana* L.	Isoetaceae	Quill wart	Orally used in spleen disorders and liver complains.	Throughout India	Singh *et al,* 2003
73.	*Lemmaphyllum carnosum* (Sm.) Presl	——	——	Fronds are pectoral diuretic and astringent. Used in urine calculus and rheumatism. Decoction of fronds used in stop haemorrhages.	N.E. Himalayas E. Himalayas	Das, 2003
74.	*Leucostigia immersa* (Wall. *ex* Hook.) Presl	Davalliaceae	——	The leaves cures constipation.	N.W. Himalayas Central and Peninsula India E. Himalayas	Dixit *et* Das, 1978. Singh *et al,* 2003

Continued .....

	Botanical name	Family	Vernacular/ English name	Uses	Distribution	References
75.	*Loxogramme involuta* (D. Don.) Presl	Loxogramm-aceae	——	The paste of rhizome applied on cuts and	N.W. Himalayas E. Himalayas N.E. and Peninsular India	Singh *et al.*, 2003
76.	*Lycopodium clavatum* L.	Lycopodiaceae	Club Moss	Plant decoction is anti-spasmatic useful in rheumatism diseases of lungs and kidneys. Spores are used as water repellant and protective dusting powder for tender skin.	Throughout India	Vasudeva 1999; Watt, 1932
77.	*Lygodium flexuosum* (L.) Sw.	Lygodiaceae	Kalijar	Fronds boiled with mustard oil useful as local application to carbuncles externally in rheumatism, sprains, scabies, ulcers and cutwounds. Rhizome cures gonorrhea and paste applied on the piles. Spores cure high fever.	Throughout India	Vasudeva, 1999; Dixit, 1975; Chopara *et al*, 1956; Singh *et al*, 2003; Dixit *et* Sinha, 2001; Manickam, 1999.

Continued .....

Botanical name	Family	Vernacular/ English name	Uses	Distribution	References
78. *L. japonicum* (Thunb.) Sw.	Lygodiaceae	Climbing fern	Plant used as expectorant. Used in cure of haematuria and An amulet of this plants is tied round the neck of patient suffering from malaria. It is diuretic or cathartic Rhizome decoction taken by ladies to regularize menstruation.	N.E. India N.W. Himalayas Peninsular India	Vasudeva, 1999; Das, 1997; Dixit, 1975; Kirn *et* Kapahi, 2000; Manickam, 1999.
79. *L. circinatum* (Burm.) Sw.	Lygodiaceae	——	Stipe chewed and applied against bite of venomous snake or insect. Fronds used in externally application for wounds. Leaves used in rheumatic pain in gynecological disorders.	N.E. India, Andaman and Nicobar N.W. Himalayas	Quisunbing, 1951; Dixit *et* Sinha, 2001.
80. *L. microphyllum* (Cav.) R. Br.	Lygodiaceae	——	Decoction of leaves used in dysentery and also useful in skin disorders. Leaves cure hiccough. Rhizome used as astringent.	N.W. Himalayas E. Himalayas Central India Gangetic plain Andaman and Nicobar	Dixit, 1975. Bouquest, 1974; Dixit *et* Sinha, 2001; Singh *et al,* 2003.

Continued .....

	Botanical name	Family	Vernacular/ English name	Uses	Distribution	References
81.	*L. longifolium* (willd.) Sw.	Lygodiaceae	——	Fronds chewed with ash salt to cure stomach ache and and diarrhea.	Peninsular India	Holdsworth *et al,* 1975.
82.	*Marsilea minuta* L.	Marsileaceae	Water clover	Decoction of leaves mixed with ginger given in bronchitis, spastic condition of leg, cough, sedative and insomnia. Fresh young leaves are in form of juice is effective in urinary disorder. Leaves extract with sugar or fishes is useful in migraine. Drops of juice are very effective eye disease.	Throughout India except Kashmir	Vasudeva, 1999; Roy *et* Gupta, 1965. Dixit, 1975 Dixit *et* Sinha, 2001. Dadhich *et* Sharma, 2000; Jha *et* Verma, 1996. Singh *et al,* 2003; Vyas *et* Sharma, 1988.
83.	*M. quadrifolia* L.	Marsileaceae	——	Whole plants is useful in bone fracture.	North India	Manickam, 1999.
84.	*Marattia fraxinea* Smith	Marattiaceae	——	Young leaves are used remedy for bronchial and intestinal catarrh.	Peninsular India	Dixit, 1975.

Continued .....

	Botanical name	Family	Vernacular/ English name	Uses	Distribution	References
85.	*Microlepia speluncae* (L.) Moore	Dennstaed-tiaceae	——	Leaves are effective in curing of high fever,	N.W. Himalayas N.E. and Central India Andaman and Nicobar	Dixit *et* Sinha, 2001. Manickam, 1999. Singh *et al,* 2003
86.	*Microsorium punctatum* (L.) Copel.	Polypodiaceae	——	Whole plant used for curing snake bite. Used as purgative, diuretic and wound healer.	E. Himalayas N.E and Peninsular India. Andaman Nicobar	Dixit *et* Sinha, 2001. Bouquest, 1974.
87.	*M. alternifolium* (Willd.) Copel.	Polypodiaceae	——	Fronds are used as strength tonic.	N.E. Himalayas, Western Ghat	Das, 2003
88.	*Nephrolepis cordifolia* (L.) Presel	Nephrolepid-aceae	Sanay	Rhizome and tubers are used in intestinal and kidney disorders. Extract of rhizome is used in permanent curing of sterility in women. Decoction used for cough. Rhizome stock is applied locally in skin disease.	Throughout India	Dixit, 1975. Jamir, 1997. Henry *et al,* 1996. Quisumbing, 1951. Singh *et al,* 2003

Continued .....

Botanical name	Family	Vernacular/ English name	Uses	Distribution	References
89. *N. auriculata* (L.) Trimen	Nephrolepid-aceae	Banu	A decoction of fresh tubers and fronds are given in cough.	N.W. Himalayas Peninsular India. N.E. India	Singh, 1993.
90. *N. biserrata* (Sw.) Schott.	Nephrolepid-aceae	——	Haemostatic treatment for wounds.	N.E. and Peninsular India.	Bouquest, 1974.
91. *N. exaltata* (L.) Scott	Nephrolepid-aceae	——	Chew the washes tubers for sinus trouble, tooth-ache, kidney and liver problems. Rhizome extract used against women sterility. Juice of fresh fronds used in checking the blood.	Throughput India	Singh *et al*, 2003
92. *Oleandra wallicii* (Hook.) Presl	Nephrolepid-aceae	——	The rhizome is reported to possess rejuvenating and beneficial to the aged.	N.W. Himalayas N.E. India	Das, 2003
93. *Ophioglossum costatum* R. Br	Ophiogloss-aceae	Sheambli	Dried tubers are crushed in the forms of power mixed with mustard oil used in skin disease.	Throughout India	Dadhich *et* Sharma, 2002; Singh *et al*, 2003

Continued .....

	Botanical name	Family	Vernacular/ English name	Uses	Distribution	References
94.	*O. gramineaum* Willd.	Ophiogloss-aceae	——	The whole plant used as astringent. Fronds are styptic used in contusions Wounds, boils and haemorrhages.	Throughout India	Singh *et al,* 2003
95.	*O. pendulum* L.	Ophiogloss-aceae	——	Spores given to babies at birth to purge me conium.	E. Himalayas, N.E. India Andaman Nicobar	Fosberg, 1942.
96.	*O. reticulatum* L.	Ophiogloss-aceae	Vanpalak	Fronds are used in preparation of tonic. It is useful in relief of headache. Paste of fresh leaves and tubers applied on boils.	Throughout India	Vasudeva, 1999. Singh *et al,* 1989; Singh *et al,* 2003
97.	*O. vulgatum* L.	Ophiogloss-aceae	Adder's Tongue	Fronds used as tonic and styptic. Antidote for snake bite Whole plants vulnerary, in cuts and wound. Decoction of fronds used in angina and as lotion for boils.	N.E. Himalayas	Vasudeva, 1999; Caius, 1935; Kaushik *et* Dhiman, 1995.

Continued .....

	Botanical name	Family	Vernacular/ English name	Uses	Distribution	References
98.	*Ophioderma pendula* (L.) Presl.	Ophiogloss- aceae	——	Fronds mixed with coconut oil prevent falling hair.	Andaman Nicobar N.E. India	Dixit *et* Sinha, 2001.
99.	*Osmunda clatoniana* L.	Osmundaceae	Interrupted fern	Rhizome used as adulteration for male fern.	N.W. Himalayas N.E. India	Vasudeva, 1999. Kaushik *et* Dhiman, 1995
100.	*O. regalis* L.	Osmundaceae	Royal fern	Plants used as tonic and styptic. Young fronds cure toothache, rheumatism and internally for intestinal gripping.	Throughout India	Vasudeva, 1999; Dixit, 1975; Chopara *et al*, 1956. Manickam, 1999.
101.	*O. hugeliana* Presl	Osmundaceae	——	Plants used as tonic and styptic Young fronds cure toothache, rheumatism and internally for intestinal gripping.	E. Himalayas Central and Peninsular India	Singh *et al*, 2003

Continued .....

	Botanical name	Family	Vernacular/ English name	Uses	Distribution	References
102.	*Onychium siliculosum* (Desv.) C. Chr.	Cryptogram-maceae	—	Fronds used for dysentery and to check/ prevent falling hairs.	N.E. Himalayas N.W. Himalayas Andaman Nicobar	Dixit *et* Sinha, 2001. Manickam, 1999.
103.	*Palhinhaea cernua* (L.) Franco. *et* Vasc	Lycopodiaceae	—	Leaves are effective in skin eruptions, beri-beri	Throughout India	Dixit *et* Sinha, 2001; Singh *et al,* 2003
104.	*Pellaea calomelanos* (Sw.) Link	Sinopteridaceae	—	Rhizome are anthel-mintic used in cold and headache.	N.W. Himalayas	Vasudeva, 1999; Caius, 1935
105.	*Phlegmariurus phegmaria* (Herter) Holub	Lycopodiaceae	—	Rhizomes are useful in dropsy, emmenogogue.	Andaman Nicobar	Manickam, 1999.
106.	*Phymatosorus longissimus* (Bl.) Pic. Ser.	Polypodiaceae	—	Fronds used as strength tonic.	N.E Himalayas Peninsular India Nicobar	Das, 2003
107.	*P. scolopendria* (Burm. f.) Pic.-Ser.	Polypodiaceae	—	Fronds used for curing chronic diarrhea, dysentery and fore gonorrhea.	Peninsular India Nicobar	Dixit *et* Sinha, 2001.

Continued .....

	Botanical name	Family	Vernacular/ English name	Uses	Distribution	References
108.	*Pityrogramma calomelanos* (L.) Link	Hemionitidaceae	——	Whole plant decoction remedy for kidney trouble. Tea prepared by its fronds useful in flu, fever and hypertension.	Throughout India	Quisumbing, 1951; Croft, 1976.
109.	*Pleopeltis macrocarpa* (Bory *ex* Willd.) Kaulf	Polypodiaceae	——	Whole plants used in form of tea to cure itching.	N.E. Himalayas Central and Peninsular India	Vasudeva, 1999. Kaushik *et* Dhiman, 1995; Singh *et al,* 2003; Manickam, 1999.
110.	*Polypodium vulgare* L.	Polypodiaceae	——	Rhizome are anthelmintic and purgative.	H.W. Himalayas	Vasudeva, 1999; Manickam, 1999.
111.	*Psilotum nudum* (L.) P. Beauv.	Psilotaceae	——	Whole plant used as purgative.	Throughout India in Hilly region.	Dixit & Sinha, 2001. Singh *et al,* 2003

Continued .....

	Botanical name	Family	Vernacular/ English name	Uses	Distribution	References
112.	*Pteris biaurita* L.	Pteridaceae	——	The paste of Stipe and stem is useful in wounds.	Throughout India	Manickam, 1999; Singh *et al*, 2003
113.	*P. criteca* L.	Pteridaceae	——	The paste of founds applied of cuts and wounds	N.E Himalayas Central and South India	Singh *et al*, 2003
114.	*P. ensiformis* Burm. f.	Pteridaceae	Laich ang khrang	Leaf juice used as an astringent to cure dysentery. It is also used as diuretic and cooling agent and to cure malaria. Decoction of fresh fronds given in the glandular swelling of neck.	N.E. Himalayas Peninsular India	Dixit *et* Sinha, 2001. Singh *et al*, 2001; Burkill, 1966; Quisumbing, 1951.
115.	*P. quardriaurita* Retz.	Pteridaceae	——	The paste of rhizomes applied to take and hasten the recovery of boils.	Central and Peninsular India	Singh *et al*, 2003
116.	*P. multifida* Poir.	Pteridaceae	——	Decoction of whole plant is used in	N.W. Himalayas Peninsular India	Das, 2003

Continued .....

Botanical name	Family	Vernacular/ English name	Uses	Distribution	References
			dysentery, also a good vermifuge. Paste of rhizome and fronds with seasame oil applied on skin affections of infants.		
117. *P. tripartica* Sw.	Pteridaceae	——	Fronds are used during parturition.	Peninsular India	Dixit *et* Sinha, 2001
118. *P. vittata* L.	Pteridaceae	Jasumba	Rhizome used as demulcent. It used in glandular swelling	Throughout India	Dixit *et* Sinha, 2001 Dadhich *et* Sharma, 2002; Singh *et al.*, 2003
119. *P. wallichiana* Agardh	Pteridaceae	Unio	Fresh leaves are crushed and applied on cuts to stop bleeding and also on wounds for healing.	N.E. Himalayas	Dixit *et al.*, 1978

Continued .....

	Botanical name	Family	Vernacular/ English name	Uses	Distribution	References
120.	*Pteridium aquilinum* (L.) Kuhn.	Pteridiaceae	Bracken fern	Rhizome have astringent and anthelminthic properties. The decoction of rhizome and fronds is given in chronic disorders arising from obstructions of viscera and spleen.	Throughout India	Dixit, 1975. Chopara *et al*, 1956. Manickam, 1999; Singh *et al*, 2003
121.	*P. aquilinum* var. *wightianum* (L.) Kuhn.	Pteridiaceae	Kakaie	Plants boiled with mustard oil is locally applied on Wounds and skins affections of cattle.	N.E. Himalayas	Kirn *et* Kapahi, 2000. Chopera *et al*, 1956.
122.	*Pyrossia adnascens* (Sw.) Ching	Polypodiaceae	——	Fronds used in skin burn and in dysentery.	N.W. Himalayas Peninsular India	Dixit *et* Sinha, 2001. Manickam, 1999; Singh *et al*, 2003
123.	*P. lanceolata* (L.) Forewell	Polypodiaceae	——	Decoction of plant used in fever, cough and cold. Fronds paste used as itch guard.	N.W. Himalayas Andaman Nicobar	Singh *et al*, 2003

Continued .....

	Botanical name	Family	Vernacular/ English name	Uses	Distribution	References
124.	*Selaginella bryopteris* (L.) Bak.	Selaginellaceae	Sanjivani	Whole plans diuretic and gonorrhea. Dried plants with Tobacoo smoked for hallucination. Plants of leaves with sugar is taken in stomachache, inflammation of urinary tract and in some venereal disease. A popular strength tonic amongst local people. Effective in spermatorrhoea and leucorrhoea.	N.W. Himalayas Andaman Nicobar	Vasudeva, 1999; Singh *et al*, 1989. Shah *et al*, 1990; Kaushik *et* Dhiman, 1995; Singh *et al*, 2003 Manickam, 1999.
125.	*S. ciliaris* (Retz.) Spring	Selaginellaceae	——	Plant extract is used in stopping bleeding from fresh	N.W. Himalayas Andaman Nicobar	Singh *et al*, 2001; Singh *et al*, 2003
126.	*S. involvens* (Sw.) Spring	Selaginellaceae	———	An infusion of plant is administrated in cases of high fever. Also used in the prolapsed of bleeding, piles and gravel aminorrhoea.	Throughout India	Das, 2003 Singh *et al*, 2003

Continued .....

	Botanical name	Family	Vernacular/ English name	Uses	Distribution	References
127.	*S. radicata* (Hook. *et* Grev.) Spring	Selaginellaceae	——	Fronds are antibacterial.	Central India	Singh *et al,* 2003
128.	*S. willdenovii* (Desv. *ex* Poir.) Bak.	Selaginellaceae	——	Whole plants is useful in backache and rheumatism.	Andaman and Nicobar	Manickam, 1999
129.	*Sphenomeris chinensis* (L.) Maxon	Lindsaeaceae	Farsley fern	Fronds paste used internally for chronic enteritis.	Throughout India	Vasudeva, 1999; Manickam, 1999; Singh *et al,* 2003
130.	*Schizaea dichotoma* (L.) Sm.	Schizaeaceae	——	Juices of fronds to cure cold and cough. It is useful in child birth.	Peninsular India	Dixit *et* Sinha, 2001.
131.	*Stenochlaena palusris* (Burm. f.) Bedd.	Stenochl-aenaceae	——	Whole plants used to stomachache and remittent fever. Cooling agent and for the treatment of burns and ulcers.	N.W. Himalayas Peninsular India	Dixit *et* Sinha, 2001. Manickam, 1999.

Continued .....

	Botanical name	Family	Vernacular/ English name	Uses	Distribution	References
132.	*Taenitis blechnoides* (Willd.) Sw.	Taenitidaceae	——	Whole plants used as tonic for postpartum prospective medicine.	Nicobar, Penninsular India	Dixit *et* Sinha, 2001.
133.	*Tectaria cicutario* (L.) Copel.	Tectariaceae	Harishakar pattor	10 gm. Rhizome along with seven fruits of black pepper mixed with cow's milk given once daily as tonic and to purified blood and also in blood dysentery. About 10 gm. of rhizome paste with *Zingiber purpureum* Rosc. and *Croton roxburghii* Balak. in equal proportion is given in menstrual problems. About 15 gm. of rhizome power with root of *Cissampelos pareira* L. and *Smilax ovalifolia* Roxb. in equal proportion is given in leucorrhoea and body temperature.	N.W. Himalayas	Brahman *et* Saxena, 1990 Brahman *et* Saxena, 1990

Continued .....

	Botanical name	Family	Vernacular/ English name	Uses	Distribution	References
134.	*T. coadunata* (Smith) Chr.	Tectariaceae	——	The decoction of rhizome is anthelmintic, Used in asthma bronchitis in getting relief from bites and stings by insect and centipedes.	Throughout India	Pande *et al*, 1994; Manickam, 1999; Vasudeva, 1999; Vyas *et* Sharma, 1988; Singh *et al*, 2003
135.	*Tectaria polymorpha* (Wall. *ex* Hook.) Copel	Tectariaceae	——	Rhizome used as anthelminthic.	N.W. Himalayas N.E. India Peninsular	Vasudeva, 1999; Kaushik *et* Dhiman, 1995; Singh *et al*, 2003
136.	*Vittaria elongata* Sw	Vittariaceae	——	Leaf juice used for curing rheumatic pain and stiffness	Throughout India in hilly region.	Dixit *et* Sinha, 2001. Manickam, 1999
137.	*Vittaria zosterifolia* Willd.	Vittariaceae	——	Whole Plant is useful in skin disease.	N.E. Himalayas Andaman and Nicobar	Manickam, 1999.

and Arora, 1999). The main bio-diversity centers are : Himalayas, Gangetic plains, Western and Southern Ghats and Thar Desert. Maximum pteridophytic diversity of the country is occupied by Northwestern and Eastern Himalayas region (Sikkim, West Bengal, Arunachal Pradesh, Assam, Nagaland, Manipur, Meghalaya, Mizorum and Tripura), Southern and Western Ghats (Nilgiri hills, Palni hills, Palghat gap and Salient Valley). In Central region Madhya Pradesh (Pachmarhi and Amarkantak), Chhattishgarh (Bailadilla hills) and Orissa (Parasnath hills) are the rich sites for Pteridophytic diversity. Gangetic plains and Aravalli hills or towards Northwest region, the percentage of Pteridophytic diversity is too poor except occurrence of few species.

The study deals with documentation of medicinally important Pteridophytes along with their uses, which are prevalent in present political boundary of the country and may be lost if not properly documented. In the lack of proper education, excessive collection, rapid industrialization and biotic interference the number of these useful species of Pteridophytes are likely to be decreasing continuously, therefore it is urgent need to formulate conservation strategies to save the medicinally rich Pteridophytic plant diversity occurring in the country.

The present paper consolidates all such information available on medicinal importance of Indian Pteridophytes in condensed tabular form to enable and access all available publication at one place (Table-1). The table provides current botanical name, family, vernacular/English name (wherever available) along with medicinal importance, distribution and the authority of **137 species** of Pteridophytes. It is hope that the workers in India may initiate the work on this neglected group of plant.

METHODOLOGY

Study is based on extensive survey cum collection practices in the rural and tribal pockets of various parts of India. The information collected was verified by present available literature and herbarium consultation as well as by cross-checking with tribal living herbal medicine practitioners of various ethnic groups though interviews, discussions, personal contacts and keen observations.

RESULTS

All known 137 Pteridophytic species from the Indian region are enumerated in a comprehensive table with botanical name, vernacular name/English, medicinal importance and distribution along with related authorities under family. The families are arranged after Pichi Sermolli (1977) classification.

DISCUSSION AND CONCLUSION

The present work is the result of intensive systematic medicinal and ethnomedicinal observation conducted throughout the country. All the 137 species of Pteridophyes of medicinal interest are recorded after critical screening with the available literature. These are recommended for further phytochemical / pharmacological investigation and nutritional analysis, which might result in the discovery of new drug molecules for human welfare.

In India about 1200 species of Pteridophytes are distributed throughout, in which numerous species are traditionally and ethomedicinally used by the local tribals. They may be lost if their medicinal aspects are not properly documented. They are facing several hurdles for their survival due to continuous threat of habitat destruction and therefore, their number have been enormously affected. At present, their population densities have been decreasing continuously at an alarming rate especially in respect of such restricted taxa, which are endemic to India and medicinally important. These plants are extracting widely along with the rhizome resulting rarity of species from their natural habitat throughout the country. Therefore it is essential to conserve these species before extinction by way of their cultivation and stopping wide extraction from the forest area.

Public awareness programme should be intensified. Publicity plays a big role in building up the opinion and participation of the local inhabitants in the conservation programme and it will further help in developing a general awareness within the society regarding the conservation. Botanists should discharge their duty in building up the public opinion in favour of conservation of the biodiversity of the country. It is hoped that the workers in India may also initiate work on this neglected group of plant (Pteridophtyes).

ACKNOWLEDGEMENTS

Grateful thanks are due to Dr. M. Sanjappa, Director, Botanical Survey of India, Kolkata and Dr. V. S. Manickam, Director, Centre of Biodiversity and Biotechnology Research, Department of Botany, St. Xavier's (Autonomous) College, Palayamkottai for the encouragement and facilities.

REFERENCES

Asolkar, L.V., Kakkar, K.K. and Chakae, O.J. (1992). *Glossary of Indian Medicinal plants with active principals.* Pt. 1.CSIR. New Delhi.

Ambasta, S.P. (Ed.) (1986). The useful plants of India CSIR. New Delhi.

Bhardwaja T.N. and Garg, A. (1984). The antifertility effect of an Australian species of the aquatic fern *Marsilia minuta* L. *Indian Fern J.*, **1** : 75-82.

Bir, S.S., Satija, C.K., Vasudeva, M. and Goyal, Pramod (1983). Pteridophytes Flora of Garhwal Himalaya Jungal. (Kishore & Co. Dehradun).

Bouquet, A. (1974). *Plantes Medicinales de la cote d' Ivoire Oratom.* Paris.

Brahaman, M.A. and Saxena, H.O. (1990). Ethonobotany of Gandhamardan Hills - Some note worthy Folk Medicinal uses. *Ethnobotany,* **2** : 71-79.

Burkill, I.H. (1966). *A Dictionary of the economic products of the Malaya Peninsula,* (Gov't of Malaysia and Singapore).

Caius, J.F. (1935). The medicinal and poisonous ferns of India. *J. Bombay Nat. Hist. Soc.,* **38** : 341-361.

Chandra, S. (2000). *The Ferns of India (Enumeration, Synonyms & Distribution),* International Book Distributors, Dehradun.

Chopra, R.N., Nayar, S.L. and Chopara, L.C. (1956). *Glossary of Indian Medicinal Plants.* (CSIR New Delhi).

Copeland, E.B. (1942). Edible Ferns *Am. Fern J.,* **32** : 121-126.

Croft, Jim (1982). *Ferns and Man in New Guinea.* Centre for plants biodiversity research Papua, (New Guinea Botany Society).

Dadhichi, L.K. and Sharma, A.P. (2002). *Biodiversity strategies for conservation* (Dr. S.K. Aggarwal Comm. Vol.) A.P.H. Publishing Corporation New Delhi. 223-235.

Das, A.K. (1997). Less known uses of plants among the aids of Arunachal Pradesh. *Ethnobotany,* 9 : 90-93.

Das, Silpi (2003). Usefulness of Pteridophytes in India with special refernce to medicin and conservation. *J. Econ. Tax. Bot.,* **27(1)** : 7-16.

Dixit, R.D. (1975). Fern—a much neglected group of medicinal plants –I. *J. Res. Indian Med.,* **10(2)** : 74-90.

Dixit, R.D., Das, Anjali and Kar, B.D. (1978). Studies on Ethnobotany-III. On some less known edible Ecomomic and Medicinal ferns of Darjeeling District. W. B. *Nagarjun,* **21(10)** : 1-4.

Dixit, R.D. (1984). *A Census of the Indian pteridophytes,* Botanical survey of India, Calcatta.

Dixit, R.D. and Sinha, B.K. (2001). *Pteridophytes of Andaman Nicobar Islands,* Bishen Singh Mahendra Pal Singh, Dehradun.

Dymock, William (1890). *Pharmacographia indica*-A History of the Principal Drugs of Vegetable Origin met within British India. (Kegan Paul, London).

Fosberg, F.R. (1942). Uses of Hawaiian Ferns. *Amer. Fern J.,* **32** : 15-23.

Henry, A.N., Hosagoudor, V.B. and Kumar, R. (1996). Ethnomedicobotany of the Southern and Western Ghats of India in S.K. Jain (Ed). *Ethnobiology in Human Welfare.* 173-86. New Delhi.

Holdsworth, D.K. and Giheno, J. (1975). A preliminary survey of highland medicinal plants. *Sci. New Guinea,* **3** : 198.

Jain, S.K. (1991). Dictionary of Indian folk medicine and ethnobotany. Deep Publication. New Delhi.

Jain, S.K. (2000). Human aspects of plants diversity. *Economic Botany,* **54(4)** : 459-470.

Jamir, N.S. (1997). Ethnobotany of Naga tride in Nagaland. Medicinal Herbs. *Ethnobotany* **9** : 101-104.

Jacobs, Marion Lee (1958). *Index of plants of North Carolina with reputed medicinal uses.* (Chapel Hill. North Carolina).

Jha, R.R. and Verma, S.K. (1996). Ethnobotany of Sauria Paharia of Santhal Paragna, Bihar-I Medicinal plants. *Ethnobotany,* **8** : 31-35.

Joshi, Pramila (1997). Ethnobotany of pteridophytes of hilly district of Uttar Pradesh, India. *Indian Fern J.,* **14** : 14-18.

Kapur, S.K. and Sarin, Y.K. (1977). Useful Medicinal Ferns of Jammu Kashmir. *Indian Durgs,* **14(7)** : 136-140.

Kaushik, P. and Dhiman, A.K. (1995). Common medicinal pteridophytes. *Indian Fern J.,* **12** : 139-145.

Kirtikar, K.R. and Basu, B.D. (1918). *Indian MedicinalPlants.* Lalit Mohan Bose. Leader Road, Allahabad.

Kirn, H.S. and Kapahi, B.K. (2001). Ethnobotanical notes of some ferns and ferns allies of Jammu Kashmir, State. India. *Indian Fern J.,* **18** : 35-38.

Lloyd, Robert M. (1964). Ethnobotanical uses of California Pteridophytes by Western Indian. *Amer. Fern. J.,* **54** : 56-140.

May, L.W. (1978). The economic uses and associated folklore of ferns and ferns allies. *Botanical Review,* **4(4)** : 491-528.

Manickam, V.S. (1999). Medicinal ferns of India. *Amruth* (Feb): 3-9.

Nayar, B.K. (1957). Medicinal Ferns of India. *Bull Nat. Bot. Garden,* **29** : 1-36.

Niranjan, A.K.S. and Dubey, J.P. (1978). Some Medicinal Ferns from Pithoragarh Hills. *J. Sci. Res.* (B.H.U.), **29(10)** : 183-189.

Pande, P.C. and Pangtey, Y.P.S. (1987). Studies on the Ethnobotany-I On less known Edible and economic ferns of Kumaun regions of W. Himal. *J. Eco. Tax. Bot.,* **2(1)** : 81-85.

Pande, P.C., Joshi, P. and Pande, H.C. (1994). Studies on type Ethnobotany-II On less known Edible and economic ferns of Kumaun regions of W. Himal. *Curr. Res. Pla. Sci.* (Eds) Sharma, T.A., Saini, S.S., Trivari, M.L. and Sharma, M. (Bishan Singh Mehandra Pal Singh. Dehradun).

Puri, G.S. and Arora, R.K. (1961). Some medicinal ferns from Western India. *Indian For.,* **87** : 179-188.

Puri, H.S. (1970). Indian Pteridophytes used in folk remedies. *Amer. Fern. J.,* **56** : 79-81.

Quisumbing, Eduardo (1951). Medicinal Plants of Philippines. *Manila Dept. Arg. Nat. Res. Bull.,* **16(2)** : 174-176.

Roy, P. and Gupta, H.N. (1965). Cataka Samhita (A scientific synopsis) *Nat. Inst. Sci. India.* 1-120. (History of Science in India publication).

Remero, John Bruno (1954). The botanical Lore of California Indians. Vantage Press. N.Y.

Scully, Virginia (1970). *A treasury of American Indian Herbs.* Crown. N.Y.

Shankar, Rama and Khare, P.K. (1992). Ethnobotanical observation on some ferns of Pachmarhi hills. *Econ. Tax. Bot. Add. Ser.*, **10** : 97-100.

Shah, N.C. and Singh, S.C. (1990). Hitherto Unreported Phototherapeutic uses from Tribals of M.P. (India). *Ethnobotany,* **2** : 91-95.

Singh, J.B. (1969). Some medicinal ferns of Sikkim Himalayas. *Indian J. Medi Res.*, **3** : 71-73.

Singh, J.B. (1969-1970). Some medicinal ferns of Pachmarhi hills (M.P.) *J. Sci. Res. BHU,* **20** : 227-230.

Singh, S. (1993). Some edible ferns and fern-allies of North West Himalayas having promising food value. *Indian. J. Fer.*, **16(2)** : 174-176.

Singh, K.K., Sahu, S. and Maheshwari, J.K. (1989). Ethnobotanical uses of some ferns among the Tribals of Uttar Pradesh. India. *Indian Fern. J.*, **6** : 63-67.

Singh, L.S., Singh, P.K. and Singh, E. Jadu (2001). Ethnobotanical uses of some Pteridophytic species in Manipur, India. *Indian. Fern. J.*, **18** : 14-17.

Singh, Shweta, Dixit, R.D. and Sahu, T.R. (2003). Some medicinally important Pteridophytes of Central India. *Int. J. For. Usuf. Magt.*, **4(2)** : 41-50.

Smith, H.H. (1924). Ethnobotany of Menomini Indians. *Bull. Pub. Muse. City Milwaukee.* Green wood Press, Westport). 4-174.

Turner, N. and Marcus, Bell. A.M. (1971). The Ethnobotany of the Coast Salish Indians of Vancouver Islands. *Economic Botany,* **25** : 63-99.

Uddin, Md. G. Mirza, M.M. and Pasha, M.K. (1998). *Bangaladesh J. Pl. Tax.*, **5(2)** : 29-41.

User, G. (1971). *A dictionary of plants used by Man.* (Hafner Press. N.Y.)

Upnof, J.C. Rh. (1968). *Dictionary of Economic Plants.* (Stechert Hafner. N.Y).

Vasudeva, S.M. (1999). Economic Importance of Pteridophytes. *Indian Fern J.*, **16** : 130-152.

Verma, P., Khan, A.A. and Singh, K.K. (1995). Traditional Phytotherapy among the Baiga Tribe of Shadol District of Madhya Pradesh. India. *Ethnobotany.*

Vyas, M.S. and Sharm, B.B. (1988). Ethnobotanicalimportant of ferns of Rajasthan, *Indigenous medicinal plants including microbes* and fungi (Ed) Kaushik P. (Todays and Tommarow Printers and Publications, New Delhi.).

Watt, J.M. and Beryer-Brandwijk, M.G. (1962). *The medicinal and poisonous plants of Southern and Eastern Africa.* (2nd ed. Edinburgh, E & S. Livingstone).

Recent Advances in Medicinal Plant Research : Vision 21st Century, 2005
Editors : S.K. Prabhuji, G.P. Rao and S.K. Patil
Satish Publication House, New Delhi

Chapter 7

IDENTIFICATION, CHARACTERIZATION AND CONSERVATION OF *EMBLICA OFFICINALIS* (AMLA)

Divya Goyal, Seema Bhadauria and Kaushal Pratap Singh
Department of Botany, Raja Balwant Singh College, Agra - 282 002 (U.P.)
e-mail : divyajrf@yahoo.com

Emblica officinalis (Amla) is an important indigenous tree of Central and Southern India. Fruit of *Emblica* is highly nutritious and therapeutically important. In early years there were three main varieties of *Emblica* - Banarasi, Chakaiya and Francis. Kanchan, Krishna, Balwant and Neelam are selected varieties from these three, which are grown in Agra. Seven varieties of *Emblica* were characterized on the basis of morphological characters- Tree shape, Height of tree, No. of fruits per tree, Yield per tree, Fruit color, Fruit shape, Fruit size, Percent seed, Percent pulp, Fruit weight, No. of Segments per fruit, Leaf area, No. of leaves/10 cm long determinate shoot, Fruit astringency and quality of juice. There were many morphological differences between these varieties. Since *Emblica* has many uses, uncontrolled collection may lead to its extinction. Thus, it needs proper identification and conservation.

INTRODUCTION

The 'Mother of herbs' - *Emblica officinalis* belongs to family Euphorbiaceae. It is one of the most useful medicines in Indian Pharmacopoeia, and is considered to be one of the strongest rejuvenatives (rasayana), particularly for the blood, bones, liver and heart. It is an exceptionally rich source of vitamin C, containing thirty times the amount found in oranges. *Phyllanthus emblica* Linn. is synonymous to *Emblica officinalis* Gaertn. *Emblica* has got various names: Emblic, Myrobalan tree, Indian Gooseberry (English), Amalkamu, Urirkai (Andhra Pradesh), Amalaki (Bengal), Amali, Ambala (Gujrat), Aonla (Himachal Pradesh, Orissa and Punjab), Amla, Onilika (Hindi), Amalka (Karnataka), Nelli (Kerala and Tamilnadu), Alathanda, Khondona, Adiphala, Dhatri, Amalaka, Shriphalam and Vrittophalam (Sanskrit). Its fruit is used for ailments like fever, liver disorder, indigestion, anaemia, heart complaints, scurvy, pulmonary tuberculosis, etc. It is a potent anti-oxidant and antimutagen. It reduces the incidence of heart diseases and blood sugar in diabetic patients. It stimulates bone marrow cells and the production of WBC. It is effective in gastric syndrome, raises protein level and increases body weight. *Emblica* is used for a wide variety of anti-inflammatory conditions, including haemorrhoids, gastritis and colitis. Its fruits are also used to make pickle, shampoo and hair dye.

Emblica is found in subtropics and tropical part of China, India, Indonesia and Malay Peninsula. It is met with wild or cultivated throughout the greater part of India– Jammu and Kashmir, Gujrat, Tamilnadu, Karnataka, Himachal Pradesh, Haryana, Punjab, Rajasthan and Andhra Pradesh. The tree of *Emblica* is found growing in plains and Sub Mountains on tracts all over the Indian subcontinent from 200 to 1500 m above sea level. The plant genus *Phyllanthus* (Euphorbiaceae) consisting of approximately 550 to 750 species.

It is medium sized deciduous tree with usually a crooked trunk, average height being 5.5 meters. Its bark is light brown to black. Bark peels small irregular flakes or strips that resemble the bark of *Anogeissus latifolia.* Average girth of main stem is 70 cm. In most cases the main trunk is divided into 2 to 7 scaffolds very near the base. Its branch and branchlets are mostly weak and drooping. Branchlets are alternate and superposed. They all face one plane. Length of branchlet is upto 40 cm. More than hundred leaves are found to be present in a branchlet. Tree bears two types of shoots on the basis of growth characteristics- long or indeterminate and short or determinate. The indeterminate shoots are longer and do not bear flowers while smaller, determinate shoots bear leaves arranged so closely that shoot appears like pinnately compound leaf. Small sized leaf measures about 1.5 cm × 0.3 cm. Leaves are linear, apiculate at tip, pale green glabrous or puberulous beneath, stipules minute, ovate, finely acute. Inflorescence is recemose type. Flowering and fruiting behaviour of this plant is quite variable. In Northern India, flowers appear on determinant shoots from February onwards while in Southern India, flowers appear twice in a year-first in February to March and then in June to July. Growth studies have revealed that the fruit remains dormant while extension growth of the shoots continue at a rapid rate. With cessation of vegetative growth, initiation of flowering and fruit growth takes place. *Emblica* does not show any symptom of fruit growth from March to July. Fruits start growing from August and continue to gain the size. Young fruitlet remains encased in a cup like structure. Fruit is oval, round or triangular in shape. Colour of fruit is whitish green to yellowish green, skin smooth to rough, segments 6-8 and seed coat hard.

Emblica is cultivated commercially mainly in Raibareilly, Sultanpur, Azamgarh, Allahabad, Bareilly, Nainital, Agra, Mathura, Lucknow and Kanpur in Uttar Pradesh. Singh and Arora (1967) studied two varieties of Amla on the basis of physical and chemical properties of mature fruits. Singh (1973) described three varieties grown in U.P., particularly in Banaras and Pratapgarh district. These are Banarasi, Chakaiya and Francis. New varieties are Balwant, Kanchan, Krishna, Amrit, Neelam, etc. Seven different *Emblica* cultivars are cultivated in Agra zone. These are Banarasi, Chakaiya, Francis, Kanchan, Krishna, NA-7 or Neelam and Balwant. *Emblica* varieties are required to be characterized morphologically and biochemically.

Identification of plant species is the first and foremost criterion in planning for sustainable utilization of resources. Proper identification of commercial plant may focus more light on its distribution and its resource generation capability. Lack of knowledge on distribution or population status may make the species extinct due to overexploitation.

METHODOLOGY

Collection of samples

Samples of seven different *Emblica* cultivars were collected from Agricultaral Research Farm, R.B.S. College, Agra; Aonla Nursery, Dauki, Fatehabad road, Agra and Keetham, Agra.

Morphological analysis of fruits and other samples of *Emblica*

Tree shape, Height of tree, No. of fruits per tree, Yield per tree, Fruit color, Fruit shape, Fruit size, Percent seed, Percent pulp, Fruit weight, No. of Segments per fruit, Leaf area, No. of leaves/10 cm long determinate shoot, Fruit astringency and quality of juice were analysed by standard methods.

RESULTS AND DISCUSSION

Morphological analysis of seven cultivars of *Emblica* has been shown in Table-1.

Table - 1 : Morphological Characterization of Seven Different Emblica Varieties

Characters	Banarasi	Chakaiya	Francis	Kanchan (NA.4)	Krishna (NA.5)	Neelam (NA.7)	Balwant (NA.10)
Tree shape	Erect	Erect	Spread	Spread	Spread	Spread	Spread
Height of tree	Moderate	Tall	Moderate	Moderate	Moderate	Tall	Moderate
No. of fruits/tree	3606	1004	1108	3809	1220	950	3110
Yield/tree	High	Less	Less	High	Less	Less	Moderate
Fruit weight (g)	65.5	35	46	31	32	42	42.5
Fruit size (cm)							
Length	5.7	4.0	4.0	3.6	3.7	4.2	5.7
Diameter	6.8	3.5	3.8	3.7	3.6	4.2	6.0
Fruit colour	Whitish green	Green	Light green	Creamy white	Light green	Yellowish green	Yellowish green
Fruit shape	Round	Triangular	Elliptical	Oval	Round flattened	Elliptical	Round flattened
Segments no./Fruit	6	6	6-8	6-8	6-8	6	6
Seed size	Medium	Small	Large	Small	Medium	Large	Medium
Percent seed	1.9	2.1	2.2	2.4	2.5	2.5	1.7
Percent pulp	98.5	94.2	96.5	88.6	92.9	97.5	96.0
Leaf area (mm^2)	38.5	36.0	39.0	40.5	38.0	32.5	38.5
No. of leaves/10 cm long deter. shoot	52	56	61	50	60	64	60
Astringency	Medium	High	Medium	Medium	Medium	High	High
Juice	Medium	Less	Medium	Medium	Medium	Medium	High

Initially, There were three varieties of *Emblica officinalis* for cultivation – Banarasi, Chakaiya and Francis. Other varieties were selected from these varieties. Kanchan was selected from Chakaiya; Krishna and Balwant from Banarasi and Neelam or NA-7 is a selection from Francis. There were morphological differences between these varieties as shown in Table-1. Chakaiya and Kanchan were tall trees. Others were moderate sized trees. Data on average yield per tree showed that the cultivar Kanchan yielded maximum fruits followed by Banarasi and Balwant. NA-7 and Chakaiya were least yielders.

Fruits of these varieties possessed major differences- Banarasi had largest and heaviest fruit; followed by Balwant, NA-7 and Francis while Kanchan had smallest fruits. Fruits of Chakaiya were triangular and that of Francis and Neelam were elliptical; all others had round fruits. Colours of fruits were whitish green to yellowish green. Balwant possessed yellowish green fruits with pink tinge unique to it and Kanchan had creamy white fruits.

Smallest leaves were of NA-7 while Kanchan got largest leaves.

Varieties were also differed in astringency, juice, seed and pulp percentage.

Now-a-days, *Emblica officinalis* is cultivated on large scale due to an increase in demand. Since it is an important medicinal plant with restricted distribution in India, indiscriminate collection may lead this species to extinction. Proper identification of *Emblica* is very important for its conservation.

ACKNOWLEDGEMENTS

Authors are grateful to Principal, R.B.S. College, Agra for providing lab facilities. The senior author is grateful to the Council of Scientific and Industrial Research, New Delhi, India for the award of JRF (Junior Research Fellowship).

REFERENCES

Singh, D.B. (1973). Studies on the Problems of Fruitfulness in Amla *(Emblica officinalis.)*, Ph.D. Thesis, Agra University, Agra

Singh, J.R. and Arora, J.S. (1967). Physiochemical differences in two varieties of Amla *(Phyllanthus.emblica)* at maturity, *Punjab Horticulture,* **7** : 145.

Websites : Amla, *http://www.haryana_online.com*
Emblica-Triphala, *http://www.ayurveda-herbs.com*

Recent Advances in Medicinal Plant Research : Vision 21st Century, 2005
Editors : S.K. Prabhuji, G.P. Rao and S.K. Patil
Satish Publication House, New Delhi

Chapter 8

EXPLORATION OF MEDICINAL AND AROMATIC PLANTS IN TRIBAL AREAS OF JHARKHAND AND WEST BENGAL FOR CHEMO-ETHNOBOTANICAL STUDIES

C.S. Raghav*, Poonam Suneja, K.C. Bhatt, J.B. Tomar and S.S. Malik
National Bureau of Plant Genetic Resources, Pusa Campus, New Delhi - 110 012
*e-mail : raghav63@yahoo.com

An exploration was conducted during August 2001 in East and West Singhbhum districts of Jharkhand as well as Bankura, Midnapur and Purulia districts of West Bengal for ethno-medicinal studies of some of the important medicinal and aromatic plants of the area. Ethnomedico-botany and natural habitat conditions of the plants were recorded during the exploration through local herbal-men in the area and the active principles of the respective plants were analysed at NBPGR, New Delhi. The areas surveyed were rich in medicinal and aromatic plants diversity and most of the local people were dependent on the crude herbs for the treatment of their common ailments. The plant species collected and studied were *Asparagus racemosus, Curculigo orchioides, Costus speciosus, Curcuma aromatica, Urginea indica, Hemidesmus indicus, Catharanthus roseus, Smilax macrophylla, Andrographis paniculata and Ocimum* species. The details of finding of ethnobotanical studies and chemical analyses of the active principles are discussed in the paper.

INTRODUCTION

Eighty percent of the world's population relies on traditional medicines. India has the ancient history of using plants for various kinds of diseases and ailments as evidenced in Vedas, Purans, Ramayan and Mahabharata etc. It was the charismatic traditional knowledge of early Indians, which cured almost all kinds of diseases without using any sophisticated diagnostic tools and equipment as are available in modern age. However, with the advent of allopathic system of medicine and modernisation and prosperity of Indian urban communities, a shift has been observed from herbal drugs to allopathic medicine. But, Indian rural communities, which consists of about 80% of country's population, are still dependent on locally available herbal medicines. At present, this trend is also being observed in urban population due to increasing awareness and knowledge of unwanted side effects of allopathic drugs (Anonymous, 2001). Therefore, the importance of studying local ethnobotanical knowledge of rural people about these plants is obvious. Further, it is a well known fact that the beneficial effect of medicinal and aromatic plants lies in their chemical constituents. One individual constituent or in combination with others act on the body system and cure the diseases. The unique multi disease action of Ayurvedic drugs works on synergistic effect theory of healing. The action of individual constituents has also been proven through enormous research work (Farnsworth and Bingel, 1977).

Therefore, study of important active principles in locally used medicinal plants is of utmost importance. This study further opens various dimensions of multipurpose utilisation of individual plant species as crude drugs, raw material for pharmaceutical industries, plant breeding lines etc. It shows that the utilisation of natural products in the drug industries is 35-50%. Almost every class of drug includes a model structure derived from natural products exhibiting the classical effect of that specific pharmaceutical category (Schultes and Reis, 1995). Jharkhand and West Bengal forests possess vast variety of medicinal and aromatic plants. Very few explorations have been conducted in the limited areas of these two states in the past. They, too, have been conducted purely for ethnobotanical studies (Chaudhary *et al.,* 1982). Further, most of the plants surveyed in our exploration have not been covered in these studies. Therefore, many parts of these two states are still unexplored for ethnobotanical cum biochemical studies. In the present paper attempt has been made to study the chemo-ethnobotnical aspects of locally available medicinal and aromatic plants.

METHODOLOGY

An exploration team comprised of taxonomist, Ayurvedic physician, Ethnobotanist and Chemist surveyed the East and West Sinhbhum districts of Jharkhand and Bankura, Purulia and Midnapore districts of West Bengal (Fig. 1) during August 2001.The detailed information on area, location (latitude, longitude, altitude), soil type, plant habitats, local use of individual plant etc. (Table 1) was recorded. The information on local uses was gathered with the help of local Vaidya and herbal-men of the locality. These people located the plants on collection site and the botanical identity was established by the taxonomist. All the samplings were done randomly. Sufficient quantity of each sample was collected. One part of it was established in the field for *ex situ* conservation and the other part was used for chemical analysis. The important active principles of the collected plant species were identified through the published literature and the quantitative analysis of the individual compound or group of compounds was done for all the samples.

RESULTS

A total of 36 accessions of 20 plants species (live plants and seeds) were collected for *ex situ* conservation. Out of these, 28 accessions of 9 plant species were analysed for active principles along with other parameters. Eleven plant species were collected as herbarium specimens. A total of 6 forest ranges/hills of 5 districts of two states were explored.

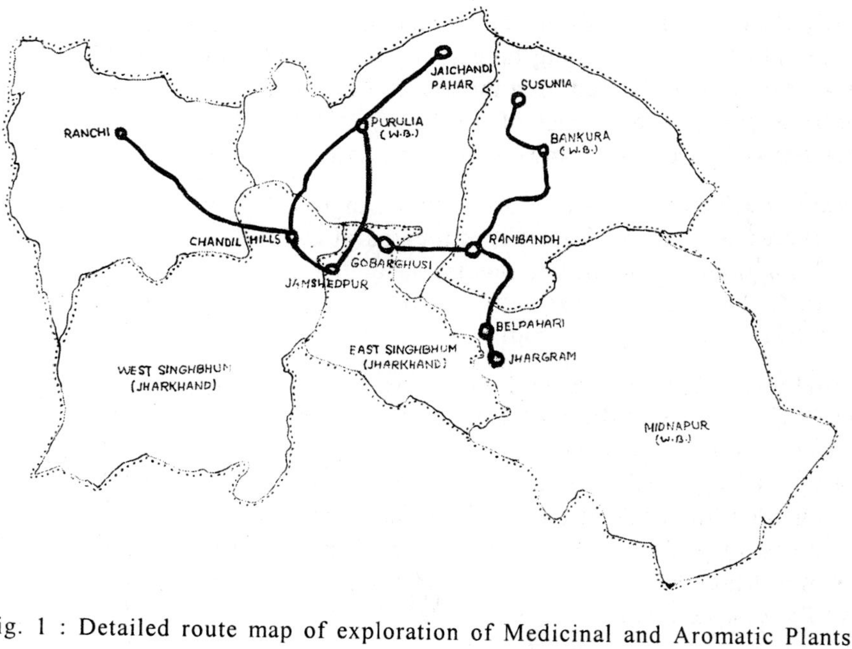

Fig. 1 : Detailed route map of exploration of Medicinal and Aromatic Plants

Geographical details of the explored area

There was a great diversity in ecology, topography, soil type, vegetation etc. Area-wise details of important features as observed during the exploration have been presented here.

Susunia Hills : Susunia hills are located at 23° 23.8' N latitude and 86° 58.4' E longitude at an elevation ranging from 150-500 m. mean sea level. The hills are dry, no water streams or reservoirs were available on the hills. The soil is very shallow, gravelous and the most part is rocky. The entire hill is disturbed by the human activities. The hills are surrounded by widely sprawled plain land. Not much medicinal and aromatic plants were available on the hills.

Chandil Hills : Chandil hills are sprawling over a very large area and are very rich in medicinal and aromatic plant resources. It is located at 22° 56.3' N latitude and 86° 02.6' E longitude at an altitude ranging from 200-500 m. The forest is very dense and some of its parts are not disturbed by human or domestic animal activities. The soil is least stony, partially gravelous and rich in organic matter.

Gobarghusi : It is located in lower hills at 22° 57.6' N latitude and 86° 46.8' E longitude at an elevation ranging from 150-250 m. The soils of the area are black, rich in organic matter, slightly gravelous at the hillocks. The area is poorly rich in plant diversity. A considerable part of the area is either fallow pasturelands or under cultivation.

Belpahari Forest : Belpahari forest range is located at 22° 37.9' N latitude and 86° 45.6' E longitude at an altitude ranging from 120-300 m. It is a non-hilly undulated area mostly covered with wood forest. Some of its part is partially and, rest of the part is totally disturbed. Soils of the forest are red in colour, sandy loam with gravels.

Ranibandh Forest : It is a large forest range of low hills, very undulated and covered with wood forest. It is located at 22° 50.7' N latitude and 86° 43.7' E longitude at an altitude ranging from 150-350 m. The area is partially disturbed. The soil is fine silty, partly gravelous, red in colour and moderately rich in organic matter.

Jaichandi Hills : Jaichandi hills are a small group of 4-5 hillocks surrounded by plain cultivated area. It is located 23^0 31.2' N latitude and 86^0 40.1' E longitude at an altitude ranging from 250 to 300 m. The hills are near a village and therefore, it is completely disturbed. There are hardly any considerable populations of important medicinal or aromatic plants. The soil is very shallow developed on large rocky surface. The hills have been encroached by domestic animals for grazing. The total area of the hill is hardly 1 square km.

Details of collected samples

Asparagus racemosus **Willd. (Liliaceae)** : It is locally known as 'satmooli'. A total of 4 samples of *Asparagus racemosus* (satawari) were collected each one from Susunia, Chandil, Gobarghusi and Belpahari hills. The availability of *Asparagus* was abundant at Gobarghusi and occasional at rest of the places. The root growth of *Asparagus* was found maximum under Goberghusi conditions. Roots up to one meter length were found in this area. The root size at other places was maximum up to two feet. Shortest roots were observed at Susunia hills.

The sapogenin content in the dry roots of these samples were 1.08, 1.19, 1.37 and 1.24 %, respectively. The roots of *Asparagus* and *Smilax macrophylla* in 3:2 ratio are taken to treat constitutional disorders. The powdered root (one spoon) with milk is prescribed twice a day as mild aphrodisiac tonic to male and galactogogue to female. The root powder and sugar (misri) in the ratio of 2:1 is given with water as coolant to young boys. It is also used with clarified butter (ghee) of cow in the ratio of 2 :1: sugar to taste to children up to 15 years of age for improving eye sight.

Curculigo orchioides **Gaertn. (Hypoxidaceae)** : It is locally known as 'Taalmuli' or 'Kali musli'. A total of 4 samples were collected from Susunia, Chandil, Gobarghusi and Belpahari hills. The availability of *Curculigo* in these hills were occasional, frequent, rare and frequent, respectively. When observed the thickness and length of roots, the roots found at Gobarghusi were thickest (up to 1.5 cm diameter) and longest (up to 25 cm). Among the roots collected from other three locations, the maximum thickness and length was up to 1.0 cm diameter and 15cm in length, respectively. The probable reason for vigorous growth and development was the black and porous soil with sufficient moisture and organic matter. The petroleum ether extract, chloroform extract and methanol extract in the above four samples were found in the range of 0.80-0.90%, 2.0- 2.3% and 5.20- 5.60%, respectively. The root powder (1/2 spoon at night) is given to treat dyspepsia. The root powder (one spoon) with curd (4: 1) is given to cure leucorrhoea in women. It is also used with cow milk and black pepper in the ratio 4: 3: 1 (one spoon twice a day) to treat seminal debility.

Costus speciosus **(Koen.) Sm. (Costaceae)** : Locally known as 'Keon' was found at four locations i.e. Susunia , Chandil, Gobarghusi and Jaichandi hills . It was frequently available at Gobarghusi, but rarely available at other places. Mostly the plants were available in population. Again, under the Gobarghusi condition i.e. high organic matter rich, pulverised black soils, the maximum growth of plant (height up to 5 feet) and fresh rhizomes (approx. 1 and ½ Kg/plant) was noticed. At other locations normally the plant height was noticed 3-4 feet and per plant fresh rhizome yield was 250 gm to 1 Kg. The root depth was similar in all cases. Even in deep pulverised soils, the rhizomes were available only in the upper 6" layer of soil. They were horizontally spreading in the upper soil layer .The sapogenin (mainly diosgenin) content of the above four samples were found to be 1.79, 1.99, 1.72 and 1.57 %, respectively. The variation in the sapogenin content was not considerable and contents were within the reported limits. The plant height at Gobarghusi was more than the reported one. People of these areas use the decoction of the rhizomes for fever. The filtered extract of rhizome and roots *of Artocarpus integrifolia L.* (kanthal) with Sugar (to taste) is taken in 10 ml doses thrice a day in jaundice. Some tribal people also use it as pickle.

***Curcuma aromatica* (Zingiberaceae) :** *Curcuma aromatica*, locally known as 'Ban haldi' was found at Ranibandh forest range, Chandil, Belpahari and Gobarghusi areas. It was frequently available at Ranibandh, occasionally at Chandil, rarely at Gobarghusi and abundantly at Belpahari area. The highest plant height (up to 1.5 m) was noticed at Gobarbhusi area as compared to plants found at all other areas where maximum height was observed 1.0 m. The rhizome size was almost the same under all conditions and locations. The essential oil contents of these four samples were 0.6, 1.0, 0.7 and 1.3%, respectively. In addition to its use as spices, it is also used to treat cuts, wounds and burns. Some tribals use it with black pepper in 3:1 ratio (¼ spoon) after each meal twice a day as carminative. Rhizome powder ½ spoon in 200 ml hot boiled rice water (starch) is taken to cure influenza.

***Urginea indica* (Roxb.) Kunth (Liliaceae):** *Urginea indica* is locally known as 'Ban-pyaz'. It was found only at two locations i.e. Belpahari and Ranibandh forest ranges.. At both the locations it was frequently available. The higher availability was noted in moist and partially shaded forest under fine soils with moderate organic matter. The scillarin content of the two samples were 0.45 and 0.52%, respectively. It is used by the local people in urinary problems. The juice of fresh bulb is taken in 5-10 ml doses thrice a day. It is also given to asthma patients.

***Hemidesmus indicus* (L.) Schult. (Periplocaceae)** *: Hemidesmus indicus,* locally known as 'Anantmool' was spotted at three locations i.e. Jaichandi, Chandil and Belpahari areas. At Jaichandi hill it was seen in the population of few plants rarely at one or two places on the entire hill. At this place the growth of the roots and the aerial part was stunted. This may be due to lack of moisture and stoniness in the soil. At Chandil hills it was frequently available while at Belpahari it was abundantly available. At Chandil hills the plants were growing mostly near the water resources under shady conditions. The length of the plant was up to two feet creeping on the ground. It was found growing in gravelous, fine sandy soil, fertile and non fertile soils. The root and plant size was almost the same under all the conditions except slightly bigger roots were found under leaf covered shady and moist area. The essential oil content of roots of three samples were 0.07, 0.1 and 0.09 %, respectively. As told by the local people and herbal men, the root powder with milk is taken to cure nervousness. The root infusion is given to treat tonsilitis. Roots and leaves are also used locally to stop bleeding and to heal wounds. The root, taken with honey, cures intestinal ulcers.

***Smilax macrophylla* Roxb. (Smilacaceae):** Locally known as 'Ram datun', *Smilax macrophylla* is available at Chandil hills and Belpahari forest range. At both the places it was abundantly available. However, Chandil hills were having most abundant plant populations. Entire area was full of *Smilax* climbers. All the plants were profusely growing covering trees and bushes. The roots were spreading up to 2-3 feet in depth. The sapogenin content in the dry roots of two samples were 0.5 and 0.7%, respectively.

Locally it is used to treat leucorrhoea. This is mixed with the roots of *Asparagus* in the decoction of *Curcuma aromatica* (2:2) to make pills. Sun dried pills are taken twice a day. Root decoction in goat milk (3:1) is also given to cure blood dysentery. It is also well known to be a teeth protector. As told by the local people, it fixes the loose teeth if it is used as chewing stick (datun) for one month.

***Andrographis paniculata* Nees (Acanthaceae) :** *Andrographis* (Kalmegh) was seen only at Chandil and Susunia hills. At both the locations, it was occasionally available. The plants were thin, small and poorly growing. Maximum plant height was observed to be 30 cm. with branchless single stem. At both the locations, the plant population was growing on few small patches with thinly populated limited plants. The andrographolide contents of two samples were 1.57 and 2.29%, respectively.

***Catharanthus roseus* (L) G. Don. (Apocynaceae) :** During the entire exploration, *Catharanthus* was seen growing at two sites at Ranibandh area. Both the sites were fully disturbed by human activities. On one site along the road, all the plants were pink flowered. They were densely growing up to 200-300 metre distance. On the other site white and pink flowered plants were seen. This site was few kilometres away from the previous site. The total alkaloid content of these two samples were o.478 and 0.392% respectively. The seed powder with decoction of black pepper in 2:1 ratio is given to cure epilepsy. The latex of fresh plant is also used to cure scabies. Some local herbal men use it to control high blood pressure.

Locally, it is used in diarrhoea, malaria and scabies. In diarrhoea, its roots are used with *Aegle marmelos* (Bael) and *Terminalia ballerica* (Bahera) roots in the form of small tablets, one tablet thrice a day. In malaria, water extract of whole plant is mixed with root extract of *Eriodendron anfraetuosum* De (Shet simul) and 10-15 ml. of this extract is taken thrice daily.

Table - 1 : Details of collected species with their active principles

Plant	Accession No.	Location	Availability in the area	Soil	Global position	Altitude (m)	Active principle (%) DWB
Asparagus racemosus	IC-336838	Susunia hills, Bankura (West Bangal)	Occasional	Black	N 23^0 23.8' E 86^0 58.4'	250	Sapogenins 1.08
Asparagus racemosus	IC-336839	Chandil hills, West Singhbhum (Jharkhand)	Occasional	Black	N 22^0 56.3' E 86^0 02.6'	210	Sapogenins 1.19
Asparagus racemosus	IC-336840	Gobarghusi, East Singhbhum (Jharkhand)	Abundant	Black	N 22^0 57.6' E 86^0 46.8'	180	Sapogenins 1.37
Asparagus racemosus	IC-336841	Belpahari, Midnapore (West Bengal)	Occasional	Red	N 22^0 37.9' E 86^0 45.6'	150	Sapogenins 1.24
Curculigo orchioides	IC-336842	Susunia hills, Bankura (West Bangal)	Occasional	Black	N 23^0 23.8' E 86^0 58.4'	250	PE Ext. 0.80 Chloroform 2.10 Methanol Ext. 5.52
Curculigo orchioides	IC-336843	Chandil hills, West Singhbhum (Jharkhand)	Frequent	Black	N 22^0 56.3' E 86^0 02.6'	210	PE Ext. 0.90 Chloroform 2.02 Methanol Ext. 5.30
Curculigo orchioides	IC-336844	Gobarghusi, East Singhbhum (Jharkhand)	Rare	Black	N 22^0 57.6' E 86^0 46.8'	180	PE Ext. 0.84 Chloroform 2.30 Methanol Ext. 5.60
Curculigo orchioides	IC-336845	Belpahari, Midnapore (West Bengal)	Frequent	Red	N 22^0 37.9' E 86^0 45.6'	150	PE Ext. 0.85 Chloroform 2.16 Methanol Ext. 5.23

Table - 1 : Continued .....

Plant	Accession No.	Location	Availability in the area	Soil	Global position	Altitude (m)	Active principle (%) DWB
Costus speciosus (White)	IC-336846	Gobarghusi, East Singhbhum (Jharkhand)	Rare	Black	N 22^0 57.6' E 86^0 46.8'	180	Sapogenins 1.79
Costus speciosus (Red)	IC-336847	Susunia hills, Bankura (West Bangal)	Rare	Black	N 23^0 23.8' E 86^0 58.4'	280	Sapogenins 1.99
Costus speciosus (Red)	IC-336848	Chandil hills, West Singhbhum (Jharkhand)	Rare	Black	N 22^0 56.3' E 86^0 02.6'	260	Sapogenins 1.72
Costus speciosus (Red)	IC-336849	Jaichandi Pahar, Purulia (West Bengal)	Rare	Red	N 23^0 31.2' E 86^0 40.1'	270	Sapogenins 1.57
Curcuma aromatica	IC-336850	Ranibandh, Bankura (West Bengal)	Frequent	Red	N 22^0 50.7' E 86^0 43.7'	270	Essential oil 0.60
Curcuma aromatica	IC-336851	Chandil hills, West Singhbhum (Jharkhand)	Occasional	Black	N 22^0 56.3' E 86^0 02.6'	240	Essential oil 1.00
Curcuma aromatica	IC-336852	Gobarghusi, East Singhbhum (Jharkhand)	Rare	Black	N 22^0 57.6' E 86^0 46.8'	180	Essential oil 0.70
Curcuma aromatica	IC-336853	Belpahari, Midnapore (West Bengal)	Abundant	Red	N 22^0 37.9' E 86^0 45.6'	140	Essential oil 1.30

Table - 1 : Continued .....

Plant	Accession No.	Location	Availability in the area	Soil	Global position	Altitude (m)	Active principle (%) DWB
Urginea indica	IC-336854	Ranibandh, Bankura (West Bengal)	Frequent	Red	N 22^0 50.7' E 86^0 43.7'	270	Scillarin 0.45
Urginea indica	IC-336855	Belpahari, Midnapore (West Bengal)	Frequent	Red	N 22^0 37.9' E 86^0 45.6	140	Scillarin 0.52
Hemidesmus indicus (narrow leaf)	IC-336856	Jaichandi Pahar, Purulia (West Bengal)	Rare	Red	N 23^0 31.2' E 86^0 40.1'	250	Essential oil 0.07
Hemidesmus indicus (Broad leaf)	IC-336857	Chandil hills, West Singhbhum (Jharkhand)	Frequent	Black	N 22^0 56.3' E 86^0 02.6'	210	Essential oil 0.10
Hemidesmus indicus (narrow leaf)	IC-336858	Belpahari, Midnapore (West Bengal)	Abundant	Red	N 22^0 37.9' E 86^0 45.6'	150	Essential oil 0.09
Smilax macrophylla	IC-336861	Gobarghusi, East Singhbhum (Jharkhand)	Occasional	Black	N 22^0 57.6' E 86^0 46.8'	180	Sapogenin 0.50
Smilax macrophylla	IC-336862	Chandil hills, West Singhbhum (Jharkhand)	Abundant	Black	N 22^0 56.3' E 86^0 02.6'	220	Sapogenin 0.70

Table - 1 : Continued .....

Plant	Accession No.	Location	Availability in the area	Soil	Global position	Altitude (m)	Active principle (%) DWB
Andrographis paniculata	IC-336863	Susunia hills, Bankura (West Bangal)	Occasional	Black	N 23^0 23.8' E 86^0 58.4'	250	Andrographolide 1.57
Andrographis paniculata	IC-336864	Chandil hills, West Singhbhum (Jharkhand)	Occasional	Black	N 22^0 56.3' E 86^0 02.6'	210	Andrographolide 2.29
Catharanthus roseus (White)	IC-336866	Ranibandh, Bankura (West Bengal)	Rare	Red	N 22^0 50.7' E 86^0 43.7'	180	Total alkaloids 0.48
Catharanthus roseus (Pink)	IC-336867	Ranibandh, Bankura (West Bengal)	Rare	Red	N 22^0 50.7' E 86^0 43.7'	180	Total alkaloids 0.39
Ocimum basilicum	IC-336873	Jaichandi Pahar, Purulia (West Bengal)	Rare	Red	N 23^0 31.2' E 86^0 40.1'	210	Essential oil 0.094

***Ocimum basilicum* L. (Lamiaceae) :** *Ocimum* was seen only at Jaichandi hills. As described earlier, the entire small hillock was fully disturbed by villagers. Therefore, the *Ocimum* plant growing there in large population seemed to be migrated from nearby village. There was only one 10-15 metre patch of small sized mature plants among all 4-5 hillocks. The essential oil content was 0.094%. *Ocimum* along with pepper is used by local people to treat cold and cough of children. A liniment of fresh leaves on forehead relieves headache. Some people soak the seeds in water overnight and take the water in early morning as purgative.

Most of the tuberous/rhizomatous plants available in the area were collected. Seed bearing plants were in their vegetative growth stage. However the seeds of various available plants could not be collected. In general the entire area is full of medicinal plants. However there was a scarcity of naturally growing important aromatic plants. Largely available main species are *Curculigo orchioides, Smilex macrophylla, Asparagus racemosus, Curcuma aromatica* etc. The plant species were varied in phenotypic as well as chemotypic traits.

CONCLUSION

The Jharkhand and the adjoining area of West Bengal is rich in medicinal plants diversity. Many parts of these areas are still unexplored from chemo-ethnobotanical point of view. More explorations may be conducted to collect, conserve and evaluate the precious plant material.

REFERENES

Anonymous (2001). Herbal Products Current Status, Vision and Action Plan. (Report of Scientific advisory committee to the cabinet (SAC-C) Govt of India). Technology Information Forecasting and Assessment Council. (TIFAC) New Delhi, Pp. 7.

Chaudhary, R.H.N., Sareen, A.M. and Molla, H.A. (1982). Some Less Known Uses of Plants From the Tribal Areas of Bankura District. West Bengal, *Indian mus. Bull.*, **14** : 71-73.

Farnsworth, N.R. and Bingel, A.S. (1977). Problems and Prospects of Discovering New Drugs from Higher Plants by Pharmacological Screening. In : *New National Products And Plant Drugs With Pharmacological, Biological or Therapeutic Activity.* Edited by H. Wagner and P. Wolff. Berlin; Springer 1-22.

Schultes, R.E. and Reis, S.V. (Eds) (1995). Ethnobotany. Evolution of a Discipline (Chapman and Hall), 340.

Wargoda, P.B. (1980). Some Questions About the Future of Traditional Medicines in Developing Countries. *Journal of Ethnopharmacology,* **2** : 193-194.

Recent Advances in Medicinal Plant Research : Vision 21st Century, 2005
Editors : S.K. Prabhuji, G.P. Rao and S.K. Patil
Satish Publication House, New Delhi

Chapter 9

CHEMISTRY AND CULTIVATION TECHNOLOGY OF *EMBLICA* – AN ALTERNATIVE LAND USE FOR ALKALI WASTELAND

Seema Bhadauria
Microbiology Lab, Department of Botany, Raja Balwant Singh College, Agra - 282 002, U.P.
e-mail : bhadauriaseema1@rediffmail.com

The 'Aonla', also known as Indian Gooseberry' is a minor sub-tropical deciduous tree belonging to the family, Euphorbiaceae. Amla is a native of India, Ceylon, Malaysia and China. It thrives well throughout tropical India and is wild or cultivated in the region extending from the base of Himalayas to Ceylon and from Malaysia to South China. In Uttar Pradesh, India it is grown as a commercial medicinal crop. In Uttar Pradesh state of India, there is 1.3 million ha of land is salt affected. Inspite of several limitations, salt affected land have good potential for cultivation of hardy medicinal tree species under specific management approaches. In this context new cultivation technology for growing *Emblica officinalis* nearby Agra district has been developed and discussed in this chapter. This consists of land development, composition of filling material, preparation of *Emblica* saplings, irrigation, weed management, pest and disease management. The chemical composition of seven important cultivars of *Emblica* has also been reported.

INTRODUCTION

Amla (*Emblica officinalis*) the "Kayakalp Vraksh" is the one of the most important medicinal fruits of India (Anonymous, 1996).

With the growing modern civilization, the everincreasing role of plants to human life has placed before us manifold task for understanding intimately the world of plants and their relationship with human being. The dependence of man on medicinal plants is next to food for life sustenance. Continuous exploitation of medicinal plants has also been causing extinction of certain species. There is, therefore, need for intensive agricultural studies leading to genetic improvement and cultivation methods for expansion of area under medicinal and aromatic plants (Anonymous – website).

The '**Aonla**' (***Phyllanthus emblica or Emblica officinalis*** Gaertn), also known as Indian Gooseberry is a minor sub-tropical deciduous tree belonging to the family, Euphorbiaceae. Amla is a native of India, Ceylon, Malaysia and China. It thrives well throughout tropical India and is wild or cultivated in the region extending from the base of Himalaya to Ceylon and from Malaysia to South China. The tree also grows wild at elevations upto 1500 m above sea level in South India. It is grown as a commercial medicinal crop in Uttar Pradesh (Khopde *et al.*, 2001). The fruits are borne on the branchlets of the tree.

Out of 329 million ha geographical area of the country about 107.4 million ha is categorized as degraded land. Among different categories of degraded land about 6.32 million ha suffers from salinity and alkalinity and in Uttar Pradesh alone about 1.3 million ha of land is salt affected. The major farming constraints of these lands are high pH, excess of calcium carbonates, chloride, sulphate and exchangeable sodium, imbalance of electrical conductivity, low infiltration rate, impeded drainage and poor fertility status. Already a large land area has gone out of cultivation in canal commands particularly in arid and semiarid regions because of the water logging and salinity build up making landscape devoid of vegetation except for a few hardy trees and grasses.

Inspite of several limitations, salt affected land have good potential for cultivation of hardy medicinal fruit type under scientific management and now-a-days these medicinal trees are considered as an integral component in various Agroforestry System (AFS) and watershed management both from production and protection point of view. Considering these facts, efforts were made to utilize the salt affected lands by medicinal Agroforestry System

(AFS), so that permanent vegetational cover may stop further degradation and also rehabilitate these lands in due course of time.

DETAILS OF EXPERIMENTAL SITE

The Department of Biotechnology, Govt. of India has supported Raja Balwant Singh College, Agra for a massive programme on Emblica fruit based agroforestry for salt affected soil in Mainpuri district of Uttar Pradesh. The area falls in sub-humid, subtropical zone between 26 ° 53' to 27°32' North latitude and 78 ° 27' to 79 ° 26' East longitude and covers an area of nearly 2759 sq. km. The climate of the district is markedly periodic and of semiarid nature which is characterized by a dry and hot summer, a warm monsoon and a cold winter. The soil of this district is divisible into three types i.e. Matiyar, Bhur and Dumat, within this, there is vast area of wasteland known as "usar" which is not fit for agriculture. Here 66752 hectares of land is usar land and the vegetation on this land is scarce.

METHODOLOGY

Quality Analysis of Emblica Fruits

Amla varieties have not been standardized, however, they are classified according to their size, colour or named after the places of growing. Amla varieties classified according to their colour are Green-tinged, Red-tinged,

Table - 1 : Physical Characteristics of Fruits of Seven Cultivars of Emblica

Cultivars	Yield/tree		Average wt. of fruits (g)	Fruit size (cm)	
	No.	Wt. (Kg)		Length	Diameter
Balwant	3115	76.6	44.5	5.2	4.7
Banarasi	3632	102.8	77.0	7.5	6.5
Chakaiya	964	33.5	36.0	4.8	3.8
Francis	1109	34.0	52.5	4.12	3.9
Kanchan	3890	93.0	37.2	4.0	3.7
Krishna	1220	43.6	35.8	3.8	3.9
NA-7	905	30.8	45.2	4.4	4.38
SE	-	-	2.1	0.08	0.06
CD 5%	-	-	3.2	0.20	0.18

Table - 2 : Chemical Characteristics of Seven Cultivars of *Emblica officinalis*

Cultivar	Ash %	Poly Phenol %	Crude fibre %	Phos-phorus mg/ 100g	Sodium mg/100g	Pota-ssium mg/ 100g	Calcium mg/100g	Protein %	Moisture of pulp %	T.S.S. %	Acidity %	Red	Sugars N.R.S.	Total	Asc-orbic aicd mg/ 100g
Balwant															
15 Oct.	2.6	2	5.6	800.7	152	635	425	3.5	86	11	2.35	2.6	3.8	6.4	300.13
30 Nov.	3.4	2.5	5.8	1214.8	145	750	620	3.6	87.8	11.2	3.19	1.82	2.57	4.39	384.26
Banarasi															
15 Oct.	2.8	2.4	6.2	1335.4	126	825	725	3.7	88.8	ss12.6	2.15	5.2	5.7	9.9	584.19
30 Nov.	3.6	2.5	6.4	1976.5	172	1150	900	4.2	89.4	12.9	2.56	3.13	4.66	7.79	609.15
Kanchan															
15 Oct.	2.2	2.8	7.1	1232.5	146	827	512	2.55	88.05	11	1.18	2.3	1.28	3.58	283.92
30 Nov.	3.1	2.4	7	1239.5	106	751	681	1.94	85.97	11.9	2.38	1.3	0.13	1.43	309.98
Krishna															
15 Oct.	2.5	1.8	6.8	1171.7	138	927	423	3.14	87.38	10.1	1.31	2.7	1.19	3.89	417
30 Nov.	2.5	1.9	7.1	1283.9	141	802	424	2.85	88.05	9.9	2.1	1.2	0.49	1.69	585.2
Chakaiya															
15 Oct.	2.3	2.4	7	1171.9	159	1000	697	1.76	87.53	7.5	1.21	0.89	0.87	2.08	425.28
30 Nov.	2.4	2.3	8.6	1343.9	194	800	686	1.5	88.34	9.7	1.87	0.32	0.35	0.67	463.45
NAA-7															
15 Oct.	2.2	2.2	6.6	1185.6	142	951	765	3.43	88.21	6.7	1.11	2.27	0.31	2.58	253.72
30 Nov.	2.2	1.98	6.8	1304	154	802	796	3.28	88.69	8	1.97	0.78	0.26	1.04	331.59
Francis															
15 Oct.	2.7	2.2	7.6	1268	141	872	512	0.33	87.15	9.6	1.11	1.12	1.4	2.52	277.64
30 Nov.	2.1	2.5	7.4	1261.8	116	608	423	0.14	88.61	9.4	2.15	1.17	1.19	1.35	430.32

Pink-tinged, White-streaked and Bansi-red. Some important varieties of Amla recommended for commercial cultivation in Northern India are Chakaiya, Krishna, Kanchan, Narendra Amla-6, Narendra Amla-7, Balwant Amla.

Emblica fruits possess attractive pale green colour. It has shining surface and size varies from 3.5 to 6.8 cm in length and 3.7 to 6.5 cm in diameter. The fruit weight varies from 36.0 to 77 gm on harvesting. But it depends on the quality of cultivars (Table-1). The Chemical composition of ***Emblica*** varieties are given in Table-2.

CULTIVATION TECHNOLOGY OF *EMBLICA* IN ALKALI SOIL

Land Development

(i) Land Marking and Levelling

Measuring of plantation site and marking by boundary channels on square grid basis was done. The layout of plantation site was planned as per designed programme. Out of the four levelling methods viz., plane method, profile method, plan inspection method and contour adjustment method, the plane method of levelling was employed.

(ii) Protection Against Browsing

The plans for afforestation programme should include measures for protection against stray animals and wild life. Much of the afforestation efforts may be of little avail if saplings are not protected against chewing, grazing and nibbling etc. by stray animals and wild life. The opted control measure against grazing was Barbed Wire Fencing. Fencing posts 2.0-3.5 m in length preferably with RCC cement fixing at about 3-3.5 m distance and 4 to 6 strands of two ply galvanized barbed wire with 4 barbs spaced at 7-10 cm apart are then fixed to the poles. Fixing of branches of thorny plants like *Acacia* can further enhance efficacy of the fence against the entry of wild animals.

(iii) Planting Method

Post auger hole planting : This method was designed to facilitate root penetration through kankar layer or hard pans found at some depth in most of the alkali soil. In this method, holes of 15-30 cm diameter are dug to the depth of 120 cm or more so it pierces beyond the kankar layer. For reducing manual labour cost and to speed up the operations, tractor driven augers are now available in the market. Main features of this method are :

- Pierces through hard canker layer.
- Encourages and trains deeper rooting.
- Does not cause coiling of roots or face constriction which adversely affect plant growth.

- Improves the penetration of water through the holes thereby, causing better water storage and its availability.
- Ensure deeper reclamation of soil.
- Makes the mixing of fertilizers possible even in deeper layer.

Performance of post auger hole planting method has been observed to be highly satisfactory in field trials. Saplings transplanted in post-auger holes have to be spot irrigated. In order to avoid spot irrigation the post-auger hole technique has been further modified to reduce irrigation cost. First, auger holes of 15-20 cm in diameter are dug and then connected with each other through an irrigation channel. Thus, the post-auger holes are finally positioned in irrigation furrows. Saplings are then planted in boreholes at a depth of 15-20 cm from original soil surface. The method is a variant of 'sub surface planting furrow irrigation method' (SPFIM) elaborated in an ensuing section.

Alternately for high value species like fruit trees, pit of 45 × 45 × 35 cm size are manually prepared and then post-augered (15-20 cm dia.) beyond the kanker layer with a tractor driven auger. The pit holes are then interconnected with an irrigation channel. Additional benefits accruing from refined post-auger hole method include :

- Ensures easy and uniform irrigation.
- Provides for larger volume of reclaimed soil for the development of feeder root during the initial year of growth.
- Curtails run off and promotes *in-situ* rainwater conservation.

(iv) Composition of Filling Material

The filling mixture for auger holes is as follows :

a) Gypsum @ 3.5 Kg per hole or on the basis of gypsum requirement of the soil excavated by auger.

b) Farmyard manure @ 10 Kg per hole along with urea and DAP @ 50 gm per hole.

c) Zinc Sulphate @ 10 to 15 g/hole. Most alkali soils are deficient in Zinc, therefore, it is obligatory to apply it.

d) BHC 30% @ One Table Spoon per hole to avoid termite attack.

e) Additional use of rice husk @ 5 kg/hole, is beneficial is highly, deteriorated alkali soils and other soils with very poor permeability.

f) Use of Iron Sulphate @ 20 g/hole, in soil having $CaCO_3$ content more than 5%, is advocated.

g) 500 g/Auger hole biofertilizer and deoiled neem cake (125 g/Auger hole).

(v) Preparation of Emblica Saplings

For the development of nursery following steps were taken-

(a) **Site selection and preparation :** The ploughing is done with a soil turning plough. Afterwards 2-3 light harrowing are needed. One pre-irrigation, 12 days before seeding, has to be done for the better germination of Amla seeds.

(b) **Land levelling :** One light cultivation and levelling is required before sowing to protect seeding from white ants, Gujia weevil and other soil insects, pests. Aldrin 5% dust should be mixed in the soil at the rate of 25 kg/ha with first cultivation.

(c) **Fertilizer application :** For the 100 sq. ft. seed bed 0.5 kg N_2, 0.5 kg P, 0.25 kg K, 0.5 kg biofertilizer and 100 kg cowdung should be used. These fertilizers were mixed properly with the soil.

(d) **Rate of Seeding :** Seeding is done in rows, (15 × 20 cm) 100 gm seeds/100 sq.ft. seed bed.

(e) **Seed treatment :** There were two different techniques taken for the better germination and growth. They were as follows :

***Treatment with Biofertilizer :** A treatment of dipping seeds in *Azospirillum* biofertilizer slurry for 24 hr. and then sowing is done.

***Treatment with G.A. (*Gibberellic* Acid) :** Seeds were soaked with 500 ppm Gibberellic acid upto 24 hr. then soaked seeds were used for sowing. This process gives better result for germination. But G.A. has not any significant role for the growth of plants. Seeds can be sowed in seed beds, polythene bags, pots and root trainer. In Poly bags and pots two seeds are sowed.

(f) **Depth and method of seed sowing :** The methods of sowing depends upon the equipment available. It is important that seed be placed in moist soil so as to ensure uniform germination. The seed was covered with not more than 5 cm soil and not below 2 cm to ensure uniform and maximum emergence.

(g) **Irrigation :** Adequate irrigation is provided throughout the growing period. Approximate timing of irrigations are as follows:

- 3 days after sowing, by sprinkling of water onto seed bed. This process was continued upto seedlings stage.
- Then irrigation was done by pump as per the schedule given in table 3, 4.

Table - 3 : Plantation Schedule

Months	Filling of bags	Seeds and Seedling Preparation	Planting /Cutting	Shifting and Gardening	Root Prunning	Thatching	Weeding	Plant Protection	Budding
January	★	★	-	★	★	★	-	-	-
February	★	★	★	★	★	-	-	-	-
March	-	★	-	-	-	-	-	-	-
April	-	-	-	*	-	-	★	-	-
May	-	-	-	-	-	-	★	-	★
June	-	-	-	★	-	-	★	-	★
July	-	-	-	-	★	-	-	★	★
August	-	-	-	-	★	-	-	★	★
September	★	★	-	-	★	-	★	-	★
October	★	★	-	★	-	-	★	-	★
November	-	-	-	-	-	-	-	-	-
December	-	-	-	★	-	★	-	-	-

- = Nil; ★ = Appropriate Time

Table - 4 : Schedule of Field Preparation

Months	Harro Laveling	Preparation of Furrows	Back Filling	Trans-Planting	Inter-Culture	Thatching	Prunning and Lopping	Plant Protection	Irrigation Schedule
January	★	★	★	-	-	★	★	-	★
February	★	★	★	★	-	-	★	-	★
March	-	-	-	★	-	-	-	★	★
April	★	★	★	-	-	-	-	-	★★
May	★	★	★	-	-	-	-	★	★★
June	★	★	★	-	-	-	-	★	★★
July	-	-	-	★	-	-	-	-	-
August	-	-	-	★	-	-	-	★	-
September	-	-	-	★	★	-	-	-	-
October	-	-	-	-	★	-	-	-	★
November	-	-	-	-	-	-	-	-	★
December	★	★	-	-	-	★	-	-	★

★ = Irrigation one a month; ★★ = Irrigation twice a month; - = No Irrigation

(h) **Weed control :**

- Weed control in seed bed is too necessary, otherwise plant growth will be affected.
- Spray of 2-4 D at the rate of 0.5 kg active ingredient per hectare has given good results, 400 to 600 litres of water was used for one hectare.
- Amount of water may be increased or decreased according to spray machine.

(i) **Budding :** Patch budding, shield budding and softwood grafting were carried out using 10 months old Amla (seeded plant) as a root stock and activated but unsprouted buds from mature shoots of cv. Balwant as scion. Treatment was carried out in mid June to mid July. Greatest success was achieved with patch budding. Patch budding was the most successful technique in each period.

Cultivation of economically important trees in such reclammed soil solves three problems by providing employment, revenue generation and soil eco-management.

(vi) Harvesting

The amla gets ready for harvesting by November-December. However, the fruits may be allow to remain on the tree till February without much fruit drops. The optimum stage of harvesting falls between the period extending from the second week of December to the third week of January in Northern plains. However, at some places in southern India the fruit is available through out the year. A fully mature amla tree may yield 2.5 to 3.0 quintals of fruits annually.

(vii) Storage

Storage is one of the most important aspects. The object of the storage of fresh fruit is to extend the period of their availability. The storage life of the fruit is dependent on the initial quality and stage of maturity on harvest. The available information on the storage behaviour of amla is very limited. Amla fruits are hard and unyielding to touch and as such well suited for long distance transportation. The fruits may be kept well in cold storage for 7-8 days at 32-35° F and RH 85-90%.

(viii) Pests/fungal diseases and their control

Emblica was found to be attacked by following insects and fungi :

(a) Bark eater insects (*Inderbela tetraonis*) : The female of this insect lays near about 2000 eggs after rainy season. Larva of this insects makes 15-20 groups. They lived groupwise on the branches of Amla. During day time they come down to shady area of the stem. At the early morning and evening their groups climb up nearly all branches and eat the pale green or greenish bark. They eat leaves also. So all the leaves of Amla can be eaten by insects in short time.

Growth of the plants retards, due to severe attack of larvae and plants could dry. The insect can be managed by garden cleanliness, keeping limited number of branches and spraying of Methyl Parathion (Metacid), Indosulphan, Virat, 0.5% solution in morning hours once a week. Three times of spraying may be necessary.

(b) Shoot Galls maker (*Betoura stylophora*) : Seeded amla plants are more affected by these insects, caterpillar invade the old branches and they enter into the stem and makes a knob like structure. They take food from branches (juice from vascular tissues). Vascular transportation is cut down. The growth of branches is completely stopped. This insect is controlled by cleanliness of garden, removal of affected parts and spraying of Monocrotophos 0.05% (June-July) two times.

(c) Aphids (*Ceraciaphis emblica*) : At the late rainy season greenish aphids attack on young branches of amla. They secrete glucose. Due to the presence of glucose fungi attack on branches and colour of the branches change into black smut. The aphids sucks the juice from the leaves and young stem. Due to severe attack the growth of plants retards.

Rogor 30 E.C. or Metacid-50 or Virat, 0.05% solution should be sprayed at morning to control the Aphides.

(d) Termites : They are soil dwellers and attacks plants roots. Severely attacked plants can be identified by the soil hips on the stem. The root of nursery plants and young planted plants are eaten by termites. So plant soon die. But in large plants they can rise on arial parts of the plants quickly they eat the bark and outer side of the wood.

Control of termites is difficult. It is better to control on the plantation time or nursery bed preparation time. Furadon 5 G, or Phoratox 10 G-25 gm can be used in auger holes.

(e) Rust of Amla (*Ravenellia*) : Rust sori are light brown and circular or semi circular in shapes. The sori are present on both side of the leaves. Fruits are also affected by this fungus, where light brown or some times dark brown spots are formed. Severe infections result in complete defoliation of leaves and growth of the plants retards. Presence of brownish patches on fruits decreases the economic value of fruits.

Sulphur dust is a good chemical means of control. Dithane M-45, Indophil-78.2% (2 gm/lt) solution also gives better control of termite.

(f) Blue Mould (*Penicillium isaudicum*) : The symptoms of this disease can be seen only in fruits. The pathogen produces rot of ***Emblica officinalis***.

The pathogen enters through stomata or lenticels. The rot areas has three colours - Bright pale yellow, violet and Blue patches respectively. For management, affected fruits should be separated. Uses of borax and NaCl can reduce the pathogen on fruits and can be safely stored.

DISCUSSION

Horizontal expansion of farming on good land is not possible due to reduction in land: man ratio, hence only alternatives seems to utilize either degraded lands or to go for intensive farming on fertile lands with assured input facilities for fulfilling the basic need of our increasing population. Regarding former approach, a sizeable land in our country is subjected to various kinds of land degradation problems and about 6.32 million ha, alone suffers from salinity and alkalinity. These lands have good potential if managed properly for sustained utilization by putting them under agroforestry land use system. In present study, the suitable agrotechniques have been detailed out by series of experimentation on farmers field in participatory mode. The site specific agrotechniques for establishment of various components, management practices and extent of participation have been suggested for efficient utilization of salt affected land through *Emblica* based agro forestry system.

Amla fruits are hard and used as a source of medicine since long for common cold scurvy, cancer and heart diseases. The fruit is a rich source of ascorbic acid. The amla sugars are composed of fructose, glucose, sucrose and myoinisitol. The fruits also contain tannins and many polyphenolic substances. Amla is a rare example of an edible material which is rich in tannins as well as ascorbic acid. The fruit also contains alkaloids (phyllantidine and phyllantine), auxins and minerals. Amla stones contain six small seeds. The seed oil of amla comprise 64.8% of linolenic acid and closely resembles to linseed oil.

ACKNOWLEDGEMENTS

Authors are Thankful to Deptt. of Biotechnology, Ministry of Science and Technology, Govt. of India, New Delhi for financial assistance and to the Principal, R.B.S. College, Agra for providing research facilities.

REFERENCES

Anonymous (1996). Indian Pharmacopia, Government of India, Ministry of Health and family welfare, the controller of Publications, Delhi. Vol. 1.

Anonymous website : http//genome.cornell.edu/botony/about phytochemdb.html.

Khopde, S.M. ***et al.*** **(2001).** Characterizing the antioxidant activity of amla (*Phyllanthus emblica*) extract. *Curr. Sci.,* **81(2)** : 185-90.

Recent Advances in Medicinal Plant Research : Vision 21st Century, 2005
Editors : S.K. Prabhuji, G.P. Rao and S.K. Patil
Satish Publication House, New Delhi

Chapter 10

PACKAGE OF PRACTICES FOR CULTIVATION OF *BALIOSPERMUM MONTANUM* (WILLD.) MUELL.-ARG.

V.B. Shimpale[1]* and S.R. Yadav[2]
Department of Botany, Shivaji University, Kolhapur - 416 004
**Present address : [1]CEMDE, School of Environmental Studies, University of Delhi, Delhi - 110 007*
[2]Department of Botany, University of Delhi, Delhi - 110 007

During 1999-2003, attempts have been made to study and find out suitable conditions for optimum growth and yield of required plant materials of *B. montanum*. It includes seed viability, seed germination, and growth performance of the species under pot and field cultivation for optimum yield. The primary package of practices for cultivation of *Baliospermum* is discussed in this chapter.

INTRODUCTION

World Health Organization (WHO) has estimated that aleast 80% of the world population depends on traditional systems of medicine for their primary health needs and these systems are largely plant based. WHO stated that over 21,000 plant species are useful in the preparation of medicines. The present global herbal market is worth about US$ 62 billion per annum and India's share in it is only 0.2%.

International trade in medicinal plant material for the production of pharmaceutical and cosmeticeutical preparations is blooming. Several medicinal plants are cultivated and domesticated; cultivation of others that are collected from the wild in vast amount is desired. However, according to one estimate, of more than 400 plant species used for production of medicine by Indian Herbal Industry, less than 35 are currently under cultivation in different parts of country. In fact, cultivation of medicinal plants is not so easy. It is rather a challenging task and very little is known about the seed biology and methods for cultivation of medicinal plants. Little efforts have been made to search for elite stock and to propagate them. There is a need for a scientific approach for propagation of medicinal plants and to collect relevant information regarding agrotechnology, genuine planting material, economics of field cultivation, high yielding varieties etc.

Baliospermum montanum (Willd.) Muell.-Arg. (Dantimul) is one of the most important medicinal plant species used in ayurvedic medicines as well as in pharmaceutical industries for preparation of different formulations. Species is used to treat several ailments like piles, anaemia, jaundice, and used as purgative (Sharma *et al.*, 1996). Leaves and seed oil are reported to have irritant property and rubefacient (Nadkarni, 1954). The roots are pungent, purgative and in over doses is highly poisonous (Anon, 1976). Plant contains montanin (a daphnane polylester), baliospermin and other tigliane poly esters, which are effective compounds (Rastogi and Mahrotra, 1993; Ogura, 1978; Ranghunathan and Mitra, 1999). Locally the seeds and roots are used as drastic purgative and oil extracted from seeds is used for rheumatism. In market it is sold under name, 'Dantimul' but many times adulterated with roots of *Ricinus communis* due to lack of availability of the species.

METHODOLOGY

Seeds of *B. montanum* were collected from natural populations of Ratnagiri. To study the effect of seed storage on seed germination, seeds stored in dry stoppered glass bottles in laboratory conditions were subjected to petriplate germination test in the first week of each month and percentage germination in 20 days was recorded to test effect of storage on seed germination, dormancy and viability. Pot culture and field cultivation experiments were done in experimental garden of the department. Earthern pots (30 cm diameter and 25 cm heights) were used as containers for each type of experiment for pot culture. Similar set of containers was used for each of the treatment. Seeds were sown at 2 cm depth and after 30 days of sowing thining was done to keep required number of plants per container to make further studies on growth performance. Roots, stem and leaves were dried separately at 60°C for 72 hours and the contribution of dry biomass by different organs to the total biomass was determined. For each condition and at each time of harvest, minimum three plants were analyzed. Results are based on average of three replicates.

RESULTS

Baliospermum montanum (Willd.) Muell.-Arg. is monoecious stout erect undershrub, upto 1.5 m tall. Leaves are variable, sinuate, toothed, often 3-5 lobed and 2-glandular at base. It produces flowers on axillary racemes with numerous male and a few female flowers. Capsule are 8-13 mm long, obovoid, usually three valved with solitary ovule in each cell. Seeds are 8 × 5 mm, ellipsoid, smooth and mottled.

It is commonly called as dantimul or jamalgota. Species is distributed throughout the Sub-Himalayan tract, North Bengal, Chota Nagpur and South India. The fruits mature by the end of January and February to March is the best season for seed collection from natural populations.

Seed germination

Storage and seed germination

Freshly collected seeds showed fairly good germination percentage (90.6%) in 1st month of storage. Again it was low (62.6%) in month of March. The germination percentage increased during April onwards and maximum germination percentage (98%) was obtained during June to July. Thereafter there was drastic reduction in germination percentage with increased storage period and in December the germination was reduced to 1.33%. Thereafter, i.e. 12 months of storage seeds failed to germinate and found to be non-viable. It indicates that the seeds remain viable only for about 11 months.

Effect of pH solutions on seed germination

To study the effect of pH on seed germination, seeds were subjected to petriplate germination tests and supplied with water of particular pH throughout the investigation (Fig. 1A). The seeds showed wide range of pH tolerance for their germination. Seeds showed fairly good seed germination at pH range between 4-9. Maximum germination percentage (81.78%) was obtained in distilled water (control) followed by 77.7% at pH 8. Minimum germination percentage was observed at pH 4.

Effect of depth of sowing on seed germination

Seeds were sown equidistantly at different soil depths in earthen pots and the pots were watered daily once with equal amount of water. The seeds showed maximum and equally good germination percentage (86%) at 1-2 cm sowing depth. The seeds sown on surface of soil showed 18% germination. Seeds showed decrease in germination percentage with increase in depth of sowing above 2 cm and its was minimum at sowing depth of 6 cm. Therefore, sowing depth of 1-2 cm was found to be ideal for proper seed germination (Fig. 1B).

Effect of soil composition on seed germination

The seeds were sown at 2 cm depth in earthen pots containing different compositions of soil and the pots were watered on alternate day. The seeds showed good germination except in sand. Maximum germination (82.67%) was found in farm Yard Manure followed by garden soil. In sand there was no seed germination (Fig. 1C).

Effect irrigation practices on seed germination

The seeds sown in garden soil and farm yard manure (GS + FYM 3:1) in earthen pots at 2 cm depth were irrigated at different rates. Seeds showed maximum percentage germination (89.3%) in pots watered daily followed by pots watered on alternate day (86%). The seeds showed 47.3%

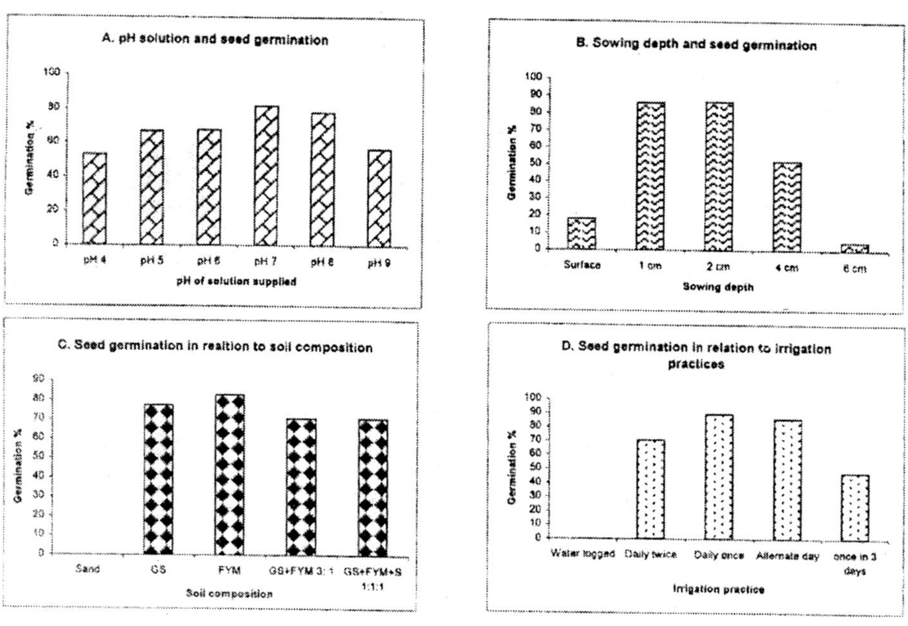

Fig. 1 : *Baliospermum montanum* : Seed germination

Habit

Seeds

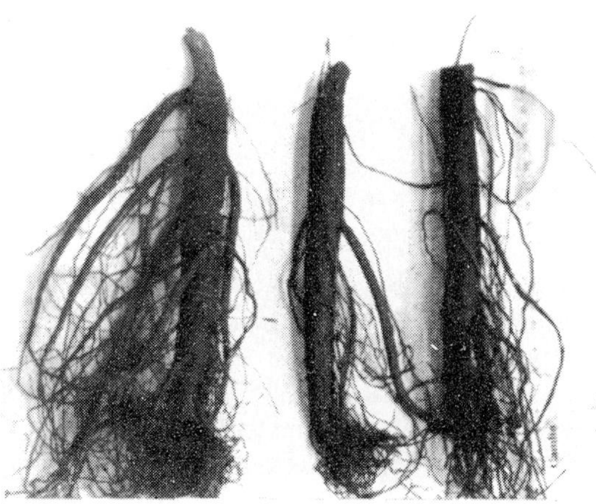

Roots

Plate I : *Baliospermum montanum*

germination in pots watered once in 3 days and failed to germinate under waterlogged conditions (Fig. 1D).

Pot culture and field cultivation experiments

(I) Growth performance in relation to soil composition under pot culture

Soils of different compositions were obtained by using sand, garden soil and farm yard manure. The following four sets of soils were used in the experiment

Set - I : Garden Soil (GS);

Set - II : Farm Yard Manure (FYM);

Set - III : Garden Soil (GS) + Farm Yard Manure (FYM) 3:1 and

Set - IV : Garden Soil (GS) + Farm Yard Manure (FYM) + Sand (S) 1:1:1.

Initially 25 seeds were sown at 2 cm depth in uniform sized earthen pots with different compositions of soil. After germination, thinning was done so as to keep only three equidistantly placed seedlings per pot. Three replicates were analyzed for each type of soil and at each time of harvest. Harvesting was done at three stages viz. pre-flowering (102 days), peak flowering (151 days) and late fruiting stage (204 days).

Maximum contribution to the total biomass came from leaves followed by roots and stem at both pre-flowering and flowering stage of harvest in all the sets of soils. At final stage of harvest maximum contribution to the total biomass was by roots followed by leaves. Plants showed best growth performance when grown in garden soil and farm yard manure in 3:1 proportion. Observations indicated that the roots are the storage organ of the species and synthates are accumulated in roots with increasing age (Fig. 2a).

(II) Growth performance in relation to light conditions under pot culture

Baliospermum montanum usually grows as undergrowth in semi-shaded places in moist and dry deciduous forests. To study the effect of light on growth performance of the species experiments were designed. Plants were grown in different light conditions such as tree shade (20,300 Lux); wall shade (48,200 Lux); diffuse light (69,600 Lux) and full sun light (85,900 Lux). Seedlings were raised in suitable soil composition (GS+FYM 3:1). Growth analysis of plants was done at the age of 101, 158 and 208 days representing vegetative (pre-flowering), flowering and fruiting stages of growth. Analysis of percentage contribution by different plant parts did not show any definite trend for light conditions used at all three stages of harvest. There was no significant different in proportion of contribution under different light conditions. In pre-flowering stage of growth, maximum contribution to the biomass was by leaves (except wall shade) followed by root and minimum contribution was by stem. Equally good contribution by leaves and roots to the total biomass was observed at flowering as well as fruiting stage of harvest and minimum by stem. In general, with increase in age there was decrease in percentage contribution by stem, root and leaves under all the four conditions of light (Fig. 2b). Results shows that partial shade light is good for better performance of the species. Increase in productivity with increasing age from vegetative to fruiting stage was observed under all the four conditions of light.

(III) Growth performance in relation to irrigation practices under pot culture

To study the effect of irrigation practices on growth performance of the species, plants were raised in earthen pots filled up with suitable soil composition (GS+FYM 3:1) and kept under tree shade. The pots were irrigated differentially as W_1 : Irrigated daily twice; W_2 : Irrigated once daily; W_3 : Irrigated on alternate day; W_4 : Irrigated once in 3 days and W_5 : Water logged condition.

Percentage contribution to the total biomass was maximum by leaves at the time of pre-flowering stage of harvest followed by roots and it was maximum by stem under all the four practices of irrigation (Fig. 2c). Increase

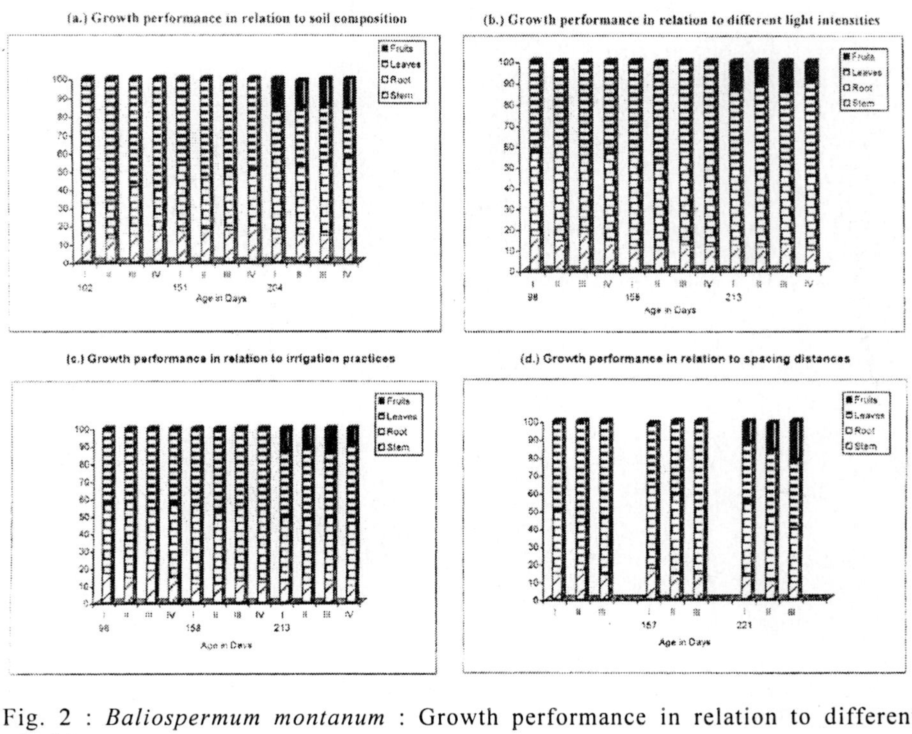

Fig. 2 : *Baliospermum montanum* : Growth performance in relation to different edaphic parameters

A- Pre-flowering stage B- Peak flowering stage C-Late fruiting stage

Plate II : *Baliospermum montanum* : Field cultivation

in percentage contribution by roots and decrease by leaf was observed in flowering stage of harvest under all the four irrigation practices. Slight increase in percentage contribution by stem was also observed at flowering phase of harvest. During fruiting stage harvest, maximum contribution to the total biomass was by leaves and roots but low as compared to two initial harvests as the synthates were diverted to reproductive growth. Plants when irrigated daily once showed better growth performance. No definite trend in percentage contribution under four irrigation practices at any harvest time was observed.

(IV) Growth performance in relation to spacing distance under field cultivation

To study the effect of spacing on growth performance of *B. montanum*, experimental plots of 2 × 2 m size were prepared in botanical garden of the department (Plate-II). After preparation of soil, seeds were sown at 10, 30 and 50 cm distance in different plots.

Under field condition, maximum biomass was contributed by leaves followed by roots and minimum by stem at initial phase of harvest at all the three spacing distances (Fig. 2d). During second harvest stage, at spacing distance of 50 cm, maximum percentage contribution to the total biomass came from leaves but it was less as compared to first harvest and there was corresponding increase in dry biomass of roots and leaves. Percentage contribution to the total biomass was maximum and equally good by leaves and roots in plants with spacing distance of 30 cm followed by fruits and stem at third stage of harvest. Plants with different spacing distance showed a common trend i.e. increase in productivity with age.

DISCUSSION

Supply of plant-based medicines to the world is an important challenge of 21st century. The international trade in medicinal plants is a major force in the global economy. Over exploitation, declining forest areas and increased demands resulted in rarity of medicinal plants. About 10,000 species of medicinal plants are threatened with extinction through world (Leaman, 1998). Cultivation of medicinal plants is viewed as a means of meeting current and future demands of plant-based drugs and for relieving pressure on wild populations (Palevitch, 1991; WHO *et al.*, 1993; FAO, 1995; Lambert *et al.*, 1997).

Knowledge of seed dormancy, viability and germination is of prime importance in establishing package of practices for cultivation of any wild plant species. The selected plant species have their own ecological nitche and differences in seed behaviour. *Baliospermum montanum* possess no dormancy mechanism. The fresh air-dry seeds germinate readily.

Seeds of *B. montanum* has non-dormant seeds. Air-dry fresh seeds showed over 90% germination in first month of storage and again there was decreased during March coinciding with dry spell of year. The species showed high percentage germination during rainy season. Thereafter, there was drastic reduction in percentage germination with increasing period of storage. The seeds stored for 12 months lost their ability to germinate. The seeds of the species remained viable only for about one year under laboratory storage conditions. Loss of ability to germinate with increasing seed storage period is but natural phenomenon and viability period of seeds differ from species to species (Robert and Lockett, 1977).

It is well known fact that species respond differentially to different soil pH depending on their genetic constitution. pH of soil is one of the important factors, which govern distribution of any species. Seed germination, survival and successful establishment of seedling are the critical stages in lifecycle of any species. Soil pH is known to play important role in seed germination and establishment of seedling. Seeds of *B. montanum* showed wide range of tolerance from 4 to 9, however, maximum seed germination was observed at pH 7 (distilled water). Many species have been reported to have narrow range of pH tolerance during seed germination (Ramakrishnan, 1963; Rao and Reddy, 1981; Yadav and Kulkarni, 1986; Bajpai, 1997).

The germination capacity of the seeds and establishment of seedlings are strongly influenced by abiotic factors such as temperature, light and edaphic factors. These factors also show interactions affecting seed germination, seedling establishment and survival of species (Rao and Reddy, 1981). Edaphic factors such as depth of soil, soil moisture, texture of soil are known to effect seed germination and seedling establishment. *Baliospermum montanum* showed best germination at 1-2 cm sowing depth. With increase or decrease in depths of sowing, there was decrease in germination of seeds. One to two cm sowing depth was found to be optimum for *Baliospermum montanum*. Depth of sowing is found to influence seed germination in number of species (Ramakrishnan, 1963; Yadav and Kulkarni, 1986; Kumar *et al.*, 1996). Ideal depth at which maximum germination is obtained varies from species to species. For example the seeds of *Erigeron linifolius* and *Euphorbia nivula* show maximum germination when sown superficially (Ramakrishnan, 1963; Agarwal, 1973), that of *Calligonum polygonoides* at 0.5 cm sowing depth (Kumar *et al.*, 1996), *Michelia champaca* at 1.5 cm sowing depth (Pyare Lal *et al.*, 1997) and *Ambrosa trifida* and *Quercus semecarpifolia* at 2 cm depth (Lalhal *et al.*, 1996).

Soil moisture plays important role in seed germination. Seed germination markedly influenced by the moisture levels of soil (Baskin and Baskin, 1970a) and the response depends upon the species (Singh, 1972; Gupta, 1973). Seeds of *B. montanum* showed maximum seed germination under daily once irrigation practices.

Texture of soil, porosity, soil moisture, water holding capacity of the soil and the microclimate depends upon the composition of the soil and all these factors together affect the seed germination and seedling establishment. *B. montanum* showed maximum germination in farmyard manure and garden soil. Differential response of seeds to germinate in different soil compositions has been recorded by various workers in different species. Singh (1972) observed 50% seed germination in *Solanum surttense* in farm yard manure. Like *B. montanum*, seeds of *Anthocephalus cadamba* showed maximum germination when sown in garden soil (Beniwal and Dhawan, 1991). Seeds of *Mallotus phillipensis* and *Hamiltonia suaveolens* showed good germination in soil composed of equal parts of garden soil, farm yard manure and sand (Bahuguna and Pyare Lal, 1996, 1997).

Cultivation of medicinal plants is a need of present day for continuous and pure supply of raw drug materials. It has numerous advantages including conservation, utilization and pure reliable supply of medicinal plants. Importance of ecological studies in establishing cultivation practices for wild medicinal plants is realized all over the world. Ecological data together with data on pot culture and field cultivation provides valuable information in establishing cultivation practices for medicinal plants for increased yield. *B. montanum* showed best performance when grown in soil composed of 3 parts of garden soil and one part of farm yard manure. Effect of soil texture and composition on growth performance of different species have been studied by Verma (1966) and Gupta (1973). Wakhloo (1961, 1963) recorded that in addition to morphological character, alkaloid content of *Ravoulfia sepentina* is influenced by soil texture.

Effect of light intensity on various plant species have been studied by many workers (Wahi, 1976; Yadav and Kulkarni, 1986). *B. montanum* showed better performance under reduced light intensities. The growth performance of species was adversely affected with increase in intensities of light. Similarly, *Crotalaria juncea* and *Crotalaria sericea* showed poor performance with decreasing light intensities (Pandey and Sinha, 1977).

Soil moisture is one of the important factors, which has influenced on plant growth. Response of species populations to moisture stress had formed a subject of numerous investigations (Bohara and Sen, 1980). Irrigation practice is one of the important aspects of cultivation of any species. *Baliospermum montanum* showed equally good performance under daily once and alternate day irrigation practice.

Density of plants is one of the important factors, which affect growth performance and behaviour of the species. With the increased pressure on available land source, maximum harvestable quality biomass production per unit area is the prime concern of high-density plantation. A suitable spacing can provide maximum material of desirable quality (Pacholi and Pandey,

1998). *Baliospermum montanum* showed good performance when sown at 50 × 50 cm spacing distances. Stem and root contribution decreased with increase in spacing distance, however, contribution by leaves and fruits increase with increased in spacing distance at late fruiting stage.

CONCLUSION

1. Seeds of *B. montanum* possess no dormancy. The seeds remain viable for about 11 months but after 8 months of storage, seeds show drastic reduction in germination and start loosing their viability.
2. Seeds of *B. montanum* can be sown directly into soil at depth of 2 cm for better seed germination and seedling establishment. Soil composed of 3 parts of garden and 1 part of farmyard manure is found to be appropriate in cultivation of the species for higher yield. Planting distance of 50 cm between plants and rows is found to be appropriate for higher yield of root and for maximum seed output per plant.
3. For higher yield of leaves, the species needs to be grown in farm yard manure under partial light shade, irrigated once in three days and harvest need to be done at pre-flowering stage. Similarly for higher yield of root biomass, the species needs to be grown in GS+FYM+S (1:1:1) or garden soil, partial shade, irrigated daily once and harvest have to be done at fruiting stage.
4. For higher seed-out put, the species has to be planted at 50 × 50 cm spacing distance however, for high root dry biomass per unit area, it needs to be planted at 10 × 10 cm distances.

The experiments conducted on seed germination, pot culture and field cultivation provide the basic information for bringing this species under cultivation to produce maximum yield.

MARKET POTENTIAL AND SYSTEMS

Although demand for medicinal plants is too high but there is a lack of authentic data for demand and supply. Market is unorganized and does not have regulatory body. Furthermore, there is lot of fluctuation in price of crude drugs.

In the present system, farmers are not getting the right price for their produce. Middleman and agents are being benefited from the present system. Cultivation produce seems to be costlier than wildly collected material. So, it is necessary to check illegal collection and trading to encourage cultivation.

REFERENCES

Agarwal, S.K. (1973). Some aspects of seed germination of *Euphorbia nivulea* Buch.-Ham. *J. Biol. Sci.,* **16** : 10-13.

Anon. (1976). Medicinal Plants of India. Vol. I. ICMR, New Delhi.

Bahuguna, V.K. and Pyare Lal (1996). Prelimiary studies on the germination behaviour of some shrub species of Himalaya. *Ind. For.,* **123(4)** : 345-347.

Bahuguna, V.K. and Pyare Lal (1996). Studies on evolving nursery practices for *Mallotus phillipensis* seeds. *Ind. For.,* **122(4)** : 298-305.

Bajpai, A. (1997). Influence of moisture stress and temperature on germination studies of *Boerhaavia diffusa* L. *Ad. Pl. Sci.,* **10(2)** : 229-234.

Baskin, J.M. and Baskin, C.C. (1970). Germination ecophysiology of *Draba verna. Bull. Torrey Bot. Club,* **97** : 209-216.

Beniwal, B.S. and Dhawan, V.K. (1991). Standardization of nursery technique of *Anthocephalus chinensis. Ind. J. For.,* **117(2)** : 105-108.

Bohara, P.N. and Sen, D.N. (1980). Behaviour of some arid zone plants under water stress. In Progress in Ecology III (eds. Agarwal, V.P. and Sharma, V.K.) Today and Tomorrow printers and publishers, New Delhi pp. 167-178.

Food and Agriculture Organization (1995). Non-Wood forest products for rural income and sustainable forestry. Non-wood forest products-7. Food and Agriculture Organization of the United Nations, Rome, Italy.

Gupta, S.K. (1973). Ecology of *Rumex* species (*R. dentatus* and *R. nepalensis* Spreng) Ph.D. Thesis. BHU, Varanasi (India).

Kumar, V., Nagavani, H.C., Kumar, D. and Singh, V.P. (1996). Propagation and growth studies of *Calligonum polygonoides* (Phog.). *Ind. For.,* **122(5)** : 377-382.

Lalhal, J.S., Jishtu, V. and Singh, O. (1996). Seed collection and nursery emergence in *Quercus semecarpifolia* Sm. *Ind. For.,* **122(1)** : 85-86.

Lambert, J., Srivastava, J. and Vietmeyer, N. (1997). Medicinal plants : rescuing a global heritage. World Bank Technical paper No. 355. The World Bank, Washingtone.

Leaman, D. (1998). Letters. *Plant talk,* **14** : 4.

Nadkarni, K.M. (1954). Indian materica medica. Popular Prakashan Pvt. Ltd. Bombay.

Ogura, M. (1978). Potential anti-cancer agents VIII constituents of *Baliospermum montanum* (Euphorbiaceae). *Planta Med.,* **33(2)** : 128-133.

Pacholi, R.K. and Pandey, O.N. (1998). Impact of varying plantation density of the biomass production and nutrient contents in *Dalbergia sissoo* Roxb. *Ind. J. For.,* **21(3)** : 228-231.

Palevitch, D. (1991). Agronomy applied to medicinal plant conservation. In : The conservation of medicinal plants (eds.) Akerele, O., Heywood, V. and Synget, H. Cambridge University Press. Cambridge, UK.

Pandey, B.N. and Sinha, R.P. (1977). Light as a factor in growth and morphogenesis. I-Effect of artificial shading on *Crotalaria juncea* L. and *C. sericea* Retz. *New Phytol.*, **79** : 431-439.

Pyare, Lal, Bisht, N.S. and Agarwal, S.K. (1997). Determination of proper depth of seed sowing in *Michelia champacha* in nursery. *Ind. For.*, **123(2)** : 118-122.

Raghunathan, K. and Mitra, R. (eds.) (1999). Pharmacognosy of indigenous drugs Vol. I-III. Central Council for Research in Ayurveda and Siddha, New Delhi.

Ramkrishnan, P.S. (1963). Seed germination studies on *Erigeron linifolius* Willd. *Proc. Nt. Inst. Sci. India*, **29B** : 561-566.

Rao, P.N. and Reddy, B.V.N. (1981). Autecological studies in *Indigofera linifolia* (L.f.) Retz. I-Germination behaviour of the seeds. *J. Ind. Bot. Soc.*, **60** : 51-57.

Rastogi, R.P. and Mehrotra (ed.) (1993). Compendium of Indian Medicinal Plants. Vol. I-III. CDRI and Publ. and Infor. New Delhi.

Roberts, H.A. and Lockett, P.M. (1977). Temperature requirements for germination of dry stored, cold stored and buried seeds of *Solanum dulcamara*. *New Phytol.*, **79** : 505-510.

Sharma, S., Ramji, S., Kumar, S. and Baphana, J.S. (1996). Randomized controlled trials of *Asparagus racemosus* (Shatavari) as a lactogogue in lactational inadequacy. *Indian Pediatr.*, **33** : 675-677.

Singh, M. (1972). Studies on seed germination of two forms of *Solanum surattense* Burm.f. *J. Ind. Bot. Soc.*, **51** : 31-44.

Verma, M. (1966). Geographical distribution and ecology of medicinal plants. *Solanum surattense* and *S. nigrum* L. Ph.D. Thesis, BHU, Varanasi (India).

Wahi, P. (1976). Growth responses of *Andrographis paniculatus* to different regimes. Ph.D. Thesis, BHU, Varanasi (India).

Wakhloo, J.L. (1961). Autecological aspects of three indigenous medicinal plants. Ph.D. Thesis, BHU, Varanasi (India).

Wakhloo, J.L. (1963). Variation in the total alkaloid content of *Rauvolfia sepentina* roots, a consideration from ecological point of view. *J. Ind. Bot. Soc.*, **42** : 214-221.

World Health Organization (1993). IUCN-the world conservation union and WWF-the world wide fund for nature. Guidelines on the conservation of medicinal plants. IUCN-the world conservation union, Gland, Switzerland.

Yadav, S.R. and Kulkarni, A.R. (1986). Ecological studies in *Psoralea corylofolia* L. I-Pod/seed and seed germination. *Geophytol.*, **16(1)** : 198-215.

Recent Advances in Medicinal Plant Research : Vision 21st Century, 2005
Editors : S.K. Prabhuji, G.P. Rao and S.K. Patil
Satish Publication House, New Delhi

Chapter 11

PROSPECTS OF *STEVIA* CULTIVATION IN INDIA: THE SUGAR OF 21ST CENTURY

S.D. Singh[1] and G.P. Rao[2]
[1]*AKS Herbal Research and Land Development Centre*
[2]*Society for Conservation and Resource Development of Medicinal Plants*
AKS House, Parha, Kasla Road, P.O. Railway Colony
Behind Riyaz Haddi Hospital, Mohaddipur, Gorakhpur - 271 012
Uttar Pradesh

Stevia is the new emerging alternative source of calorie free sweetener having no carbohydrate and fat. It is 20 to 30 times more sweetener than cane and beet sugar, highly nutritious, delicious, non-toxic and non-additive sugar. It also enhances the flavour, helpful in digestion, weight reduction, anti oxidant, prevent dental caries and having antimicrobial and anti-plaque properties, increases mental alertness, increase energy levels but does not affect the blood sugar level, therefore key-source sweetener for diabetic world. Besides, *Stevia* can be used in hypertension, hypoglycemic, helpful in skin toning and healing, tobacco and alcohol cravings and a tonic for pancreas. It can also be used as alternative source of sugar for food confectioneries, bakeries, fruit juices, jams, biscuits, chocolates, vegetables and other food stuffs. The recent researches alongwith future prospective of this new emerging medicinal plant has been discussed in this chapter.

INTRODUCTION

Stevia is a native plant of Paraguay (South America). It is indigenous to the Rio Monday Valley of the Amambay Moutain Region where it grows as a perennial at an altitude between 200-500 metres having a mean temperature of 23 °C (range -6 °C + 43 °C) and rainfall ranging from 1500-1800 mm per annum. The native Guarani tribe had known for centuries the unique sweetening power of its leaves and other medicinal properties. They called the plant "kaa he-he" which translates as "sweet herb" and used it as sweetener for their green herbal tea "mate" and other domestic purposes as a flavour enhancer. In due course, it was introduced to settler. By now, Stevia is being consumed in Japan, Brazil, USA, Argentina, China, Canada, Paraguay and Indonesia.

Characteristics

Stevia rebaudiana a member of the Compositate/Asteraceae family and as such is related to sunflower, marigolds and chrysnthemums. Stevia has an alternate leaf arrangement and herbacious growth habit (Fig. 1a) with flowers arranged in indeterminate heads (Plate 1g). The flower are small and white with a pale purple throat. The pollen can be highly allergic.

Source of Sweetness

Stevia derives its sweetness from diterpene glycosides – eight in number (steviol glycosides). These are synthesized, at least in the initial stages, using such the same pathway as gibberellic acid, an important plant hormone. The four major steviol glycosides are:

- stevioside
- rebaudioside A
- rebaudioside C
- dulside A

Propogation

In the wild, Stevia regenerates from seed, from the rooting of plant stems touching the ground (Plate 1c) and from regeneration at the base of the plant (crown division). Seed germination is notably very poor, commonly due to infertile seed. Some plant varieties/selections produce virtually no viable seed due to their self-incompatibility.

Under cultivation, Stevia can be propagated by seed, by tissue culture and by vegetative cutting (and plant separation) (Plate 1c). Since germination rates are poor and seedlings are very slow to establish, it is best grown as an annual or perennial transplanted crop. Clonal propagation is practical for small scale production, but is probably not economically viable for large scale Stevia production in those regions where labour costs are high.

Plate 1 : (a) Morphology of *Stevia* plant, (b) Land and furrow preparation for *Stevia* planing, (c) Cultivation of *Stevia* shoot on ridges, (d) Young grown *Stevia* plants, (e) Network of irrigation in *Stevia* field, (f) Stevia leaves ready for harvesting, (g) Flowering stage of *Stevia*

Cultural Practices

A planting density ranging from 40,000 plants/acre have been found suitable for maximum production and is recommended by our centers. Leaf yield increased with increasing density up to 83,000 and 1,11,000 plants per ha for the first year of production (Plate 1d). In Canada, a plant population of 1,00,000 per hectare is recommended. Densities of 80-100,000 plants per hectare on row spacing of 45-65cm are generally recommended.

The concentration of glycoside in the leaves of Stevia increases when the plants are grown under long days. Since glycoside synthesis is reduced at or just before flowering, delaying flowering with long days allows more time for glycoside accumulation.

On the whole plant level, steviol glycosides tend to accumulate in tissues as they age, so that older lower leaves have more sweetener that younger upper leaves, Since chloroplasts are important in precursor synthesis, those tissues devoid of chlorolphyll, like roots and lower stems contain no or trace amounts of glycosides. Once flowering is initiated glycoside concentrations in the leaves become low. Therefore, harvesting is to be done just prior to flowering when steviol glycoside content in the leaves is at its maximum.

Nutrition and Fertilization

Stevia plants do best in red soil, alluvial soil or sandy loam with a pH range of 5-7.5. Since the feeder roots tend to do be quite near the surface, it is a good idea to add compost for extra nutrients if the soil is sandy. Besides being sensitive to cold during their developmental stage, the roots can also be adversely affected by excessive levels of moisture. So care should be taken not to over water them and to make sure the soil in which they are planted drains easily (Plate 1e). Most organic fertilizers would work well, since they release nitrogen slowly.

Scientists of AKS Herbal Research and Land Development Centre has worked out that humic, fulvic and humate contents are more important in comparison to inorganic fertilizers containing N, P, K alone or in combinations as per general recommendation. Stevia cultivation requires 120 Kg of humic contents and 20 Kg of fulvic contents per acre basis. These contents are in any existing fertilizer, so far available in India. Keeping this in view, our scientists formulated a unique pure organic fertilizer containing approx. 12% humic acid and 2% fulvic components, named AKS BIOCROP, which is under process of patent registration. The AKS BIOCROP has been recommended at fertilizer dose of 100 liters per acre of Stevia cultivation in alluvial soils to fulfill its nutritional and fertilizer requirement.

Since Stevia is very slow to establish and does not compete well with weeds, it would be essential to de-weed the plot prior to planting and control weed growth thereafter to produce ample yield and a clean crop.

Stevia is known to be free from pest attacks, its inherent sweetness in fact acting as an insect repellant. Therefore, insecticides are not required at an essentially basis as in other crops. This also helps in getting organic *Stevia.* Two fungal diseases, *Septoria steviae* (leaf spot) and *Sclerotinia sclerotoirum*, have been reported in Canada. Neem based pesticides could be used as a precautionary measure. Recently, scientists of our center have developed a pure botanical pesticide, named AKS BIOCARE, which has been found, as more suitable for IPM of Stevia cultivation. AKS Research center has recommended 2 lit/acre of AKS BIOCROP for better cultivation of disease/pest free Stevia.

Harvesting

As stated above, harvesting must be commenced before flowering begins. Stevia can grow upto a height ranging from 60-75 cms (Plate 1f). Heights upto one meter have also been recorded. When the time does come to harvest, the easiest technique is to cut the branches off with pruning shears before stripping the leaves.

In warmer climates of India, we can harvest Stevia leaves 4 to 5 times a year with the first harvesting at 90 days of the crop.The parent crops can be maintained with good yields up to 5 years . In Paraguay/Brazil, three harvests a year are normal. In mid –summer harvest intervals can be less than two months. In Indonesia up to seven harvests have been possible per year.

In winter days only one harvest per planting is possible, which will reduce the total yield . Our research center has evolved a new technique for better production of Stevia yield in winters by applying polythene + hasien cloths during night when temperature goes down to below 5 °C. Again by covering the harvested crowns of the plants (with polythene + hasien cloth etc.), plants can survive over winter and production for a second season can also be possible.

Crop Yields

Experimental yields suggest that systems with multiple harvests a year will give higher yields and, where ratoon crops are grown, harvests in the second and third years are likely to be higher than for crops requiring replanting each year.

Yields would decline after three years needing replacement of plant material. However, our research center is confident that with good management practices, yields can be stretched upto five year and more.

Post Harvest Operations

Once all the leaves have been harvested, they have to be dried. This can be accomplished on a screen or net. The drying process is not one that requires excessive heat, more important is good air circulation. If sun dried, it would take 6-8 hours depending on the temperature. Longer drying times will lower the stevioside content of the final product. A green dried leaf colour is desirable and represents good quality. Crushing the dried leaves is the final step in releasing stevia's sweetening power. This can be done either by hand or for greater effect, in a coffee grinder or in a special blender for herbs.

AKS R&D Centre has developed a new technique for the drying and harvesting of Stevia leaves, by applying weather-turf using electric motor giving a uniform heating effect harvested leaves spread on a wire-mesh. This way we can save time and labour in comparison to sun drying and quality would also be improved.

Stems of Stevia plants contain little or no sweeteners, although it is suggested that they may contain some flavour enhancers, odourisers and other agents of potential use for improving food-stuffs or alcoholic beverages.

Forms of Stevia Usage

Stevia can be used in many forms. The choice depends upon the amout of sweetness required (white extract powders are the sweetest).

(a) Fresh Stevia Leaves

This form of Stevia is the herb in its most natural, unrefined state. A leaf picked from a Stevia plant and chewed will impart an extremely sweet taste sensation reminiscent of licorice that lasts for quite a while. To have a more practical application as a tea or sweetener, the leaves must be dried or put through an extraction process, which makes the sweet taste even more potent.

(b) Dried Leaves

For more of the flavour and sweet constituents of the Stevia leaf to be released, drying and crushing is necessary. A dried leaf is considerably sweeter than a fresh one, and is the form of Stevia used in brewing herbal tea. Dried Stevia leaf may come in bulk or packaged like tea bags. It can also be finely powdered. In this form, it is 15-20 times sweeter than sugar (one tea spoon of Stevia is approximately equivalent to one cup of sugar). It has a greenish colour and can be used in a wide variety of foods and beverages, including coffee, sweets, confectionaries and several edible dishes. It can also be used to make an herbal tea blend. Its distinctive flavour is reminiscent of licorice, which will blend very well with different aromatic spices, such as cinnamom and ginger. AKS Research center has planned to release the Stevia powder in market in suitable packing.

(c) Stevia Extracts

The form in which Stevia is primarily used as a sweetener in Japan is that of a white powdered extract. In this form it is approximately 200 to 300 times sweeter than sugar (by weight). This white powder is an extract of the sweet glycosides (natural sweetening agents) in the Stevia leaf.

Not all Stevia extract powders are the same. The taste, sweetness and cost of the various white Stevia powders will likely depend on their degree of refinement and the quality of the Stevia plant used. AKS have also been planning to isolate the active compound of sweet glycosides in purest form for different commercial and medicinal uses.

(d) Liquid Concentrates

These come in several forms. These are a syrupy black liquid (that results from boiling the leaves in water), which can enhance the flavour of many foods. Steeping Stevia leaves in distilled water or a mix of water and grain alcohol makes another type. A liquid can also be made from the white and preserved mixed with water, and preserved with grapefruit seed extract. Our scientists worked that Stevia liquid concentrate can be used as a best source for filler of Ayurvedic medicinal syrups, which were previously prepared with jaggery source as filler.

USES OF STEVIA

Stevia as a Flavour Enhancer

Traditionally, Stevia has been used as a flavour enhancer in addition to its use in herbal tea as sweetening agent. A good quality leaf is estimated to be 30 times sweeter than cane sugar or sucrose. The active constituents of stevia are considered by the world's leading food scientists as the "sweeteners of the future."

Therefore, every new development in the area of Stevia research is anxiously awaited and thoroughly analyzed when it appears. Countries in which the currently used artificial sweeteners are on the brink of being banned are desperately trying to find new, sage, non-caloric sweeteners. AKS Herbal Research and Development Centre has planned to launched the world largest project of 'Organic Cultivation of Stevia' with the support of Union Bank of India in India. And in other countries, firms that hold exclusive rights to currently used sweeteners are extremely fearful of the advent of new, sager sweeteners, over which they will have no control. The emergence of a totally natural, non-patentable sweetener would undoubtedly spell the death knell check the manufacturers of artificial sweeteners like aspartame and saccharin.

In India by 2020 it is estimated that more than 40% population will be affected by diabetic health problems. Existing artificial sweeteners are found harmful to fulfill the desire of diabetic consumers. Our scientists have suggested that Stevia sugar can serve as best alternate source of sugar for diabetic patients in future by fulfilling their need of sweet taste and desire.

Use of the whole leaf is an easier way to obtain better taste than through efforts aimed at trying to improve the taste of certain specific constituents. Research is now started in many countries to improve the taste of individual steviosides or rebaudiosides. Since the white crystalline powder exhibits a quite persistent bitter and astringent aftertaste, its use as a commercial sweetener often backfires. Thus, most manufacturers who use the isolated constituents of stevia usually have to combine it with other kinds of typical sugars. Since, rebaudiosides taste better, methods are constantly being sought to synthetically convert steviosides to rebaudiosides. But even the rabaudiosides must be combined with other kinds of sugars to obtain necessary sweetness. Finally, in the ultimate irony, there are processes currently under development for improving the taste of stevioside by combining it in various ways with other substances obtaind directly from Stevia.

MEDICINAL AND THERAPEUTIC USES

Stevia users believe there are many very legitimate reasons for using Stevia as a medicinal food. In spite of the prominence Stevia has obtained as a flavour enhancer, it contains a variety of constituents besides the steviosides and rebaudiosides, including the nutrients specified above and a good deal of sterols, triterpenes, flavonoids, tannins, and an extremely rich volatile oil comprising rich proportions of aromatics, aldehyde, monoterpenes and sesquiterpenes. These and other as yet unidentified constituents, probably have some impact on human physiology and may help explain some of the reported therapeutic uses of Stevia.

(a) Hypoglycemic Action

It is believed that the presence of the steviosides themselves that has produced dozens of empirical and semi controlled reports of hypoglycemic action. Paraguayans say that stevia is helpful for hypoglycemia and diabetes because it nourishes the pancreas and thereby helps to restore normal pancreatic function. In semi-controlled clinical reports one also encounters this action. A 35.2 per cent fall in normal blood sugar levels after 6-8 hours was reported following the ingestion of a Stevia leaf extract. Other workers has reported similar trends in humans and experimental animals. These kind of results have led physicians in Paraguay to prescribe Stevia leaf tea in the treatment of diabetes; similarly, in Brazil, Stevia tea and Stevia capsules are officially approved for sale for the treatment of diabetes.

However, it is important to note that Stevia does not lower blood glucose levels in normal subjects. In one study, rats were fed crude extracts of Stevia leaves for 56 days at a rate of 0.5 to 1.0 gram extract per day. Another team of scientists replicated these procedures. Neither group observed a hypoglycemic action. The experimental research on the effects of stevia on blood sugar levels in human patients with diabetes and hypoglycemia is in progress in many countries.

The general feeling in the scientific community is that the mild acting nature of the plant and its total lack of toxic side effects obviate the need for extensive and expensive research programme.

(b) Cardiovascular Action

A good deal of experimental work has been done on the effects of Stevia and stevioside on cardiovascular functioning in man and animals. Some of this work was simply looking for possible toxicity, while some was investigating possible therapeutic action. In neither case have significant properties been found. There is almost always a slight lowering of arterial blood pressure at low and normal doses, changing to a slight rise in arterial pressure at very high doses. The long-term use of stevia would probably have a cardio tonic action, that is, it would produce a mild strengthening of the heart and vascular system.

(c) Antimicrobial Action

Stevia is known to possess the ability to inhibit the growth and reproduction of bacteria and other infectious organisms. This is important in at least two respects. First, it may help explain why users of stevia-enhanced products report a lower incidence of colds and cough, and second, it has fostered the invention of a number of mouthwash and toothpaste products. Research clearly shows that *Streptococcus mutants*, *Pseudomonas aeruginos*, *Proteus vulgaris* and other microbes do not thrive in the presence of the non-nutritive Stevia constituents. This fact, combined with the naturally sweet flavour of the herb, makes it a suitable ingredient for mouthwashes and for toothpastes. Stevia has even been shown to lower the incidence of dental caries. In India, toothpaste using *Stevia* and *Aloe vera* has been commercialized.

Digestive Tonic Action

In the literature of Brazil, stevia ranks high among the list of plants used for centuries by the "gauchos" of the southern plains to flavour the bitter medicinal preparations used by that nomadic culture. For example, it was widely used in their "mate" herbal tea. Through much experimentation, these people learned that Stevia made a significant contribution to improved digestion, and that it improved overall gastrointestinal function. Likewise, since its introduction in China, Stevia tea made in either hot or cold water, is used as a low calorie, sweet-tasting tea, as an appetite stimulant, as a digestive aid, as an aid to weight management, and even for staying young.

Effects on the Skin

One of the properties of a liquid extract of stevia that has not yet been investigated experimentally is its apparent ability to help clear up skin problems. The Guarani and other people who have become familiar with stevia report that it is effective when applied to acne, seborrhea, dermatitis, eczema, etc. when placed directly on cuts and wounds, more rapid healing, without scarring, is observed. This treatment may sting for a few seconds, but a significant lowering of pain follows this. Smoother skin, softer to the touch is claimed to result from the frequent application of stevia poultices and extracts.

Contraceptive Action

There is a small controversy surrounding Stevia that still receives attention from time to time in the popular press and even by many scientists. This is about Stevia impacting fertility. It seems that in 1968 a paper appeared that claimed that certain tribes of Indians in Paraguay (the Matto Grosso) used Stevia tea as a contraceptive with apparently very good results. In subsequent experimental work, utilizing on rats, these researchers found that the treatment was supposed good for periods of up to two months. Subsequent work has repeatedly failed to replicate the earlier study. Furthermore, at least one attempt to locate tribes in northeastern Paraguay that used stevia to control fertility failed to confirm the story.

On the contrary effect of Stevia on reproductive physiology that appears to be valid, but this needs further study before definite conclusions can be drawn, is a healing effect on the processes underlying prostate disease. Just how important this finding is must await further research.

Safety Concerns

One of the most obvious indications of the safety of Stevia is that there have never been any reports of ill effects in over 1500 years of continuous use by Paraguayans. A similar indication of safety is the observation that despite over 10 years of widespread use of stevioside as a sweetening agent in Japan, years in which literally scores of tones of stevioside were ingested, not a single report of side effects of any kind has been reported. This may be compared with the track record of aspartame, which is the number one source of consumer food complaints made to the FDA in the USA. Nevertheless, in spite of the record of safety several official laboratory tests have taken place.

The first official investigation of possible toxicity from Stevia was performed in 1930 by Pomaret and co-workers in South America. Their tests were negative. They observed that stevioside passes through the human alimentary canal without being altered by digestive processes. That is, it goes out in exactly the same form as during eating. The issue is important because some of the metabolites of stevioside, as opposed to the whole leaf, have been shown to be toxic and researchers have cautioned against the use of stevioside for human consumpOtion until it is known for certain that stevioside is not metabolized in the human body. A typical statement from a report published in 1974 says:...the long-term effects of ingestion of stevioside would have to be investigated carefully before it could be considered for human use as a sweetener in the United States ... It remains to be proved that stevioside does not split to form any steviol in the human digestive tract". This challenge is still unattended. Perhaps that is why the United States resists all efforts to seriously explore the possible use of stevia as a sweetener. No further progress on the issue has been made since 1974. It appears that Pomaret's observations still hold.More elaborate safety tests were performed by the Japanese during their evaluations of Stevia as a possible sweetening agent. Few substances have ever yielded such consistently negative results in toxicity trials as has Stevia. Almost every toxicity test imaginable has been performed on Stevia extract or stevioside at one time or another. The results are always negative. The only related effect ever observed was the inhibition cell respiration (oxidative phosphorylation) in certain isolated cell components, but never in whole cells. What this showed was a reduction in toxicity due to a substance known as atractylignin, a poison that attacks cells of the liver. This result suggests that stevia could be used as an antidote in rare cases of poisoning by that chemical. The overall result of this action of stevia, then, turns out to be positive.

Another example of a good toxicology trial was one performed in 1985 by Yamada and co-workers. They administered stevioside and rebaudioside A to rats for two years at the rate of 0.3 – 1 per cent of their diet. The animals were than sacrificed, and the researchers conducted biochemical, anatomic, pathological and carcinogenic tests on 41 organs following autopsy. In addition they performed ongoing haematological and urine tests on the same animals. Each of the animals was matched to a control animal that experienced exactly the same treatment except for the Stevia. In the end, the symptoms and alterations noted by the research staff did not vary at all between the groups, and no dose-response effects were noted, even at the highest dose (1 per cent), which is equivalent to 125 times the average daily dose of sweeteners that a normal human would require.

Similar series of tests carried out by the National Ministry of Health and Welfare in Japan also failed to find any form of toxicity whatsoever.

SPECIFIC PRODUCTS/PURPOSES OF USE

On the basis of its various qualities, Stevia has found its use for the following purposes:

- Table top sweetener – for tea, coffee etc.
- Soft drinks, fruit juices, ice creams, yogurts.
- Cakes, biscuits, pastries, pies, baking.
- Jams, sauces, pickles, jellies, desserts.
- Chewing gum, candies, confectioneries.
- Sea-foods, vegetables.
- Weight-watcher diets and Diabetic diets.
- Skin creams.
- Toothpaste, mouthwashes – plaque retardant/cavities prevention
- Hypertension treatment and blood pressure control.
- Alcoholic beverage enhancer (aging agent and catalyst).
- Tobacco additive and flavourant.

CONCLUSION

Stevia is the new emerging alternative source of calorie free sweetener having no carbohydrate and fat. It is 20 to 30 times more sweetener than cane and beet sugar, highly nutritious, delicious, non-toxic and non-additive sugar. It also enhances the flavour, helpful in digestion, weight reduction, anti oxidant, prevent dental caries and having antimicrobial and anti-plaque properties, increases mental alertness, increase energy levels but does not affect the blood sugar level, therefore key-source sweetener for diabetic world. Besides, *Stevia* can be used in hypertension, hypoglycemic, helpful in skin toning and healing, tobacco and alcohol cravings and a tonic for pancreas. It can also be used as alternative source of sugar for food confectioneries, bakeries, fruit juices, jams, biscuits, chocolates, vegetables and other food stuffs. The recent researches alongwith future prospective of this new emerging medicinal plant has been discussed in this chapter.

NABARD has included Stevia in the list of 40 Medicinal and Aromatic Plants whose financing by banks is eligible for refinance on a three-year basis vide its circular dated August 18, 2003.

Climate conditions of India favour Stevia cultivation in various parts of the country providing 3-5 harvests per annum. It can provide the following benefits:

ANNEXURE - I

PROPOSED PROJECT IN 1000 ACRES FOR ORGANIC CULTIVATION OF STEVIA (SUGAR OF 21ST CENTURY)

As per Stevia Project

Yield of Dry leaf	1st Year	3000 kgs	average yield 3600 kg per aere
	2nd Year	3500 kgs	
	3rd Year	4000 kgs	
	4th Year	4000 kgs	
	5th Year	3500 kgs	

Total Expected Yield in 1000 Acres

3600 kgs × 1000 Acre. = 3,60,0000 Kg = 3600 MT

3600 Mt × Price Rs. 150 per Kg = 54.0 cr

Stevia leaf Powder is 20 time Sweeter than CaneSugar

Thus Cane Sugar Average Price is Rs 15 per kg

Rs 15 Per kg × 20 times sweeter = Rs 300 per Kg

Thus Stevia leaf Powder will be 50% cheaper than Cane Sugar

Having so many merits and Advantages

For 1,000 Acres of Stevia Cultivation

Inputs Requirement

A1	Planting Material 40cr Plants @ Rs 5/per plant	**20.0 cr.**
2	1,00,000 Ltrs AKS BIOCROP @ Rs 1500 per Ltr	**15.0 cr**
		35.0 cr
B3	Average Yield of Dry Leaves 3600kg per Acre	
4	Total Yield in 1000 Acres 3600 MT	
5	Cost of 3600MT @ Rs 150 Per kg	**54.0 cr**

Annexure - II

Economics of Stevia Cultivation (Only Organic Cultivation) on 1 Acre basis

S. No.	Details	1st Year Rs.	2nd Year Rs.	3rd year Rs.
1.	Cost of Cleaning the Land Preparation of Land Labelling and Preparation of Beds	2,000	0	0
2.	Cost of Planting Material @ Rs 8/- per plants 30,000 plants (including transportation cost of Rs. 5000/-)	2,45,000	0	0
3.	Cost of Plantation	1,000	0	0
4.	Cost of Organic Fertilizer "AKS BIOCROP" 100 ltrs @ Rs 1250/- per ltr.	1,25,000	1,25,000	1,25,000
5.	Cost of Irrigation (including Electricity, Labour and Diesel Charges)	12,000	10,000	10,000
6.	Plant Protection and Application	1,000	1,000	1,000
7.	Hand Weeding	2,000	2,000	2,000
8.	Gap Filling, Plant Cost and Plantation	_	25,000	25,000
9.	Harvesting Expanses	2,500	2,500	2,500
10.	Plant Harvest Expenses	1,000	1,000	1,000
11.	Contingencies	2,000	2,000	2,000
12.	Transportation and Misc.	1,500	1,500	1,500
	Total	**3,85,000**	**1,70,000**	**1,70,000**
	Total Cost of Organic Cultivation of Stevia on 1 Acre basis	**3,85,000**	**1,70,000**	**1,70,000**

INCOME

While agriculture is totally nature risk based activity, if all goes favourable the aspected yield should be calculated as in

1st Year	2nd Year	3rd Year
1800 kg Leaves	2500 kg Leaves	2500 kg Leaves

INCOME

1st year Dry Leaves 3000Kgs @ Rs. 150/ per kg.	Rs. 4,50,000/-
2nd year Dry leaves 3500Kgs @ Rs. 150/per kg.	Rs. 5,25,000/-
3rd year Dry Leaves 4000kgs @ Rs. 150/per kg.	Rs. 6,00,000/-
4th year Dry leaves 4000kgs @ Rs 150/-per kg	Rs. 6,00.000/-
5th year Dry leaves 3500kgs @ Rs. 150/-per kg	Rs. 5,25,000/-
	Rs. 27,00,000/-
Total Output for 5 years	Rs. 27,00,000/-
Total Expanses for 5 ykears	Rs. 11,85,250/-
Total Gain for 5 years	Rs. 15,14,750/-

Fund Requirement

Fund Requirement for 1st Year	Rs. 4,20,250/-
2nd Year onward	Rs. 1,91,250/-

Fund Support Expected from Bank

1st Year	Total requirement	Rs. 4,20,250
	Farmers Contribution 5% i.e.	Rs. 21,125/-
	Bank loan	Rs. 3,99,125/-
2nd Year and onward	Total requirement	Rs. 1,91,250/-
	Farmers Contribution 5% i.e.	Rs. 9,562/-
	Bank Loan	Rs. 18,11,688/-

Repayment of Bank Loan

In 1st Year (within 18 months)

50% of the Bank Loan Amount	Rs. 2.0 Lac
Interest 8% for 18 month	Rs. 0.48 Lac
	Rs 2.50 Lac

In 2nd Year (within 18 months)

50% Balance of 1st year	Rs. 2.0 lac
Interest on Balance for 18 month	Rs. 0.24 lac
Current Years 50% of loan amount	Rs. 0.90 lac
Interest @ 8% on Principal loan amount For 18 months	Rs. 0.11 lac
	Rs. 3.25 lac

In 3rd year and onward

Balance of Previous Year	Rs. .90 lac
Interest of Previous Year	Rs. .11 lac
50% of the Current Year	Rs. .90 lac
Total	Rs. 2.15 lac

In 4th Years same as above	Balance Nil
In 5th Years Dues Clearance Within 18 months	

For One Acre

Summary. Total Cultivation Cost in 1st year	**Rs. 4,20 lac**
In 2nd Year and onward	**Rs. 1,91 lac**
Profit in 5 years	**Rs. 15.14 lac**

*1US$ = Rs. 46.00 (Approx.)

- Primary producers – enabling greater diversification opportunity and returns per acre of investment.
- Rural sector – additional employment if a commercial extraction plant is established.
- The consumer – potentially reduced health risk than if consuming artificial sweetener.
- The economy – through import substitution of current imports of artificial sweeteners.

AKS Herbal Research and Land Development Center has already approached to National Medicinal Plant Board, New Delhi to provide subsidy to farmers who will take initiative in cultivation of Stevia in India. AKS Herbal Research and Land Development Center has also proposed a project in 1000 acres for organic cultivation of Stevia in India and worked out the economy and budget for organic cultivation Stevia under Indian conditions (Annexures I and II).

Recent Advances in Medicinal Plant Research : Vision 21st Century, 2005
Editors : S.K. Prabhuji, G.P. Rao and S.K. Patil
Satish Publication House, New Delhi

Chapter 12

ENDOPHYTIC BACTERIA MEDIATED MANAGEMENT OF STEM BLIGHT OF *PHYLLANTHUS AMARUS*

S. Mathiyazhagan[1], K. Kavitha[1], P. Renukadevi[1], W.G.D. Fernando[2], G. Chinnasamy[3], S. Nakkeeran[1], V. Sendhilvel[1] and G. Chandrasekar[1]
[1]Department of Plant Pathology, Tamil Nadu Agricultural University, Coimbatore - 641 003, India
[2]Department of Plant Science, University of Manitoba, Winnipeg, MB, Canada R3T 2N2
[3]Cereal Research Centre, Agriculture and Agri-Food Canada, 195 Dafoe Road, Winnipeg, Manitoba, Canada R3T2M9

Stem blight of *Phyllanthus amarus* caused by *Corynespora cassicola* is reported for the first time in cultivation of *P. amarus*. Pathogen produced toxin that induced stem blight symptoms as that of pathogen. Studies on the management of stem blight through antagonistic bacterial endophytes (*Pseudomonas fluorescens*) and non endophytes (*P. chlororaphis* PA23, *Bacillus subtilis* CBE4) suppressed mycelial growth of *C. cassicola in vitro*. *B. subtilis* (CBE4) and endophytic *P. fluorescens* degraded the phytotoxin of stem blight pathogen. Infiltration of the toxin degraded by biocontrol agents resulted in many fold decrease of electrolytic leakage than toxin treated one These strains induced the accumulation of plant defense enzymes such as peroxidase, polyphenol oxidase, chitinase, glucanse, phenylalanine ammonia lyase and phenols. Native PAGE analysis for PO and PPO indicated that the level of induction of both enzymes were more with the isolate CBE4. Application of CBE4 and ENPF1 through seedling dip coupled with foliar spray gave maximum control of stem blight both *in vitro* and *in vivo.*

INTRODUCTION

Phyllanthus amarus is an important medicinal plant used for controlling human diseases such as dropsy, urinogenital problems, dysentery, asthma and jaundice (Bagchi *et al.*, 1992). It contains phyllanthine, hyphophyllanthine (Bikram *et al.*, 1989), alkaloids, bioflavonoid (quercetin), terpenoids, nirphyllin, phyllnirurin and triacontanol (Syamasunder *et al.*, 1985). Though *P. amarus* is bestowed with several medicinal principles, intensive cultivation of it has resulted in the out break of several plant diseases. Among them stem blight caused by *Corynespora cassiicola* threatens the cultivation of *P. amarus*. Though fungicide application offers a certain degree of protection against the pathogen their adverse effects on beneficial soil microorganism and the environment cannot be ignored. Therefore, exploitation of biocontrol agents offers a practical and economical alternative for the management of this disease.

Endophytic bacteria colonize an ecological niche similar to that of plant pathogens, especially the vascular wilt pathogens. This phenomenon might favour them as biocontrol agents. In addition, intensive work on biocontrol agents of rhizosphere has recently revealed six rhizobacteria, which induced systemic resistance in cucumber by both external and internal root colonization (Kloepper *et al.*, 1992b). Exploiting an additional microbial habitat for biocontrol purposes might enhance overall disease control and its consistency, since the control agent could avoid unfavourable conditions in one habitat by escaping into the other habitat. Endophytic bacteria, isolated from potato tubers, showed *in vitro* antibiosis against *Fusarium avenaciarum, F. sambucinum* and *F. oxysporun*. The antibiosis of the isolates decreased progressively with depth of their site of isolation from the tuber surface. Possibly, the plants might have adopted bacteria as part of a disease suppressive response to pathogenic attack (Sturz, *et al.*, 1999). Studies on induced resistance of cotton plants against *Verticillium dahliae* by endophytic and rhizosphere bacteria showed that the endophytic bacteria had more potent antagonistic activity than the rhizosphere bacteria. The mobility of bacteria in cotton vascular tissue positively corresponded to their ability to induce resistance against *V. dahliae* (Xia, *et al.*, 1996). Endophytic bacteria isolated from rice seeds, when applied as seed treatment, colonized the stellar region of the root and exhibited strong anti fungal activity against *Rhizoctonia solani, Pythium myrotylum, Guamanomyces graminis* and *Heterobasidium annosum* (Mukhopadhyay *et al.*, 1996). *Bacillus subtilis*, isolated from xylem fluid of chestnuts, suppressed the growth of chestnut blight pathogen, *Cryphonectria parasitica* and reduced the lesion areas on stems, when applied 3 days prior to fungal challenge, in an *in vitro* bioassay (Wilhelm, *et al.*, 1998).

The beneficial effects of bacterial endophytes vary but appear to operate through similar mechanisms, as described for plant growth promoting rhizobacteria (Kloepper *et al.* 1991). However, because of the colonization at different habitats, endophytes after another tool for developing biological control strategies. By combining the use of bacterial endophytes with rhizosphere and foliar antagonists, a holistic biological control system may be developed that can work against multiple pathogens living in entire plant.

The present investigation enlightens the efficacy of endophytes and fluorescent pseudomonads in plant growth promotion and disease management.

METHODOLOGY

Isolation of pathogen

P. amarus affected by stem blight disease was collected from the farmer's field pertaining to Avinashi Taluk of Coimbatore district. The pathogen was isolated by tissue segment method (Rangaswami, 1958) on potato dextrose agar (PDA) and pathogenicity was proved. The pathogen was identified as *Corynespora cassiicola* based on its spore characters.

Screening of endophytic isolates against *Corynespora cassiicola in vitro*

Isolation of endophytic bacteria was made from root and stem of healthy *Phyllanthus* seedlings. Three isolates of *Bacillus subtilis* (CBE4, ENBS2, ENBS3) and three isolates of *Pseudomonas fluorescens* (ENPF1, ENPF2, ENPF3) were isolated from *P. amarus*. *P. chlororaphis* (PA23) was obtained from the lab of Dr.W.G. Dilantha Fernando, Department of Plant Science, University of Manitoba, Winnipeg, Canada. All the above isolates were screened by dual culture technique in PDA medium. The best bacterial antagonist selected through screening was used for testing its efficacy under field conditions.

Isolation of *Corynespora cassiicola* toxin

Erlenmeyer flasks (250 ml) containing 50ml of Richard's medium (50 g of sucrose, 10g potassium nitrate, 5g potassium dihydrogen phosphate, 2.5g magnesium sulphate, 0.02g ferric chloride, and 1000 ml of distilled water, pH 7) were inoculated with a 4 mm disc of pathogen (*C. cassiicola.*) grown in PDA medium. After 15 days of growth under stationary condition at laboratory temperature (28 ± 2°C), the culture filtrate was obtained by filtration through three layers of cheesecloth. Richard's medium not inoculated with the fungus served as a control.

The culture filtrates were concentrated to 10 % of their volume *in vacuo* at 50 °C. An equal volume of methanol was added and the solution was stored overnight at 5 °C. Precipitates were removed by filtration through Whatman No.1 filter paper. The precipitates were then dissolved in sterile water (10ml) and their ability to induce disease symptoms was assessed by the detached stem and electrolytic leakage bioassay. After removal of the precipitate by filtration, the filtrate containing methanol and water was evaporated to dryness *in vacuo* at 40 °C. It was dissolved in 10 ml of sterile water, and bioassayed as described above. This water preparation was treated with activated charcoal at 3 % w/v, cooled to 4 °C, stirred for 12 h and filtered through Whatman No. 1 filter paper. The filtrate, after charcoal treatment, was evaporated to dryness, dissolved in 10 ml of sterile water, and bioassayed as described above. The activated charcoal remaining in the filter paper was washed with 100 ml of methanol and collected in a beaker. The methanol was evaporated *in vacuo* at 40 °C. The residue was dissolved in 10 ml of sterile water and bioassayed.

The water fraction after the removal of activated charcoal by filtration through Whatmen No.1 filter paper showed toxicity. This fraction was partitioned with three volumes of ethyl acetate, chloroform, hexane, carbon tetrachloride and petroleum ether using a separating funnel. Both water and solvent fractions were evaporated to dryness *in vacuo* separately. The residues were dissolved in 10 ml of sterile water and the ability of the individual fraction to induce disease symptom was assayed. The absorption spectrum of the crude toxin was recorded using spectrophotometer (varion cary model 50) at a scanning speed of 750 nm/min from 200-300 nm.

Toxin bioassay

a. Detached stem bioassay

Stem bits of 60 day old *Phyllanthus* plants were detached and cut into 5 cm pieces. Each piece was placed on a glass slide and ends were fixed with gum tape. The slide was kept inside a Petri dish lined with wet blotting paper. The toxin extracted from *C. cassiicola* as above-mentioned was placed on a 4mm diameter filter paper disc. It was placed on the pin pricked stem bits placed on the glass slide. The Petri dish was incubated for 5 days under laboratory conditions (25 ± 2°C; 12 h light and 12 h darkness) and the development of symptom was assessed.

b. Bioassay with live plant

Phyllanthus seedlings (30 days old) were dipped in a test tube containing

5 ml of toxin solution for 24 h. The seedlings dipped in sterile distilled water served as control. The symptoms of wilting were recorded 24 h after incubation at room temperature.

c. *Electrolytic leakage*

Stem bits of 60 days old *P. amaruss* were cut into small pieces (1cm long) and samples (100 mg) were enclosed in washed cheesecloth and placed in scintilling vials. The stem bits were infiltrated with 50 µl of toxin *in vacuo* for 30 min. After the infiltration of the toxin the stem bits were placed in vials with 4 ml of distilled water and incubated on a reciprocal shaker at 28 ± 2°C. Conductance of ambient solutions was measured at 0, 10 and 20 min after infiltration with an Orion model 120 conductivity meters. Stem bits infiltrating with water (or) non-inoculated Richards medium served as control.

Biodegradation of toxin and its bioassay

Fifty ml of the toxin equivalent to 5 g glucose /l was distributed in 250 ml conical flasks and sterilized. Twenty-four hours old culture of *P. fluorescens* (ENPF1), *P. chlororaphis* (PA23), and *B. subtilis* (CBE4) were inoculated separately into each flask. The flasks were incubated at room temperature (28±2 °C) in a shaker at120 rpm. After 36 h of incubation, the culture filtrate degraded by bacterial antagonists were filtered through sterile filter paper and collected in a sterile flask. Degraded culture filtrates by different bacterial antagonists, crude toxin and the culture filtrate of ENPF1, CBE4 and PA23 grown in non toxin amended medium were tested for its ability to induce electrolytic leakage in *P. amarus* stem bits. In addition, the UV absorption spectra of the degraded toxin were assayed calorimetrically at 200, 220, 240,260, 280 and 300nm. Non-degraded crude toxin was used as check to compare the UV absorption spectra with the degraded toxin.

Talc based formulation of bio control agents

A loop full of *P. fluorescens* and *B. subtilis* was inoculated into the King's B broth and Nutrient broth respectively and incubated in a rotary shaker at 150 rpm for 72 h at room temperature (28 ± 2 °C). After 72 h of incubation the broth containing 9 x 10 ^{8}cfu/ml was used for the preparation of talc based formulation. To 400 ml of bacterial suspension, 1 kg of the talc powder (sodium ammonium silicate), calcium carbonate 15 g (to adjust the pH to neutral) and 10 g carboxy methyl cellulose (adhesive) were mixed under sterile condition following the method described by Vidhyasekaran and Muthamilan (1995). The mixture was shade dried to reduce the moisture content to 20 % and then packed in polypropylene bags and sealed. At the time of application, the population of bacterium in talc formulation was around 2.5 to 3 × 10 ^{8} cfu/g.

Efficacy of bacterial antagonists under greenhouse condition

To study the biocontrol potential of the endophytic isolates CBE4, ENPF1 and PA23 an experiment was conducted under green house condition. The talc based formulation of the above mentioned bacterial antagonists were delivered as seedling dip (2%) and foliar spray (2%). The efficacy of the antagonists was compared with foliar application of mancozeb (0.25%). One gram of each formulation consists of 10^8 cfu/g of the product. The humidity in the green house was maintained around 80% RH and temperature was maintained at 25 °C (day) and 20 °C (night). The pathogen was mass multiplied on its own host and was incorporated to the pots at the rate of 5% inoculum level to the soil. Ten days old seedlings with different treatments were transplanted to pots of 30 cm width × 40 cm depth. In each pot 10 plants were maintained and replicated thrice for each treatment. The observation on stem blight disease was recorded 30 days after transplantation. Growth parameter like shoot length and root length were also recorded, 30 days after bacterization.

Enzyme extraction

Seven-day-old plants were uprooted intact with root and the samples were analyzed for defense enzymes such as phenylalanine ammonia–lyase (PAL), peroxidase (PO), polyphenol oxidase (PPO), chitinase, β-1, 3 glucanase, phenol and protein.

Phenylalanine ammonia – lyase (PAL)

Five hundred milligram of leaf was homogenized in 5 ml of cold 25mM borate HCl buffer (pH 8.8) containing 5mM mercaptoethanol (0.4 ml/l). The homogenate was centrifuged at 15,000g for 15 min and the supernatant (enzyme extract) was used.The assay mixture consisted of 0.2 ml of enzyme extract, 1.3 ml water and 0.5 ml borate buffer. The reaction was initiated by the addition of 1 ml of 0.1M L-Phenylalanine. The reaction mixture was incubated for 1 h at 32 ^{0}C. The reaction was stopped by the addition of 0.5 ml of 2N HCl. A blank was run in which phenylalanine was added after adding 2N HCl. The absorbance was measured at 290 nm. The enzyme activity was expressed as n mol of trans cinnamic acid/ min/ mg of protein (Dickerson *et al.*, 1984).

Peroxidase activity (PO)

One gram of fresh leaf tissue was ground in 1 ml of 0.1M phosphate buffer (pH 7.0) in a pre cooled pestle and mortar. The homogenate was centrifuged at 15,000 g at 4° C for 15 min. The supernatant was used as enzyme source. The reaction mixture consists of 1.5 ml of 0.05M pyrogallol, 0.1 ml of enzyme extract and 0.5 ml of one per cent H_2O_2. The change in absorbance of the reaction mixture was recorded at 420 nm at 30 sec interval for 3 min. The boiled enzyme preparation served as blank. The enzyme activity was expressed as change in the absorbance of reaction mixture per min per mg of protein (Hammerschmidt *et al.,* 1982).

Polyphenol oxidase (PPO)

One gram of fresh leaves was ground in 1 ml 0.1 M sodium phosphate buffer (pH 6.5). The homogenate was centrifuged at 15,000g for 15 min at 4^0 C and the supernatant was used as the enzyme source. The reaction mixture consisted of 1.5 M sodium phosphate buffer (pH 6.5) and 0.1 ml of the enzyme extract. The reaction was started when 0.2 ml of 0.01M catechol was added and the activity was expressed as change in absorbance at 495 nm at 30 sec interval for 3 min. The enzyme activity was expressed as change in absorbance per min per mg of protein (Mayer *et al.*, 1965).

Chitinase

One gram of fresh leaves was homogenized in 5 ml of 0.1M sodium citrate buffer (pH 5.0). The homogenate was centrifuged for 10 min at 10,000g at 4 °C and the supernatant was used in the chitinase assay. Colloidal chitin was prepared as per Roberts and Selitrennikoff (1988). Assay mixture consists of 10 ml of 1M sodium acetate buffer (pH 4.0)., 0.4 ml of enzyme extract and 0.1 ml colloidal chitin (1mg), which were pipetted into a 1.5 ml eppendorff tube and incubated at 37°C for 2 hr. After 2 hr the reaction was stopped by centrifugation at 1,000g for 3 min. An aliquot of the supernatant (0.3ml) was pippetted into a glass reagent tube containing 30 µl of 1M potassium phosphate buffer (pH 7.0) and incubated with 20 µl of 3 % (w/v) desalted snail gut enzyme for 1h. Desaltation of the snail gut enzyme was done as per Boller and Mauch (1988). After one hour the reaction mixture was brought pH 8.9 by addition of 70 µl of 0.1M sodium borate buffer (pH 9.8). The mixture was incubated in the boiling water bath for 3 min and then rapidly cooled in an ice water bath. After addition of 2 ml of DMAB, the mixture was incubated for 20 min at 37°C and immediately thereafter the absorbance was measured at 585 nm. *N* - acetyl glucosamine (GlcNAc) was used as a standard. The enzyme activity was expressed as nmol GlcNAc equivalents per min per mg of protein (Boller and Mauch, 1988).

β-1, 3 - glucanase

One gram of the fresh leaves were homogenized in 1ml of 0.1M phosphate buffer (pH 7). The homogenate was centrifuged at 15,000g for 15 min at 4^0 C and the supernatant was used as the enzyme source. Crude enzyme extract of 62.5 µl was added to 62.5 µl of laminarin and then incubated at 40 ^{0}C for 10 min. The reaction was stopped by adding 375 µl of dinitro salicylic acid and heated for 5 min. on boiling water bath. The resulting solution was diluted with 4.5 ml distilled water and the absorbance was read at 500 nm. The crude extract preparation with laminarin with zero time incubation served as blank. The enzyme activity was expressed as μg equivalent of glucose per min per mg of protein (Pan *et al.*, 1991).

Phenols

One gram of the fresh leaves was ground in a pestle and mortar in 10 ml of 80 % ethanol. The homogenate was centrifuged at 10,000 g for 20 min. The supernatant was evaporated to dryness and the residue was dissolved in 5 ml of distilled water. From this 0.2 ml was taken and the volume was made up to 3 ml with distilled water. To this 3ml solution 0.25 ml of Folin-Ciocalteau reagent was added. After 3 min, 1 ml of 20 % sodium carbonate was added and mixed thoroughly. Then the tubes were placed in boiling water for 1 min and cooled. The abs⌐ ɔance was measured at 650 nm against a reagent blank. The phenol activı.y was expressed in μg of catechol per g of plant tissue. (Zieslin and Ben Zaken, 1993)

PO, PPO isoforms

The isoform profiles of PO and PPO were examined by native polyacrylamide gel electrophoresis (native - PAGE) (Laemmli, 1970). Plant samples were collected on the 4th day after challenged inoculation with pathogen. The protein extract was prepared by homogenizing 1 g of fresh leaves in 1 ml of 0.1M sodium phosphate buffer pH 7.0 and centrifuged at 16,000g for 20 min at 4 °C. The protein content of the sample was determined (Bradford, 1976). Sample (50 mg protein) were loaded onto 8 % polyacrylamide gels (Sigma, USA). After electrophoresis, PO isoforms were visualized by soaking the gels in staining 0.05% benzidine (Sigma, USA) and 0.03% H_2O_2 in acetate buffer (20mM, pH 4.2) (Nadlony and Sequira, 1980). For assessing PPO isoforms , the gels were equilibrated for 30 min in 0.1% p-phenylenediamine followed by addition of 10mM catechol in the same buffer (Jayaraman *et al.*, 1987).

Statistical analysis

The statistical analyses were done using IRRISTAT 92 computer software of International Rice Research Institute Biometrics unit, Philippines by using DMRT.

RESULTS

Symptomatology and pathogenicity

The disease was initially characterized by the appearance of water soaked dark brown coloured elongated spots on the basal portion of the stem. Later the spots turned necrotic and subsequently the spots spread over the entire stem. As the disease severity increased, the spots coalesced and resulted in blighted stem. The foliage of the blighted plants turned yellow followed by premature defoliation. This disease occurred right from seedling stage (25 days old) to maturity of the crop. The pathogen responsible for the disease was isolated from the infected portion and identified as *Corynespora cassiicola* (Indian Type Culture Collection (ITCC) I.D.No. 5572.03).

Toxin

The crude toxin induced characteristic stem blight symptom in *P. amarus* as that of pathogen. Bioassay on detached stem bits indicated that methanol filtrates produced blighted symptom but the methanol precipitate did not produce any such symptom. This revealed that the toxic principle is present in the methanol filtrate and not in methanol precipitate. When the methanol filtrate was treated with charcoal, the toxicity could be detected only in filtrate, indicating that the toxin was not absorbed by activated charcoal.

Table - 1 : Characteristics of crude toxin isolated from culture filtrate of *C. cassiicola*

Treatment	Fraction	Electrolyte leakage (mS/g)^x	Symptom bioassay^y
Methanol	Precipitate	52	0
	Water fraction	524	2.30
Activated charcoal	Adsorbed fraction	42	0
	Non adsorbed fraction	478	2.50
Ethyl acetate	Solvent fraction	48	0
	Water fraction	562	2.13
Chloroform	Solvent fraction	54	0
	Water fraction	532	2.13
Hexane	Solvent fraction	39	0
	Water fraction	482	2.62
Petroleum ether	Solvent fraction	39	0
	Water fraction	453	2.46
Carbon tetrachloride	Solvent fraction	64	0
	Water fraction	502	2.71
Non inoculated Richards medium	—	45	0
Water (control)	—	48	0

^{x}The data is the mean of three replications.

^{y}Intensity of the symptom was graded into five categories : 0 = no symptoms, 1 = 1 to10%, 2 = 11 to 25 %, 3 = 26 to 50 % and 4 = 51 to 100 % stem bits area affected, respectively. The data are the average of 20 stem bits samples.

The toxic component was always found in the water fraction after portioning with organic solvents (Table 1).

In bioassay with live plants, *Phyllanthus* seedlings immersed in crude toxin showed the wilting symptom within 6h. Control seedlings showed no wilting symptom in the same period. Studies on the toxin degradation by antagonistic bacteria revealed that the crude toxin of the pathogen recorded the highest electrolyte leakage of 380 µS at 0 min and it gradually increased to 780 µS at 20 min. The electrolyte leakage of the stem bits was several fold lower when infiltered with the toxin degraded by the endophytes (Fig 1).

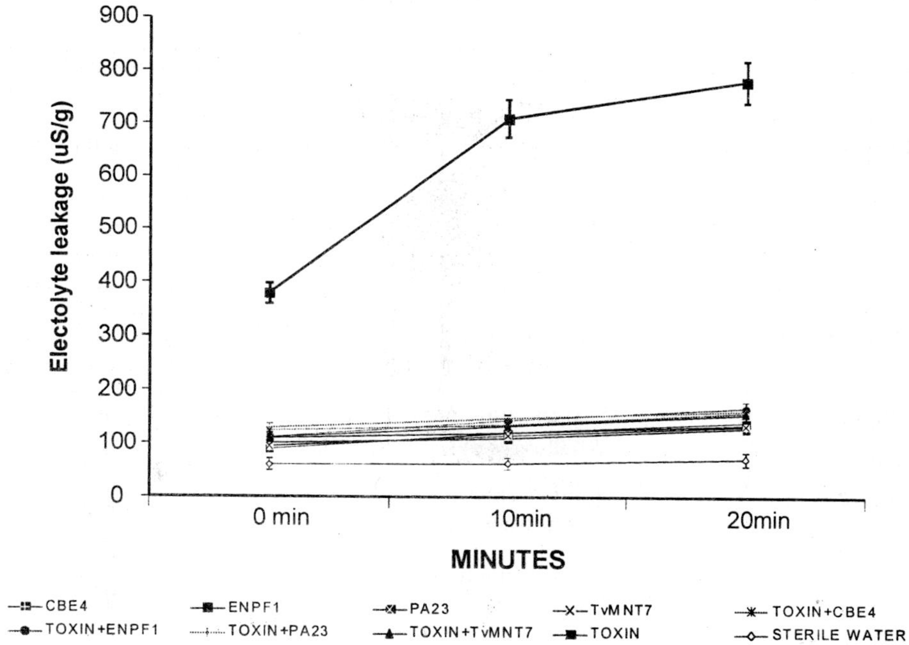

Fig. 1 : Bio efficacy of biocontrol agent to degrade *Corynespora cassiicola* toxin

The UV absorption of crude toxin and the degraded toxin indicated that the optical density (OD) was 2.2 at 260nm for crude toxin, while it was four times lower with the toxins degraded by bacterial antagonists (Fig. 2).

In vitro antagonism of endophytes

Three isolates of *B. subtilis* and *P. fluorescens* were effective *in vitro* in inhibiting *Corynespora cassiicola.* Among them CBE4 recorded the highest inhibition zone of 22 mm followed by *P. chlororaphis* (18.0 mm) and ENPF1 (16 mm) (Table 2). These three isolates were selected for further screening under glass house condition.

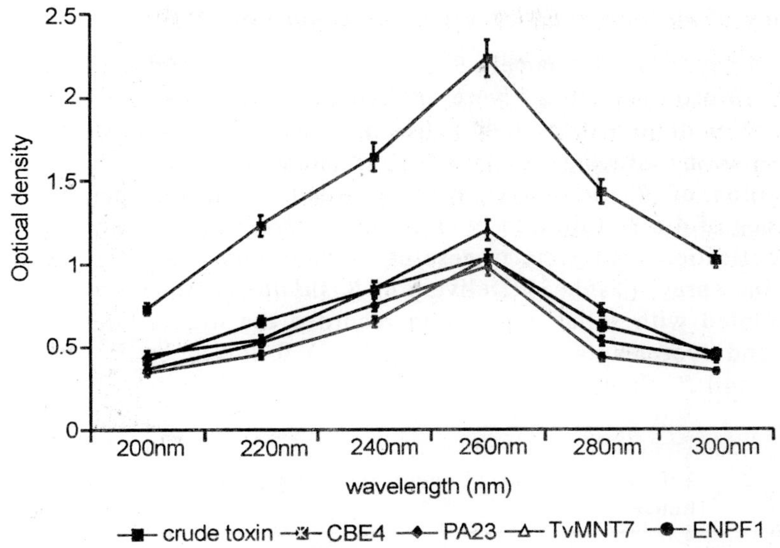

Fig. 2 : UV absorption spectra of biodegraded *Corynespora cassiicola* toxin.

Table - 2 : *In vitro* antagonism of bacterial antagonists

Bacterial antagonists	**Stem blight pathogen (*C. cassiicola*)**	
	Mycelial growth (mm)	**Inhibition Zone (mm)**
PA23	65.0^{c}	15.00
CBE4	58.0^{d}	22.00
ENPF 1	66.0^{c}	14.00
ENPF 6	70.0^{b}	10.00
ENPF 5	71.0^{b}	9.00
ENPF 8	72.0^{b}	8.00
ENPF 10	71.0^{b}	9.00
ENBS 2	73.0^{b}	7.00
ENBS 3	72.0^{b}	8.00
ENBS 4	69.0^{b}	11.00
ENBS 7	70.0^{b}	10.00
ENBS 9	73.0^{b}	7.00
Control	90.0^{a}	-

PA23 – *Pseudomonas chlororaphis*; ENPF- Endophytic *Pseudomonas fluorescens;* ENBS- Endophytic *Bacillus subtilis*

Efficacy of endophytes in greenhouse condition

The endophytic isolates namely *B. subtilis, P. fluorescens* (CBE4 and ENPF1) and *P. chlororaphis* PA23 were applied as seedling dip, foliar spray and combination of both. Among these treatment seedling dip and soil application was better in reducing the stem blight incidence. Seedling dip and soil application of *B. subtilis* (CBE4) recorded the lowest percent disease incidence of 4.6 % followed by 6.2% in *P. chlororaphis* (seedling dip and soil application) and 8.2% in case of *P. fluorescens* ENPF1 (seedling dip and foliar spray) (Table 3). Delivery of *B. subtilis* (CBE4) through seedling dip coupled with foliar application recorded the highest shoot length of 47cm and root length of 18 cm (Table 3) followed by *P. chlororaphis* (PA23) and *P. fluorescens* (ENPF1).

Table - 3 : Effect of biocontrol agent against stem blight incidence in green house

Treatment	Per cent disease incidence	Percent reduction over control	Root length (cm)	Shoot length (cm)
CBE4 (SD)	18.23^{b}	56.81	14.5^{bc}	42.6^{c}
CBE4 (FS)	16.21^{bc}	61.59	16.2^{bc}	44.5^{b}
CBE4 (SD+FS)	15.60^{bc}	63.04	18.2^{a}	47.5^{a}
ENPF1 (SD)	16.22^{bc}	61.57	12.3^{ef}	32.4^{hi}
ENPF1 (FS)	16.60^{bc}	60.67	13.6^{cde}	34.5^{fg}
ENPF1 (SD+ FS)	15.73^{bc}	62.73	14.4^{bc}	36.6^{e}
PA23 (SD)	18.30^{b}	56.65	12.5^{def}	30.5^{jk}
PA23 (FS)	16.46^{bc}	61.00	14.3^{cd}	33.4^{gh}
PA23 (SD+ FS)	16.20^{bc}	61.62	16.2^{b}	40.3^{d}
Mancozeb (FS)	13.18^{c}	68.77	10.8^{fg}	28.5^{l}
Control	42.21^{a}	—	10.2^{g}	22.5^{m}

Values are mean of three replications.
SD - Seedling dip, FS - Foliar spray
Values in parenthesis are arcsine transformed
In a column, means followed by a common letter are not significantly different at the 5 % levels by DMRT
ENPF - Endophytic *P. fluorescens*, ENBS - Endophytic *B. subtilis*, BCA - *P. chlororaphis*, CBE4 - *Bacillus subtilis*

Induction of systemic resistance against *P. amarus* stem blight

Phenylalanine ammonia lyase

In *P. amarus*, seedling dip and foliar application of PA23, CBE4 and ENPF1 induced the plant to synthesise phenylalanine ammonia lyase (PAL), where as an additional increase in the PAL synthesis was observed in PGPR isolates challenge inoculated with *C. cassiicola*. The activity of PAL reached a maximum level on 4th day (338.56 nmol of trans cinnamic acid/min/mg of protein) after challenge inoculation in *B. subtilis* (CBE4) and thereafter the activity declined slowly throughout the experimental period up to 10th day. In plants treated with pathogen alone, increased activity of PAL was observed for a period of 2-4 days and thereafter the activity declined (Fig 3).

Peroxidase

Activity of peroxidase (PO) was induced on *P. amarus* plants treated with PGPR strains alone. But an additional increase in the activity was observed in PGPR pretreated plants challenge inoculated with *C. cassiicola*. The treatment with *B. subtilis* (CBE4) recorded higher PO activity (1.324 changes in absorbance/min/mg of protein) on 4th day after challenge inoculation and thereafter the activity declined slowly throughout the experimental period up to 10 days. In plants treated with pathogen alone, the activity of PO was lower when compared to PGPR treated plants challenge inoculated with PGPR strains (Fig 3).

Polyphenol oxidase

P. amarus plants expressed higher activity of polyphenol oxidase (PPO) when challenge inoculated with pathogen irrespective of various PGPR treatments. Although an increase in PPO activity began from 2nd day after challenge inoculation, the activity increased further and remained significantly higher from 4th day to 6th day in *B. subtilis* (CBE4) treated plants. There were no marked change of PPO activity in untreated as well as in PGPR treated plants which was not challenge inoculated with pathogen (Fig 3).

Chitinase and β-1, 3 Glucanase

Studies on the induction of chitinase and β-1,3 glucanase by PGPR isolates revealed that the expression of chitinase and β-1, 3 glucanase increased irrespective of the PGPR treatments. The activity of chitinase and β-1, 3 glucanase increased up to 595 nmol of GluNAc $min^{-1}mg^{-1}$ of protein and to 585 μ mol equivalent of glucose $h^{-1}mg^{-1}$ of protein respectively on 4 days after challenge inoculation with *C. cassiicola* in *P. amarus* plants pretreated with CBE4. Subsequently the activity of both defense enzymes declined slowly up to 10 days after challenge inoculation. In addition, comparison of the activity of chitinase and β-1, 3 glucanase in *P. amarus* plant inoculated

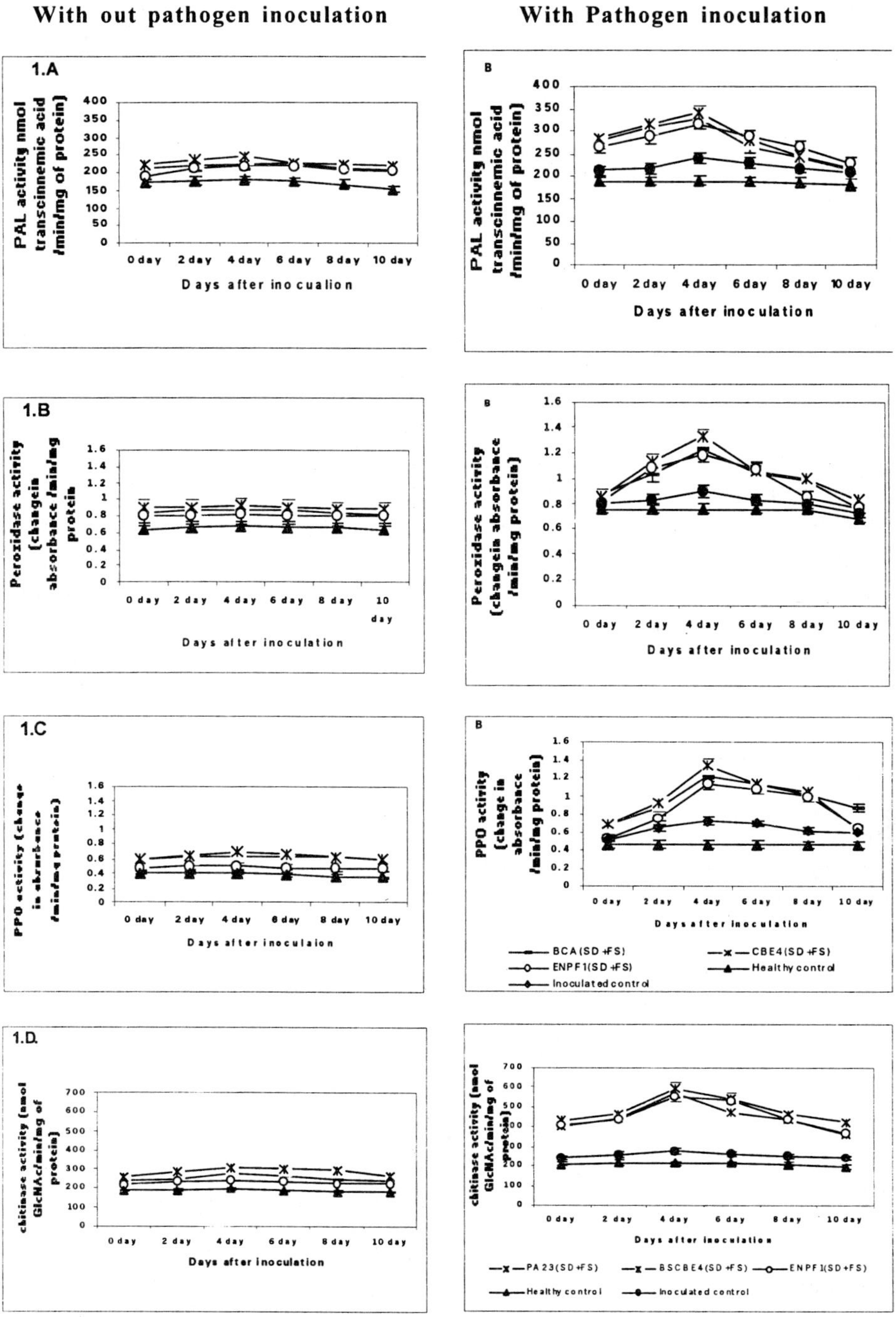

Fig. 3. Induction of defense related enzymes (PAL, PO and PPO) in *P.* amarus challenged with *C. cassiicola.*

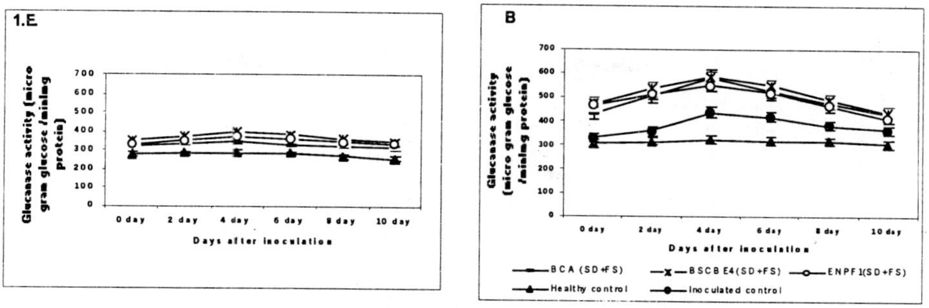

Fig. 3 : (contd). Induction of Chitinase and β1,3 glucanase in *P. amarus* challenged with *C. cassiicola.*

with pathogen alone reflected that the activity was lower than the plants pretreated with CBE4 (Fig 3).

Phenol

Higher accumulation of phenols was observed in PGPR treated *P. amarus* plants challenge inoculated with *C. cassiicola.* Accumulation of phenols started after 2nd day of challenge inoculation with pathogen. The maximum phenol content was observed on 4th day after challenge inoculation in *B. subtilis* (CBE4) treatment (3.846 μg of catechol /mg of protein). In plants inoculated with pathogen alone, the phenolic content declined to the initial level on 10thday after inoculation. Plants treated with PGPR strain had an increased content of phenols when compared to untreated plants (Fig 3).

Isoforms of PO and PPO

Peroxidase (PO)

Studies on the induction of PO isoforms by bacterial antagonists explained that PO2 was expressed in pathogen inoculated and healthy control. The bacterial antagonists *B. subtilis* (CBE4) induced 3 isoforms PO1, PO2 and PO3 while only PO2 and PO3 were expressed in plants treated with *P. chlororaphis* (PA23) challenged against pathogen (Fig 4).

Polyphenol oxidase (PPO)

Studies on the expression of various isoforms of PPO through native gel electrophoresis indicated that 4 different isoforms (PPO1-PPO4) were induced in all the treatments including pathogen inoculated and healthy control. But the intensity of expression of the isoforms was very less in inoculated control and healthy control. Comparison between biocontrol agents indicated that the intensity of expression of PPO4 was very high with PA23. But the expression of PPO1, PPO2 and PPO3 was more in CBE4 treated plants than all other treatments. Plants treated with ENPF1 and challenged with pathogen, induced all 4 isoforms of PPO (Fig 5).

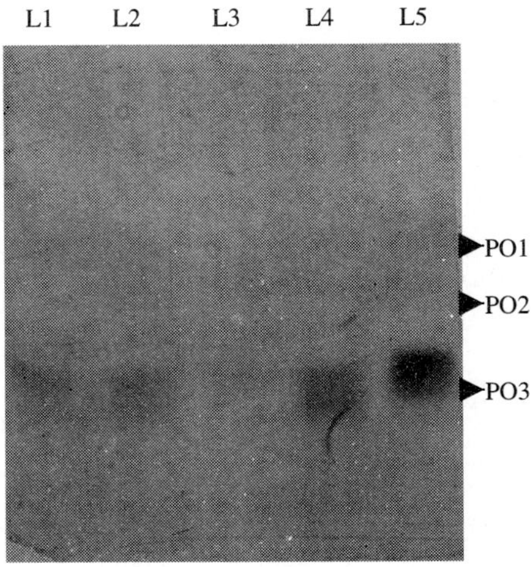

Fig. 4 : Induction of PO isoforms by bacterial antagonists
L1 - Inoculated control, L2 - Healthy control, L3 - ENPF1 challenged against pathogen, L4 - PA23 challenged against pathogen, L5 - BSCBE4 challenged against pathogen.

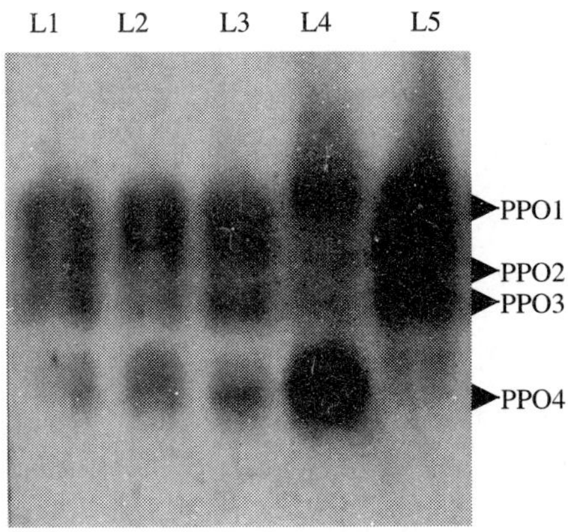

Fig. 5 : Induction of PPO isoforms by bacterial antagonists
L1 - Inoculated control, L2 - Healthy control, L3 - ENPF1 challenged against pathogen, L4 - PA23 challenged against pathogen, L5 - BSCBE4 challenged against pathogen.

DISCUSSION

Stem blight is reported for the first time in *P. amarus*. The blight symptom on cucumber caused by this pathogen was reported by Blazquez (1967). It also infects crops like cotton (Johnes, 1961) cowpea (Olive, 1949), egg plant (Onesirosan *et al.*, 1974), sesame (Stone and Jones, 1960) and tomato (Blazquez, 1972).

Studies on the *in vitro* antagonism of bacterial antagonists indicated that *B. subtilis* CBE4 and endophytic *P. fluorescens* resulted in the suppression of the mycelial growth of stem blight pathogen (*C. cassiicola*). Several strains of *Pseudomonas* spp. and *Bacillus* spp. have been shown to produce wide array of antibiotics which include lacton, 2,4 diacetylphloroglucinol, hydrogen cyanide, oligomycen, oomycen A, phenazine, pyroluteorin, pyrolnitrin, pyocyanin, iturin, surfactin and several other uncharacterized molecules (Kim *et al.*, 1989; Keel and Defago, 1997; Whipps, 1997; Nielson *et al.*, 1998)

To prove the involvement of toxin in disease development, the toxin should be able to produce typical symptoms of the disease. Toxin produced by the stem blight pathogen *C. cassiicola* induced stem blight symptoms that were indistinguishable from those induced by the pathogen.

Cassiicolin, a host selective toxin was produced by *C. cassiicola* responsible for leaf fall disease of rubber. Similarly, in the present study also, the toxin of *C. cassiicola* induced typical symptom of stem blight as that of the pathogen. Infection of *C. cassiicola* in mint produced t_1 toxin which was highly toxic and induced wilting of *Mentha* twigs. The toxin produced characteristic, necrotic and chlorotic lesions similar to that of natural disease symptoms in case of *Mentha arvensis*. In the current study, the toxin produced by *C. cassiicola* induced typical symptoms as that of the pathogen, suggesting that the toxin might be host selective.

Most microbes including bacteria and fungi are known to degrade the phytotoxin produced by plant pathogens. In the present study we explored the possibility of *B. subtilis* (CBE4), endophytic *P. fluorescens* and *P. chlororaphis* (PA 23) capable of degrading the phytotoxin produced by the stem blight pathogen. The biocontrol agents were able to grow on the culture filtrate utilizing it as a sole carbon source. It suggested the ability to degrade the toxin produced by stem blight pathogen. In addition, infiltration of the toxin degraded by biocontrol agents resulted in many fold decrease of electrolytic leakage than toxin treated one.

Several *P. fluorescens* strains and *T. viride* isolates were screened for degradation of RS toxin produced by *R. solani* causing sheath blight of rice and a *T. viride* isolate, TvMNT7 was found to degrade the toxin when provided as sole carbon source. Fu et al. (1999) reported that endophytic bacteria degrade the toxin produced by *Verticillium dahliae* causing wilt disease in cotton. Shanmugam et al. (2001) reported the involvement of 17 kDa proteins as accessory factor acting in relation with 110 kDa a-glucosidase of Tv-MNT7 grown solely on sheath blight toxin (SBTox) to degrade the toxin in culture condition.

Plant growth promoting rhizobacteria and bacterial endophytes play a vital role in the management of various fungal diseases. But one of the major hurdles experienced with biocontrol agents is the lack of appropriate delivery system. In the present study application of CBE4 and ENPF1 through seedling dip coupled with foliar spray gave maximum control of stem blight both *in vitro* and *in vivo.* Similarly, seed treatment and soil application of *P. fluorescens* reduced root rot of black gram caused by *Macrophomina phaseolina* (Jayashree *et al.,* 2000; Shanmugam *et al.,* 2001) panama wilt of banana (Raguchander *et al.,* 1997). Seed and foliar application of *P. fluorescens* reduced sheath blight of rice (Nandakumar *et al.,* 2001.) The efficacy of biocontrol agents to reduce stem blight under glass house condition might be due to the production of siderophore, antibiotics, lytic enzymes and the induction of defense related enzymes like PO, PPO, PAL, chitinase, β-1,3 glucanase and phenol. The increase in biomatter production by ENPF1 and CBE4 may be due to the production of plant growth promoters or through indirect stimulation of nutrient uptake and by producing siderophore or antibiotics to protect plant from deleterious rhizosphere organisms. There has been an increase in the root length and shoot length of tomato, cucumber, lettuce and potato as a result of bacterization with *Pseudomonas* strain. Besides direct antagonistic activity by the production of various bacterial metabolites, induction of systemic resistance by fluorescent pseudomonas against diseases has been established as a new mechanism by which the plants defend themselves from pathogen attack (Van Peer *et al.*, 1991, Van Loon, *et al.*, 1998). Prior application of fluorescent pseudomonas as seed treatment induces various defense mechanisms in the plants (Chen *et al.,* 2000).

Plant has endogenous defense mechanisms that can be induced in response to attack by insects and pathogens (Bostock *et al.*, 2001; Heil, 2001). It is well known that the defense genes are inducible genes and appropriate stimuli or signals are needed to activate them. Inducing the plant's own defense mechanisms by prior application of a biological inducer is thought to be a novel plant protection strategy. Induced resistance by inducing agents in several crops is associated with enhancement of lignification and also increased activities of enzymes involved in phenyl propanoid pathway and PR protein synthesis (Boller and Mauch, 1988; Hammerschmidt and Kuc, 1995). Recent studies imply that prior application of fluorescent pseudomonas strengthens the host cell wall structures resulting in restriction of pathogen invasion in plant tissue (Benhamou *et al.*, 2000; Chen *et al.*, 2000).

Peroxidase represents another component of an early response in plants to pathogen attack and plays a key role in the biosynthesis of lignin which limits the extent of pathogen spread (Bruce and West, 1989). The products of this enzyme in the presence of hydrogen donor and hydrogen peroxide have antimicrobial activity and even antiviral activity (Van Loon and Callow, 1983). Increased activity of cell wall bound peroxidases has been elicited in different plants such as cucumber (Chen *et al.*, 2000), rice

(Reimers *et al.*, 1992) tomato (Mohan *et al.*, 1993) and tobacco (Anlgoy *et al.*, 1992) due to pathogen infection. In bean rhizosphere, colonization of various bacteria induced the peroxidase activity (Zdor and Anderson, 1992).

Radjacommare (2000) reported that *P. fluorescens* Pf1 induced the activities of PPO in rice against *R. solani.* Similarly, Meena *et al.* (2000) reported that the *P. fluorescens* induced the activities of PPO in response to infection by *C. personata* in groundnut. Tea plants treated with *P. fluorescens* at weekly intervals increased activity of PPO under field condition. Similarly significant increase in the activity of PPO was observed in greengram plants treated with *P. fluorescens* along with chitin (Saravanakumar, 2002). In the present study, PPO activity increased from 2nd day after challenge inoculation, but maximum activity was recorded on 4th day after challenge inoculation with pathogen in all the PGPR treated *Phyllanthus* plants. In addition the various isoforms of PO were induced by CBE4 after challenge inoculation by pathogen. The intensity of expression of PPO was comparatively more in CBE4 treated plants challenged with *C. cassiicola.* The induction of PO and PPO might have resulted in cell wall thickening and would have suppressed the penetration of *C. cassiicola* into *P. amarus* as seen from the cross section of infected stems.

In general fungal cell contains chitin and glucan as their cell wall constituents. The main mode of antagonistic activity of antagonistic bacteria is the production of lytic enzymes (chitinase and β-1, 3 glucanase) which act on cell walls of organisms which have chitin and glucan as their cell wall component (Singh *et al.*, 1999) and also through induced systemic resistance (ISR) in plant system. Maurhofer *et al.* (1994) reported that PR proteins, viz., PR-1a, 1b, 1c, endochitinase and β-1,3 glucanase were induced in the intercellular fluid of tobacco plants grown in the presence of *P. fluorescens* strain CHA0.

Earlier Viswanathan and Samiyappan (1999) reported that ISR by fluorescent pseudomonas which was associated with induction of chitinase, as a promising technology for the management of red rot of sugarcane. In the present study, the activity of chitinase and β-1, 3 glucanase was higher in PGPR treated plants challenge inoculated with *C. cassiicola* indicating that it might have resulted in the lysis of invading pathogen.

Latest developments in cellular, molecular, biochemical, genomic and proteomic techniques will be helpful to address the molecular mechanisms behind the interactions between the plant, pathogen and antagonist.It would be rewarding to characterize various isoforms of PO and PPO identified in this study using cutting edge peptidomics technologies. Further studies could be extended to explore the expression level and pattern, structure, function and localization of various proteins and their isoforms involved during the interaction of plant, pathogen and biocontrol agents.

REFERENCES

Ahlgoy, P., Felix, G., Metraux, J.P. and Meins, J.R. (1992). Resistance to disease in the hybrid *Nicotiana glutinosa* x *Nicotiana debneyi* mediated by peroxidase and polyphenol oxidase. *Physiology and Molecular Plant Pathology,* **41(1)** : 11-21.

Bagchi, G.D., Srivasthava, G.N. and Singh, G.C. (1992). Distinguishing features of medicinal herbaceous species of Phyllanthus occurring in Lucknow district (U.P) India. *International journal of pharmacognosy,* **30** : 160-168.

Benhamou, N., Gagne, S., Ie Quere, D. and Dehbi, L. (2000). Bacteria-mediated induced resistance in cucumber: beneficial effect of the endophytic bacterium *Serratia plymuthica* on the protection against infection by *Pythium ultimum. Phytopathology,* **90 :** 45-56.

Bikram, S., Agarwal, P.K., Thakur, R.S. (1989). A new lignan and a new neolignan from *Phyllanthus niruri. Journal of Natural Products,* **52 :** 48-51.

Blazquez, C.H. (1972). Target spot of tomato. *Plant Disease Reporter,* **56** : 243-245.

Blazquez, C.H. (1967). *Corynespora* leaf spot of cucumber.In: Proceeding of the Florida state Horticultural society IFAS, University of Florida, *Gulf Research.* Education Centre. Brandenton, USA **80** : 177-182.

Boller, T. and Mauch, F. (1988). Colorimetric assay of chitinase. *Methods of Enzymology,* **161** : 430-435.

Bostock, R.M., Karban, R., Thaler, J.S., Weyman, P.D. and Gilchrist, D. (2001). Signal interactions in induced resistance to pathogens and insect herbivores. *European Journal of Plant Pathology,* **107** : 103-111.

Bradford, M.M. (1976). A rapid and sensitive method for quantification of microgram quantities of protein utilizing the principle of protein dye binding. *Annals of Biochemistry,* **72** : 248-254.

Bruce, R.J. and West, C.A. (1989). Elicitation of lignin biosynthesis and iso peroxidase activity by pectic fragments in suspension cultures of castor bean. *Plant Physiology,* **91(3)** : 889-897.

Chen, C., Belanger, R.R., Benhameu, N. and Paullitz, T.C. (2000). Defense enzymes induced in cucumber roots by treatment with plant growth promoting rhizobacteria. *Physiology and Molecular Plant Pathology,* **56** : 13-23.

Dickerson, D.P., Pascholati, S.F., Hagerman, A.E., Butler, L.G. and Niholson, R.L. (1984). Phenylalanine ammonia lyase and hydroxyl cinnamate : CoA ligase in maize mesocotyls inoculated with *Helminthosporium maydis* or *Helminthosporium carbonum. Physiol. Plant Pathol.,* **25** : 111-123.

Fu, Z.Q., Xia, Z.J., Wu, A.M, Yang, Y.H, Zheng, Q. and Gu, B.K. (1999). The mechanism for controlling cotton wilt (*Verticillium dahliae*) by endophytic bacteria Jiangsu. *J. Agric. Sci.,* **15 :** 211-215.

Hammerschmidt, R. and Kuc, J. (1995). Induced resistance to diseases in plants. Klewer Academic Publishers, Dordsechtl. *The Netherlands,* 182 p.

Hammerschmidt, R., Nuckles, E.M. and Kuc, J. (1982). Association of enhanced peroxidase activity with induced systemic resistance of cucumber to *Colletotrichum lagenarium*. *Physiol. Plant Pathol.*, **20** : 73-82.

Heil, M. (2001). The ecological concept of induced systemic resistance (ISR). European *Journal of Plant Pathology*, **107** : 137-146.

Jayaraman, K.S., Ramanuja, M.N., Vijayaraghavan, P.K. and Vaisyanathan, C.S. (1987). Oxidative enzyme in pearl millet. *Food Chemistry*, **24** : 203.

Jayashree, K., Shanmugam, V., Raguchander, T., Ramanathan, A. and Samiyappan, R. (2000). Evaluation of *Pseudomonas fluorescens* (Pf1) against blackgram and sesame root rot diseases. *J. Biological Control*, **14** : 55-61.

Johnes, J.P. (1961). A leaf spot of cotton caused by *C. cassiicola*.. *Phytopathology*, **51** : 305-308.

Keel, C. and Defago, G. (1997). Interaction between beneficial soil bacteria and root pathogen: mechanisms and ecological impact. In: *Multitrophic interactions in terrestrial system* Gange, A.C. and Brown, V.K., (Eds.). Oxford: Blackwell science, pp 27-47.

Kim, J.I., Choi, D.R., Cho, C.K. and Han, S.C. (1989). Influence of plant parasitic nematode in occurence of *Phytophthora* blight on hot pepper and sesame. *Crop Protection*, **31** : 27-30.

Kloepper, J.W., Wei, G. and Tuzun, S. (1992b). Rhizosphere population dynamics and internal colonization of cucumber by plant growth-promoting rhizobacteria which induce systemic resistance to *Colletotrichum orbiculare*.. *In* Biological control of plant diseases. *Edited by* E.S. Tjamos. Plenum Press, New York. pp. 185-191.

Kloepper, J.W., Zablotowiz, R.M., Tipping, E.M. and Lifshitz. R. (1991). Plant growth promotion mediated by bacterial rhizosphere colonizers. *In* The rhizosphere and plant growth. *Edited by* D.L. Keister and P.B. Cregan. Kluwer Academic Publishers, Dordrecht, the Netherlands. pp. 315 -326.

Laemmli, H.K. (1970). Cleavage of structural proteins during the assembly of the head of bacteriophage T_4. *Nature*, **227** : 680-685.

Maurhofer, M., Hase, C., Maurwly, D., Metranx, J.P. and Defago, G. (1994). Induction of systemic resistance of tobacco to tobacco necrosis virus by the root colonizing *Pseudomonas fluorescens* strain CHAO : influence of the *gac* A gene and of pyoverdine production. *Phytopathology*, **84(2)** : 139-146.

Mayer, A.M., Harel, E. and Shaul, R.B. (1965). Assay of catechol oxidase a critical comparison of methods. *Phytochemistry*, **5** : 783-789.

Meena, B., Ramomoorthy, V., Marimuthu, T. and Velazhahan, R. (2000). *Pseudomonas fluorescens* systemic resistance against late leaf spot of groundnut. *J. Mycol. Plant Pathol.*, **30** : 151-158.

Mohan, L., Vijayan, P. and Kolattukudy, P.E. (1993). Development and tissue specific expression of a tomato anionic peroxidase (*tap* 1) gene by a minimal promoter with wound and pathogen induction by an additional 5' flanking region. *Plant Molecular Biology*, **22** : 475-490.

Mukhopadhyay, K., Garrison, N.K., Hinlon, O.M., Bacon, C.W., Khush, G.S., Peck, H.D. and Datta, N. (1996). Identification and characterizalion of baclerial endophylcs of rice. *Mycopathologica*, **134** : 151-159.

Nadlony, L. and Sequira, L. (1980). Increase in peroxidase activities are not directly involved in induced resistance in tobacco. *Physiological Plant Pathology*, **16(1)** : 1-8.

Nandakumar, R., Babu, S., Viswanathan, R., Raguchander, T. and Samiyappan, R. (2001). Induction of systemic resistance in rice against sheath blight disease by plant growth promoting rhizobacteria. *Soil Biology and Biochemistry*, **33** : 603-612.

Nielson, M.N., Sorensen, J., Fels, J. and Pdersen, H.C. (1998). Secondary metabolite and endochitinase dependent antagonism towards plant pathogenic microfungi of *Pseudomonas fluorescens* isolates from sugar beet rhizosphere. *Applied Environmental Microbiology*, **64** : 3563-3569.

Olive, L.S. (1949). Target spot of cowpea and soybeans. *Mycologia*, **41(3)** : 355.

Onesirosan, P.T., Arny, D.C. and Durbin, R.D. (1974). Host-specificity of Nigerian and North American isolates of *Corynespora cassiicola. Phytopathology*, **64(10)** : 1364 -1367.

Pan, Q., Te, Y.S. and Kuc, J. (1991). A technique for detection of chitinases, b-1, 3 glucanase and protein patterns after single separation using PAGE or isoelectric focusing. *Phytopathology*, **81(9)** : 970-974.

Radjacommare, R. (2000). *Pseudomonas fluorescens* mediated systemic resistance in rice against sheath blight disease and leaf folder insect. *M. Sc., (Ag.) Thesis*, Tamil Nadu Agricultural University, Coimbatore, India. p 122.

Raguchander, T., Jeyashree, K. and Samiyappan, R. (1997). Management of *Fusarium* wilt of banana using antagonistic microorganisms. *Journal of Biological Control*, **11** : 101-105.

Rangaswami, G. (1958). An agar blocks technique for isolating soil micro organisms with special reference to Pythiaceous fungi. *Science and Culture*, **24** : 85.

Reimers, P.J., Guo, A. and Leach, J.E. (1992). Increased activity of a cationic peroxidase associated with an incompatible interaction between *X. oryzae* pv *oryzae* and rice (*Oryza sativa*). *Plant Physiology*, **99(3)** : 1044-1050.

Roberts, W.K. and Selitrennikoff, C.P. (1988). Plant and bacterial chitinases differ in antifungal activity. *Journal of General Microbiology*, **132** : 169-176.

Saravanakumar, D. (2002). Rhizobacteria induced systemic resistance against Biotroph (*Exobasidium vexans*) and Necrotroph (*Macrophomina phaseolina*) pathogens in tea and greengram. *M.Sc., (Ag.) Thesis*, Tamil Nadu Agricultural University, Coimbatore, India.

Shanmugam, V., Sriram, S., Babu, S., Nandakumar, R., Raguchander, T., Balasubramanian, P. and Samiyappan, R. (2001). Purification and characterization of an extracellular a-glucosidase protein from *Trichoderma viride* which degrades a phytotoxin associated with sheath blight disease in rice. *Journals of Applied Microbiology*, **90** : 320-329.

Singh, P.P., Shin, Y.C., Park, C.S. and Chung, Y.R. (1999). Biological control of *Fusarium* wilt of cucumber by chitinolytic bacteria. *Phytopathology*, **89** : 92-99.

Stone, W.J. and Jones, J.P. (1960). *Corynespora* blight of Sesame. *Phytopathology*, **50(4)** : 263-266.

Sturz, A.V., Christie, H.R., Matheson, B.G., Arsenault, W.J. and Buchman, N.A. (1999). Endophytic bacterial communities in the periderm of potato tubers and their potential to improve resistance to improve resistance to soil borne plant pathogens *Plant Pathol.*, **48 :** 360-369.

Syamasunder, K.V., Sikram Singh Thakur, R.S., Husain, A. and Kiso, Y. (1985). Antihepatotoxic principles of *Phyllanthus niruri* herbs. *Journal of ethnopharmacology*, **14** : 41-44.

Van Loon, L.C. and Callow, J.A. (1983). Transcription and translation in the diseased plant In: *Biochemical Plant Pathology* Callow, J.A. (Ed,) John wiley and Sons, Chichester, U.K.

Van Loon, L.C., Bakker, P.A.H.M. and Pieterse, C.M.J. (1998). Systemic resistance induced by rhizosphere bacteria. *Annual Review of Phytopathology*, **36** : 453-483.

Van Peer, R., Neimann, G.J. and Schippers, B. (1991). Induced resistance and phytoalexin accumulation in biological control of *Fusarium* wilt of carnation by *Pseudomonas sp.* stain WCS417r. *Phytopathology*, **81** : 728-734.

Vidhyasekaran, P. and Muthamilan, M. (1995). Development of formulations of *Pseudomonas fluorescens* for control of chickpea wilt. *Plant Disease*, **79(8)** : 780-782.

Viswanathan, R. and Samiyappan, R. (1999). Induction of systemic resistance by plant growth promoting rhizobacteria against red rot disease caused by *Colletotrichum falcatum* Went in sugarcane. *Proceeding of Sugarcane Technology Association*. India. **61** : 24-39.

Whipps, J.M. (1997). Developments in the biocontrol of soil-borne plant pathogens. *Advances in Botanical Research*, **26** : 1-134.

Wilhelm, E., Arthofer, W., Schafleitner, R. and Krebs, B. (1998). *Bacillus subtilis* an endophyte of chestnut (*Castanea sativa*) as as antagonist against chestnut blight (Cryphonectria parasitica). *Plant Cell Tissue Organ Cult.*, **52 :** 105-108.

Xia, Z.J., Gu, B.K. and Wu, A.M. (1996). Studies on induced resistance of cotton plants against *Verticillium dahliae* by endophytic and rhizosphere bacteria. *Chinese J. Biol. Control.*, **12 :** 7-10.

Zdor, R.E. and Anderson, A.J. (1992). Influence of root colonizing bacteria on the defense response in bean. *Plant and Soil*, **140(1)** : 99-107.

Zieslin, N. and Ben-Zaken, R. (1993). Peroxidase activity and presence of phenolic substances in peduncles of rose flower. *Plant. Physiol. Biochem.*, **31** : 333-339.

Recent Advances in Medicinal Plant Research : Vision 21st Century, 2005
Editors : S.K. Prabhuji, G.P. Rao and S.K. Patil
Satish Publication House, New Delhi

Chapter 13

STUDIES ON DAMPING-OFF, WEB BLIGHT AND STEM BLIGHT OF *PHYLLANTHUS AMARUS* (SCHUM & THONN)

S. Mathiyazhagan[1], S. Nakkeeran[1], G. Chandrasekar[1], K. Kavitha[1], P. Renukadevi[1], G. Chinnasamy[2] and W.G.D. Fernando[3]
[1]Department of Plant Pathology, Tamil Nadu Agricultural University, Coimbatore - 641 003, India
[2]Cereal Research Centre, Agriculture and Agri-Food Canada, 195 Dafoe Road, Winnipeg, Manitoba, Canada R3T2M9
[3]Department of Plant Science, University of Manitoba, Winnipeg, MB, Canada R3T 2N2

Commercial cultivation of *P. amarus* for its medicinal value has resulted in the outbreak of soil-borne diseases such as damping-off (*Pythium aphanidermatum*), web blight (*Rhizoctonia solani*) and stem blight (*Corynespora cassicola*). Histopathological studies revealed extensive damage of cortical cells with toxin deposition and vascular plugging in stem blight infected plants, which led to complete collapse of the plant. Bacterial antagonists, *Bacillus subtilis* (BSCBE4), *Pseudomonas fluorescens* (ENPF1) and *P. chlororaphis* (PA23) completely suppressed the damping – off pathogen *P. aphnidermatum* under *in vitro*. Analysis of the biocontrol efficacy of damping off in greenhouse indicated that delivering of BSCBE4, ENPF1 and PA23 through seed and soil controlled damping-off in addition to plant growth promotion. Bacterial antagonists (BSCBE4, ENPF1 and PA23) produced both catecholate and hydroxamate type of siderophores *in vitro*.

INTRODUCTION

Cultivation of medicinal plants plays a vital role in increasing the economy of farming community. The World Health Organization has compiled more than 20,000 medicinal plants used in different parts of the world. Among the medicinal plants more than one hundred botanicals have larger potential for commercial exploitation and could be marketed in the world drug markets. *Phyllanthus amarus* (Schum and Thonn), commonly known as keelanelli (Tamil), bhuamalki (Hindi) belonging to the family Euphorbiaceae occupies a prime position among the commercially cultivated medicinal plants. It has a long history of folk use in drug industry for the treatment of dropsy, urinogenital problem, dysentery, diabetes, skin ulcer, dyspepsia, fever, asthma, bronchial infections, tumour and Hepatitis B.

Hepatitis B is the very important viral disease affecting human beings and for curing of this disease *P. amarus* plays an important role. *Phyllanthus* spp, consists of medicinal principles like alkaloids, isobubbialine, epibubbialine, phyllanthine, securinine, norsecurinine, endothelinlignana phyllanthin, hyphophyllanthin, nirtetralin, terpenoids, nirphyllin, phyllnirurin, triacontanal etc. Though *Phyllanthus* is bestowed with several medicinal principles to cure several human disorders, the crop is susceptible to a number of devastating diseases that affects the quality and quantity of the medicinal principles.

Phyllanthus amarus has been reported to be affected by several diseases like rust, powdery mildew etc. A total of 23 species and 5 varieties of rust fungi belonging to 7 genera comprising of *Coleosporium, Phakopsora, Puccinia, Ravenelia, Scopella, Uredo* and *Uromyces* were recorded from Maharastra (Patil, 1991). Outbreak of powdery mildew in *Phyllanthus* was reported during the year 1971 by Narayanasamy and Ramakrishnan from Coimbatore district of Tamil Nadu.

With the rise in demand for plant based drugs tin global market, commercial cultivation of *Phyllanthus* has gained momentum during the last decade. This has led to the occurrence of stem blight, damping off and web blight diseases in severe form recently in and around the Coimbatore district of Tamil Nadu state. Complete failure of the crop was noticed within two to three days after infection. Since *Phyllanthus* is an export oriented medicinal herb, the current research on the etiology and management of the soil borne diseases will be a boon to the farming community.

METHODOLOGY

Isolation of pathogen

P. amarus seedlings affected by stem blight, damping off and web blight disease were collected from the field. The pathogen was isolated by tissue segment method (Rangaswami, 1958). The infected stem bits were cut into small bits (0.5cm in length) and surface sterilized with 0.1% mercuric chloride for 1 min and rinsed twice in sterile distilled water and blot dried with sterilized filter paper. The stem bits were placed on potato dextrose agar plates and incubated at 28 ± 2°C for 4 days. The mycelium radiating from the stem bits was sub cultured on PDA slants and the pathogen was identified based on its morphological characters.

Pathogenicity

P. aphanidermatum was multiplied in sand maize medium (Riker and Riker, 1936). Sand and maize powders were mixed at the rate of 19:1, moistened with 200 ml of sterilized distilled water /500 g of sand maize mixture. It was autoclaved at 1.4 kg/cm^2 for 2 h. *P. aphanidermatum* isolate was inoculated in the medium and incubated at 28±2°C for 10 days for conducting pathogenicity test. Pots of 22 cm diameter were filled with garden soil and sterilized in autoclave at 1.4 kg/cm^2 pressure for 2 h. Inoculum of *P. aphanidermatum* was added 10 days before sowing of seeds to the soil in pots at 5 per cent (w/w). Two hundred seeds of *P. amarus* were sown in each pot. The incidence of damping off was recorded after 7 days of germination. The pathogen was reisolated to prove Koch's postulates.

To prove the pathogenicity of *R. solani*, causing web blight disease the inoculum multiplied in sand maize medium was added to the pots at the rate of five per cent soil weight as in the case of *P. aphanidermatum*. The seedlings were observed for the expression of symptom after 30 days of germination.

Anatomy of stem blight

Microtome sectioning of the stem bits infected by *C. cassicola* was done as described below.

Fixation and washing

The stem bits were fixed in 5 parts of 30 per cent formalin, 5 parts of glacial acetic acid and 90 parts of 70 per cent ethanol for 29 h at 4 °C. The stem bits were then transferred to 50 per cent, 60 per cent and 70 per cent ethanol each at one-hour interval respectively.

Dehydration

The stems were gradually dehydrated in tertiary butyl alcohol (TBA) series as shown in the table. Later the samples were washed in running tap water for 10-15 min. followed by washing in ethanol 10%, 50%, 70% and 100% for 5 min. respectively.

Later the samples were transferred to xylene I, xylene II for 5 min. each and stained with toluidine blue (1 g/100ml). The slides were mounted in DPX and left undisturbed for 24 h for observations and finally the slide was photographed and the histopathology of the blight affected stems were studied.

Alcohol (%)	TBA (ml)	Ethanol (95%) (ml)	Distilled water (ml)	Duration (hr)
50	1.0	4.0	5.0	2-3
70	2.0	5.0	3.0	12
85	3.0	5.0	1.5	12
95	4.5	5.0	1.0	12
100	7.5	2.5	0.0	12

***In vitro* screening of bacterial antagonist**

Antagonism of *P. chlororaphis* (BCA), *B. subtilis* (CBE4), five bacterial endophytes of *B. subtilis* (ENBS2, ENBS3, ENBS4, ENBS7 and ENBS 9) and five endophytic strains of *P. fluorescens* (ENPF1, ENPF6, ENPF5, ENPF8 and ENPF10) were screened against damping off pathogen (*P. aphanidermatum*) of *P. amarus in vitro*. The bacterial antagonist was streaked 24 h earlier at 1 cm away from the periphery of Perti plate containing PDA medium. A 5 mm mycelial disc of five-day-old *P. aphanidermatum* culture was placed at opposite end of Perti plate. The growth of the pathogen towards the bacterial colony and inhibition zone was recorded 96h after incubation of *P. aphanidermatum*.

***In vitro* screening of *T. viride* isolates**

Isolates of *T. viride* were obtained from Biocontrol lab, Department of Plant Pathology, Tamil Nadu Agricultural University, Coimbatore. The efficacy of *T. viride* isolates was tested by dual culture technique (Dennis and Webster, 1971) using PDA medium. A mycelial disc of the pathogen *C. cassiicola*, *P. aphanidermatum*, and *R. solani* (9 mm dia.) was placed at one end of the Petri plate and at the opposite end, a mycelial disc of *T. viride* (TV4,TV6, TV7,TV8, UV6, UV 10, MNT5, MNT 7, MG6 and MG 9) (9 mm dia.) were placed separately in different Petri plates. The plates were incubated at 28 ±2 °C. The linear growths of the colonies (antagonist and pathogen) and over growth if any by antagonist on pathogen were measured after incubation.

Testing of bacterial antagonist for siderophores production

Productions of siderophores by bacterial antagonist were assayed by plate assay. The tertiary complex Chrome azural S (CAS) / Fe^{3+} / hexadecyl trimethyl ammonium bromide served as an indicator. Forty eight-hour-old cultures of the endophytes were streaked on to the succinate medium (Succinic acid-4.0g, K_2HPO_4- 3.0g, $(NH4)_2 SO_4$. $7H_20$- 0.2 g, Distilled water- 1 litre, pH- 7.0) amended with indicator dye. The resultant dark blue liquid was observed for the formation of bright zone with yellowish fluorescent colour in the dark colored medium. It was the indication of production of siderophore. The result was scored either positive or negative to this test (Schwyn and Neilands, 1987).

Detection and nature of siderophore

Preparation of cultures

The isolates of *P. fluorescens* and *B. subtilis* were inoculated in 10 ml of King's B broth and nutrient broth respectively. It was incubated in a rotary shaker at 120 rpm for 48 hours. The bacteria multiplied in the broths were used as the sample for the determination of the nature of siderophore.

Hydroxamate nature

It was examined by tetrazolium salt test. Instant appearance of a deep red colour by addition of siderophore sample to tetrazolium salt under alkaline conditions indicated the presence of hydroxamate (Snow, 1984).

Carboxylate nature

It was detected by Vogel's chemical test where the disappearance of pink colour on addition of Phenolphthalein to siderophore sample under alkaline condition indicated carboxylate nature (Vogel, 1987)

Preparation of talc based formulation of bio control agents

A loop full of *P. fluorescens* and *B. subtilis* were inoculated into the King's B broth and Nutrient broth and incubated in a rotary shaker at 150 rpm for 72 h at room temperature (28 ± 2 °C). After 72h of incubation the broth containing 9×10^{8}cfu/ml was used for the preparation of talc based formulation. To 400 ml of bacterial suspension, 1 kg of the talc powder (sodium ammonium silicate), calcium carbonate 15 g (to adjust the pH to neutral) and carboxy methyl cellulose (CMC) 10 g (as adhesive) were mixed under sterile condition following the method described by Vidhyasekaran and Muthamilan (1995). The product was shade dried to reduce the moisture content to 20 per cent and then packed in polypropylene bags and sealed. At the time of application the population of bacterium in talc formulation was checked to 2.5 to 3 × 10 ^{8} cfu/g.

The commercial formulation of *T. viride* was prepared as per the procedure described by Jayarajan and Ramakrishnan (1991). A mycelial disc of *T. viride* was inoculated to 100 ml molasses yeast medium in 250 ml conical flasks and incubated at room temperature (28 ± 2° C) for 5 days. The mycelial mat was homogenized and blended with talc powder in 1:2 ratio and shade dried. To this, carboxy methylcellulose was added at the rate of 0.5 per cent level as sticker.

Screening of biocontrol agents against damping off disease under glasshouse conditions

The efficacies of fungal and bacterial antagonists against damping-off were assessed under green house condition. The pots containing field soil was incorporated with *P. aphanidermatum* (5% w/w) multiplied on sand maize medium. Two hundred seeds were sown per pot (30 cm diameter, 30 cm height). The seeds were soaked overnight with 1% solution of talc based formulation of *T. viride, P. chlororaphis*, *B. subtilis*, and endophytic *P. fluorescens*. One gram of bacterial formulations consists of 10^8 cfu /gram. Ridomil (4 g/kg) was included as standard seed treatment for comparison. Application of both fungal and bacterial antagonists to soil was done at the rate of one per cent of the soil weight. For comparing the efficacy of various treatments both healthy control (without pathogen inoculation) and pathogen inoculated control were maintained through out the study. Watering was done regularly and damping-off incidence was recorded after 25 days of sowing. The root length and shoot length were also measured by randomly selecting ten plants in each pot. The humidity in the poly house was maintained around 80 per cent relative humidity and the temperature was maintained at 26°C (day)/ 20°C (night). Each treatment was replicated thrice.

RESULTS

Symptomatology and pathogenicity

Stem blight

The disease was initially characterized by the appearance of water soaked dark brown coloured elongated spots on the basal portion of the stem. Later the spots turned necrotic and subsequently the spots spread over the entire stem. As the disease severity increased, the spots coalesced and resulted in blighting of the stem. The foliage of the blighted plants turned yellow followed by premature defoliation. This disease occurred right from seedling stage (25days old) to maturity of the crop The pathogen responsible for the disease was isolated from the infected portion and identified as *Corynespora cassiicola* (Berk and Curt) Wei (ITCC, I.D.No. 5572.03).

In agar culture, colonies of C. *cassiicola* appeared as grey/pale brown; mycelium mostly immersed in the substratum, composed of branch, sub hyaline to pale brown, septate to smooth walled hyphae of 2-6 mthicknesses. Conidiophores were found to arise from immersed hyphae. The conidiophores were erect, simple, occasionally branched, straight, pale to dark brown with or without bulbous base, and slightly swollen at the tip with septations, conidia formed singly or in chains of 2-6. The conidia were obclavate or cylindrical in size, straight or slightly curved or smooth, pale olivacious brown, with 4-20 septa or pseudosepta, with a conspicuous hilum. The pathogenicity was proved by cut stem and lives plant bioassay.

Web blight

The disease was characterized by the appearance of water soaked lesions on the collar region near the ground level. As the disease progressed, the lesion spread upwards on the entire stem and the seedlings got rotten and collapsed. During advanced stages of infection, cottony growth of mycelium originated from the soil and grew over the stem. The mycelial growth of the pathogen over the stem gives a web like appearance, and hence known as web blight. Subsequently, the web like mycelial growth colonized the entire plant including leaves and seeds. It resulted in drying and defoliation of leaves followed by shedding of seeds. The pathogen was identified as *Rhizoctonia solani* (Kuhn) based on its morphological characters.

Damping off

The pathogen infected young seedlings after seedling emergence. Infection on seedlings appeared as minute pale green with brownish water soaked lesions at the basal portion of the stem. The lesions girdled the stem, affected tissues rot and the seedlings collapsed. The seedlings died within a day after infection. The pathogen was identified as *Pythium aphanidermatum* (Edson) Fitz.

Anatomy of stem blight

Studies on the anatomy of stem blight infected *P. amarus* plants revealed marked differences in healthy and infected stem sections. In the healthy stem sections, epidermal cells were normal. In cortex and hypodermis the cells were orderly arranged and cytoplasms were intact. The cell walls were of uniform thickness and the vessels in the vascular bundles were normal. In the central pith region the cells were round to polygonal in shape and found intact.

In the infected stem section, the epidermis and two to three layers of cortex turned pink when stained with saffranin stain. This revealed that the infected cells absorbed the saffranin stain. The cytoplasm became dense and granular in the cells due to pathogenic infection (*C. cassiicola*). The cortex layer adjacent to the endodermis showed discolouration due to pathogenic infection. The infected cells exhibited irregular thickening of cell wall and cytoplasm became dense, granular and occupied about 1/10th of the cell space and the remaining portions were empty. Xylem blockage was noticed in the infected cells and mycelial growth was observed in the xylem vessels. In the pith region, most of the cells were dislocated, deranged and large cavities were formed, occupying about one third to one half of the pith region. Derangement of cells and small cavities were also noticed in parenchyma.

The infected stem portion showed pink discoloration, which indicated the accumulation of phenols and other toxins due to the infection. In the heavily infected stem section, the entire cortex region was absent in the stellar region (Plate 1 and 2).

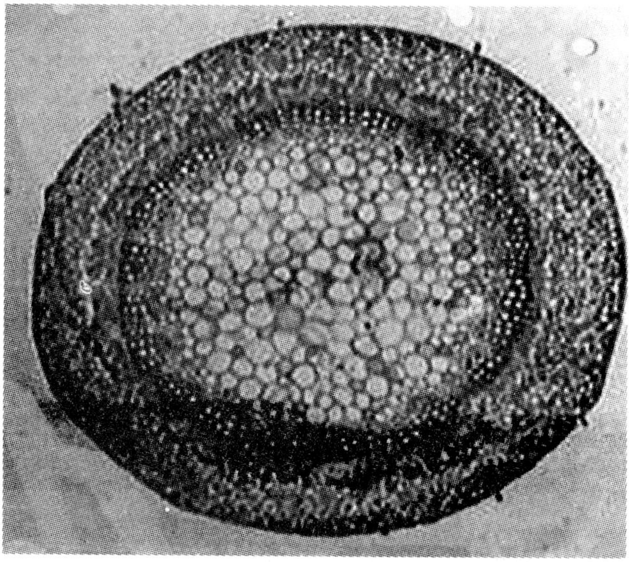

Plate 1 : Anatomy of healthy stem of *P. amarus*

Siderophore production

The endophytic bacteria isolated were tested for the production of siderophore under *in vitro*. All the isolates produced yellowish fluorescent pigments which was the indication of siderophore production. Among the different endophytes, surface area production of siderophore was higher with isolates ENPF1 (10.00cm^2) (Table 1).

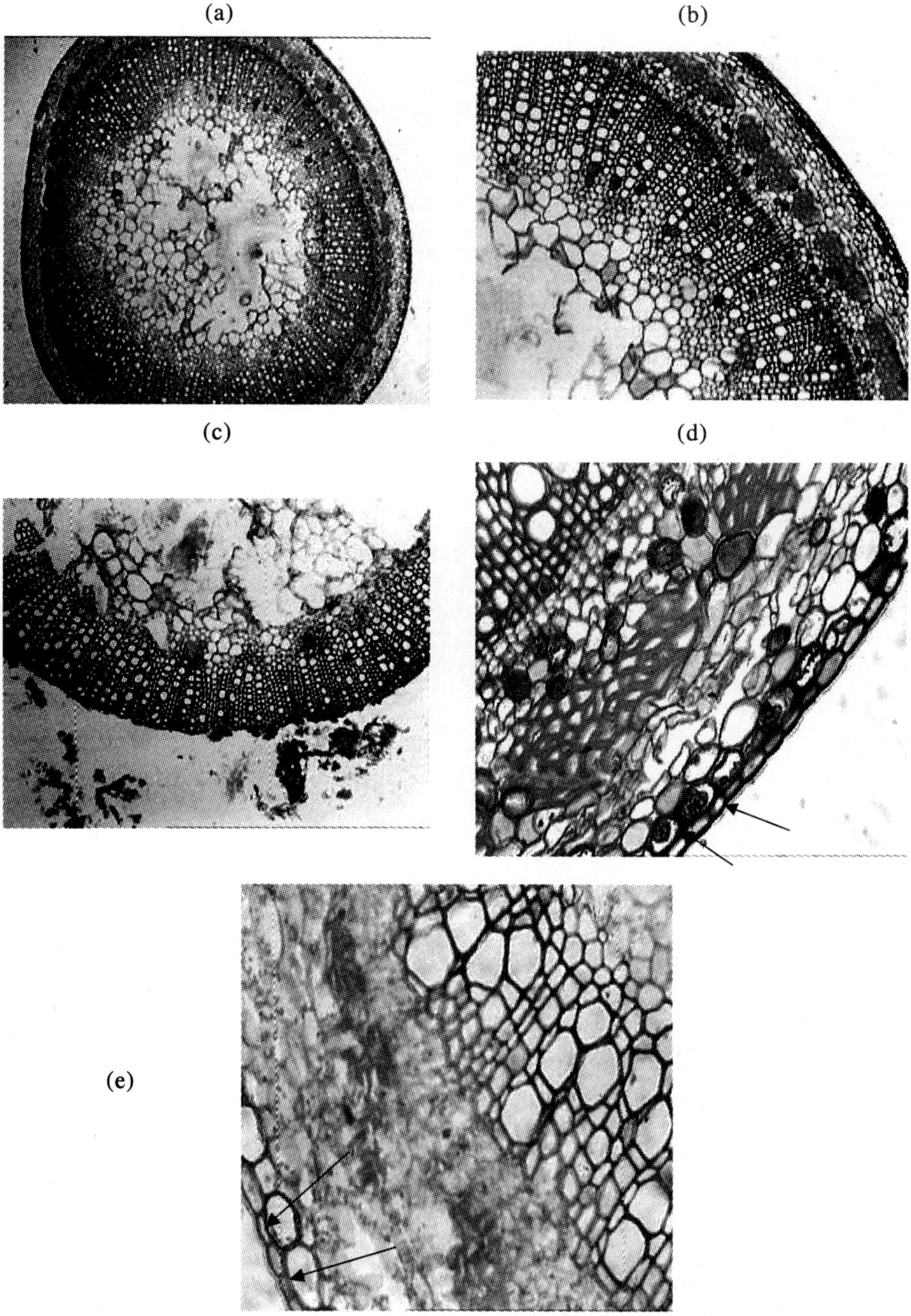

Plate 2 : Anatomy of stem blight infected by *C. cassicola* (a) Section of infected stem; (b) Blocking of xylem vessels; (c) Cortical layer damage; (d) Infected cells with granulation; (e) No granulation in healthy cells

All the bacterial isolates produced hydroxamate type of siderophores. But the comparison of isolates for the intensity of production of hydroxamate type siderophores indicated that *B. subtilis* (BSCBE4), *P. chlororaphis* (PA23), ENPF1, ENPF6, ENPF5, and ENPF8 produced higher intensity of hydroxamate types. Studies on carboxylate type of siderophore indicated that all the isolates produced carboxylate type siderophore, except ENPF6 and ENBS7 in addition to hydroxamate type (Table 1).

Effect of biocontrol agents and fungicides against damping off in green house

The biocontrol agents and fungicides were tested in pot culture for the management of damping off of *P. amarus* seedlings. Standardization of delivery system of bacterial and fungal antagonists for the management of damping off revealed that delivering of bacterial and fungal antagonists

Table - 1 : Testing of bacterial endophytes for production of siderophore

Bacterial endophytes	Siderophore production	Surface area of siderophore production after 24h (cm^2)	Hydroxa-mate type	Carboxy-late type
ENPF1	+ ve (Yellowish)	10.00^{a}	++	+
ENPF6	+ ve (Yellowish)	8.00^{bc}	++	-
ENPF5	+ ve (Yellowish)	2.50^{ef}	++	+
ENPF8	+ ve (Yellowish)	9.20^{ab}	++	+
ENPF10	+ ve (Yellowish)	4.20^{d}	+	+
ENBS2	+ ve (Yellowish)	4.20^{de}	+	+
ENBS3	+ ve (Yellowish)	8.40^{bc}	+	+
ENBS4	+ ve (Yellowish)	7.20^{c}	+	+
ENBS7	+ ve (Yellowish)	2.20^{f}	+	-
ENBS9	+ ve (Yellowish)	1.60^{g}	+	+
BCA	+ ve (Yellowish)	8.20^{bc}	++	+
CBE4	+ ve (pink)	4.20^{d}	++	+

Values are mean of three replications.
In a column, means followed by a common letter are not significantly different at the 5% levels by DMRT
++ Higher intensity; + Low intensity; - Negative

through seed treatment and soil application were found effective in suppressing the damping off rather than individual application of antagonist either as seed treatment or as soil application.

Among the biocontrol agents tested, delivering of CBE4 as seed treatment and soil application recorded the lowest percent disease incidence (15.00 %). It was followed by the application of PA23 delivered through seed

Table - 2 : Effect of biocontrol agents and fungicides against damping off in *P. amarus* under glasshouse conditions

Treatment	Per cent Disease Incidence	Per cent reduction over control	Shoot length (cm)	Root length (cm)
BCA (ST)	18.00^{cd} (25.09)	57.14	12.8^{abc}	5.20^{ef}
BCA (SA)	20.00^{bc} (26.54)	52.38	13.0^{abc}	5.80^{b-e}
BCA (ST+SA)	16.00^{de} (23.55)	61.90	14.3^{b}	6.20^{ab}
CBE4 (ST)	16.00^{de} (23.55)	61.90	13.0^{abc}	5.40^{cf}
CBE4 (SA)	18.00^{cd} (25.09)	57.14	12.5^{bc}	6.00^{bc}
CBE4 (ST+SA)	15.00^{e} (23.57)	64.28	14.8^{a}	6.70^{a}
ENPF1 (ST)	20.00^{bc} (26.54)	52.38	13.0^{abc}	6.00^{ab}
ENPF1 (SA)	22.00^{b} (26.54)	47.61	13.12	5.80^{bcd}
ENPF1 (ST+SA)	18.00^{cd} (25.09)	57.14	14.0^{ab}	6.40^{ab}
Tv MNT 7(ST)	16.00^{de} (23.55)	61.90	12.0^{bc}	5.50^{cf}
Tv MNT 7 (SA)	18.00^{cd} (25.09)	57.14	12.5^{bc}	5.60^{bcd}
Tv MNT 7 (ST+SA)	16.00^{de} (23.55)	61.90	13.0^{abc}	6.00^{bc}
Ridomil	14.00^{f} (21.96)	66.66	11.3^{cd}	4.70^{def}
Thiram	15.00^{e} (23.57)	64.28	11.6^{cd}	4.60^{def}
Healthy control	0.00^{g} (0.19)	100	11.5^{d}	4.80^{fg}
Inoculated control	42.00^{a} (40.39)	—	7.8^{e}	4.00^{g}

ST - Seed treatment; SA - Soil application
Values are mean of three replications.
Values in parenthesis are arcsine transformed
In a column, means followed by a common letter are not significantly different at the 5 % levels by DMRT

treatment and soil application. In general the growth parameters like root length and shoot length were higher when the biocontrol agents were delivered through both seed and soil. Studies on the usage of fungicides indicated that drenching Ridomil was comparable with CBE4 delivered through both seed and soil (Table 2).

DISCUSSION

Intensive cultivation of *P. amarus* has resulted in the outbreak of damping-off, web blight and stem blight. The *Corynespora* pathogen could survive up to two years in crop debris. The primary mode of field spread of this disease was through conidia disseminated through air. Similarly continuous monocropping of *P. amarus* for its high medicinal value might have resulted in the build up of inoculum, which led to disease severity. Histopathology of the stem blight infected plants indicated the extensive damage of cortical cells with toxin deposition and vascular plugging.

Production of siderophores like pseudobactin and pyoverdine which chelate the available iron in the soil results in the death of pathogen due to lack of iron for pathogen survival. (Kloepper *et al.,* 1980 a, b). Since iron being a compound of cells, its deficiency can cause growth inhibition, decrease in nucleic acid synthesis, inhibition of sporulation and change in cell morphology (Chincholkar *et al.*, 2000). In addition also regulates metabolic processes such as TCA cycle, electron transport chain, oxidative phosphorylation and photosynthesis (Chincholkar *et al.*, 2000). The deficit of available iron to pathogens might have resulted in death of *C. cassiicola.* Similar suppression of fungal pathogens by siderophores has been reported by Leong (1986). In this study, the bacterial antagonists *B. subtilis* (BSCBE4), ENPF1 and PA23 produced hydroxamate and carboxylate type of siderophores leading to iron deficit condition. Similarly Kloepper (1988) reported the production of fluorescent siderophores by *P. fluorescens* which contributed to its antagonistic action against *P. ultimum*.

REFERENCES

Chinchlokar, S.B., Chaudhar, B.L., Talegaonkar, S.K. and Kothari, R.M. (2000). Microbial Iron chelators, a sustainable tool for the biocontrol of plant diseases In: *Biocontrol potential and its exploitation in sustainable Agriculture* Vol 1 (Eds) Upadhay, R.K. Mukerji, K.G. and Chamoler, B.P. Kluwer Academic Plenum publisher, Newyork. pp 49-69.

Jayarajan, R. and Ramakrishnan, G. (1991). Efficacy of Trichoderma formulation against root rot disease of grain legumes *Petria giornale di patologia delle piante,* **1** : 137.

Kloepper, J.W., Leong, J., Teintze, M. and Schroth, M.N. (1980a). Enhanced plant growth by siderophores produced by plant growth promoting rhizobacteria. *Nature*, **64** : 286-886.

Kloepper, J.W., Leong, J., Teintze, M. and Schroth, M.N. (1980b). *Pseudomonas* siderophores a mechanism explanting disease suppressive soils. *Curr. Microbiol.*, **4** : 317-320.

Kloepper, J.W., Hume, D.J., Scher, F.M., Singleton, C., Tipping, B., Laliberte, M., Frauley, T., Kuutchaw, C., Simonson, C., Lifshitz, R., Zaleska, I. and Lee, L. (1988). Plant growth promoting rhizobacteria on canola (rape seed). *Plant Dis.*, **72** : 42-46.

Leong, J. (1986). Siderophores: their biochemistry and possible role in the biocontrol of plant pathogens. *Annu. Rev. Phytopathol.*, **24** : 187-209.

Narayanasamy, P. and Ramakrishnan, K. (1971). Powdery mildews of Coimbatore, Madras state. *J. Madras. Univ.*, **38** : 84-99.

Patil (1991). Rust fungi from Maharastra. *Indian phytopath.*, **44** : 433-439.

Rangaswami, G. (1958). An agar block technique for isolating soil micro organisms with special reference to Pythiaceous fungi. *Science and Culture,* **24** : 85.

Riker and Riker (1936). Introduction to research on plant diseases. John Swift (ed.). New York. 117 p.

Schwyn, B. and Neilands, J.B. (1987). Universal chemical assay for the detection and determination of siderophore. *Annl. Biochem.*, **169** : 47-56.

Snow, G.A. (1984). Mycobactin. A growth factor for *Mycobacterium johnei* II degradation and identification of fragments. *J. Chem. Soc.*, 2588-2596.

Vidhyasekaran, P. and Muthamilan, M. (1995). Development of formulations of *Pseudomonas fluorescens* for control of chickpea wilt. *Plant Dis.*, **79** : 782-786.

Vogel, A.E. (1987). Class reactions (reactions for functional groups) In: *Elementary Practical Organic Chemistry.* CBS publisher, New Delhi pp 190-194.

Recent Advances in Medicinal Plant Research : Vision 21st Century, 2005
Editors : S.K. Prabhuji, G.P. Rao and S.K. Patil
Satish Publication House, New Delhi

Chapter 14

ECO-CLIMATIC ZONE WISE DOCUMENTATION OF MEDICINAL PLANT DISEASES IN THE FOREST NURSERIES OF UTTARA KANNADA OF KARNATAKA

G.K. Sudarshan[1], V. Suryanarayana[1] and S.K. Singh[2]
[1]*Department of Forest Protection, College of Forestry, Sirsi - 581 401, Karnataka*
[2]*Department of Plant Pathology, Agarkar Research Institute, Pune, Maharastra*

Uttara Kannada is the core of Western Ghat part of Karnataka and is the main treasure of medicinal plants. During 2001-02, a survey for documenting seedling diseases was conducted in 15 nurseries, five each falling under plains, hills and coastal regions of Uttara Kannada. After proving Pathogenicity, the pathogens were characterized with the assistance of Agarkar institute, Pune. Major air borne diseases with respective pathogens and PDI levels were, Leaf and twig blight of *Lagerstroemia lanceolata* caused by *Pestalotiopsis mangiferae* (60.65%), leaf spot of *Sapindus emarginatus* by *Phoma prunicola* (55.65%), leaf spot of *Pongamia pinnata* by *Phyllosticta sp.* (55.32%), leaf spot of *Ficus benghalensis* by *Alternaria alternata* (42.55%), powdery mildew of *Tamarindus indica* by *Oidium sp.* (41.27%), leaf spot of *Garcinia indica* (36.51%), and leaf spot of *Vateria indica* by *Phomopsis sp* (33.45%), leaf spot of *Terminalia bellirica* by *Khuskia oryzae* (29.44%), leaf spot of *Syzygium cuminii* by *Pestalotiopsis versicolor* (28.53%), leaf spot of *Mangifera indica* caused by *Pestalotiopsis mangiferae* (28.31%), leaf spot of *Artocarpus heterophyllus* by *Phyllosticta artocarpi* (26.09%), leaf spots incited by *Alternaria alternata* on *Grewia tiliifolia* (19.84%), *Thespesia populnea* (15.30%) and *Cordia machleoldii* (13.24%), leaf spots of *Ougeinia dalbergioides*, leaf rust of *Emblica officinalis*. Major soil borne diseases was seedling blight of *Bambusa arundinacea* caused by *Rhizoctonia solani*. Symptomatology and influence of diseases on grown up trees on seedling diseases of same species were also documented. The study is first of its kind in this region.

INTRODUCTION

To shorten the gap between demand and supply of various forest products, the forest nurseries have a pivotal role to play in supplying quality seedlings. In real terms the quality implies mainly to health, a condition of freeness from a pool of diseases. Health is generally impaired basically by the menace of soil borne and air borne diseases. So, some of the diseases problems impair the nursery network, if they remain unchecked. To frame an effective management strategy, it is imperative to have nursery-wise and eco-climatic zone-wise database of seedling diseases together with the associated pathogens. Later, the incidence of the disease can be monitored for a period of time to understand its level of severity, so that control measures may be worked out economically. Uttara Kannada has a set up of 54 forest nurseries distributed in different eco-climates as hills, coastal and plains with a total production of 50 lakhs seedlings belonging to 75 species.

Uttara Kannada is the main core of Western Ghat part of Karnataka known for its richest heritage of bio-diversity with a potential treasure of medicinal plants. It has the credibility of crowning as one of mega bio-diversity centers with its own uniqueness of species distribution. In this region forest nurseries are not just concentrating the production of seedling stock of tree species of timber value, but also tree species of medicinal importance. In recent years, large-scale afforestation and rehabilitation of degraded areas have been taken up to cater the needs of the forest-based domestic requirements and the industries. Emphasis has been given in the nurseries to increase their potentials of raising large populations of quality seedlings stock. Inspite of its greater relevance, the study on drawing a nursery diseases scenario was untouched. By keeping the economic importance of diseases at nursery level and to generate the exhaustive information to fill the lacuna in getting large populations of healthy seedling stock, an effort has been made to generate valuable information.

HISTORICAL REVIEW

Raising and maintenance of healthy nursery seedling stock is the basic need of any plantation programme. The occurrence of diseases is the main constraint in getting healthy seedlings stock.

In this region no studies have been carried out with regard to diseases of nursery seedlings. However several workers have done documentation of diseases in forest nurseries. In Karnataka, Naik *et al.*, (1998) documented the diseases in forest nurseries surrounding Sirsi division and were, leaf spot of *Syzigium cumini*, root rot of *Casuarina equisitifolia*, tar spot of *Dalbergia latifolia*, anthracnose of *Mangifera indica*, powdery mildew of *Bauhinia purpurea* and *Acacia auriculiformis* and leaf spot and twig blight of *Lagerstroemia lanceolata,* seedling blight of *Terminalia chebula, T. paniculata* and *T. alata,* rust of *Dendrocalamus strictus* and *Emblica officinalis*, leaf spot of *Vateria indica* and powdery mildew of *Tamarindus indica,* bacterial leaf spot of *Mangifera indica*, damping-off *of Pterocarpus marsupium* and bacterial wilt of *Tectona grandis*. Besides this, powdery mildew on seedlings of *Acacia auriculiformis*, *A. eburnea* and *A. nilotica* has been documented by Bhat and Hegde (1990).

The survey conducted by different workers from Uttar Pradesh revealed the seedling diseases such as, collar rot of *Azadirachta indica* (Kamla Uniyal, 1999); leaf spot and blight of *Syzigium cumini* and *Michelia champaca* (Arushi Mehrotra and Mehrotra, 2000 and Mehrotra, 1992); collar rot of *Azadirachta indica* (Kamla Uniyal, 1999). In Kerala the diseases documented collar rot and leaf blight of *Gmelina arborea* (Florence and Sankaran, 1987); web blight on *Azadirachta indica* (Sankaran *et al*, 1986); nursery diseases of *Gmelina arborea* and some *Eucalyptus* species from Kerala (Sharma *et al*., 1985). The reports from other places were, damping-off of *Eucalyptus* (Jamaluddin *et al.*, 1981)

METHODOLOGY

Selection of nurseries

An intensive survey work was taken up during 2001-2002 to know the incidence and severity of different seedling diseases in the nurseries under three eco-climatic zones of Uttara Kannada district, *viz.*, coastal, hilly and plains. Coastal zone comprises of Bhatkal, Honnavar, Kumta, Ankola and Karwar talukas. Hilly zone comprises of Sirsi, Siddapur, Yellapur and Joida talukas. Whereas the parts of Mundgod and Haliyal talukas were under plains. Five nurseries were selected in each eco-climatic zone. For the documentation work, Harwada central nursery, Kasarkod central nursery,

Kaajubag, Gopshitta and Agsur nurseries were selected in coastal zone. From hilly zone, Shivaram Karanth nursery, Kalve, Kogilbana, Alageri and Subgeri nurseries were selected and Bachanki, Oralagi, Baghawathi, Karlakatta and Kurigadde nurseries were selected in plains (Fig. 1). The survey work was conducted during April-May, 2001 as this period was peak period during which maximum number of seedlings were raised in the nurseries for out planting before the onset of Southwest monsoon.

Disease incidence estimates

The disease incidence was estimated by counting number of seedlings affected out of total number of seedlings raised in a seedling lot. Percent

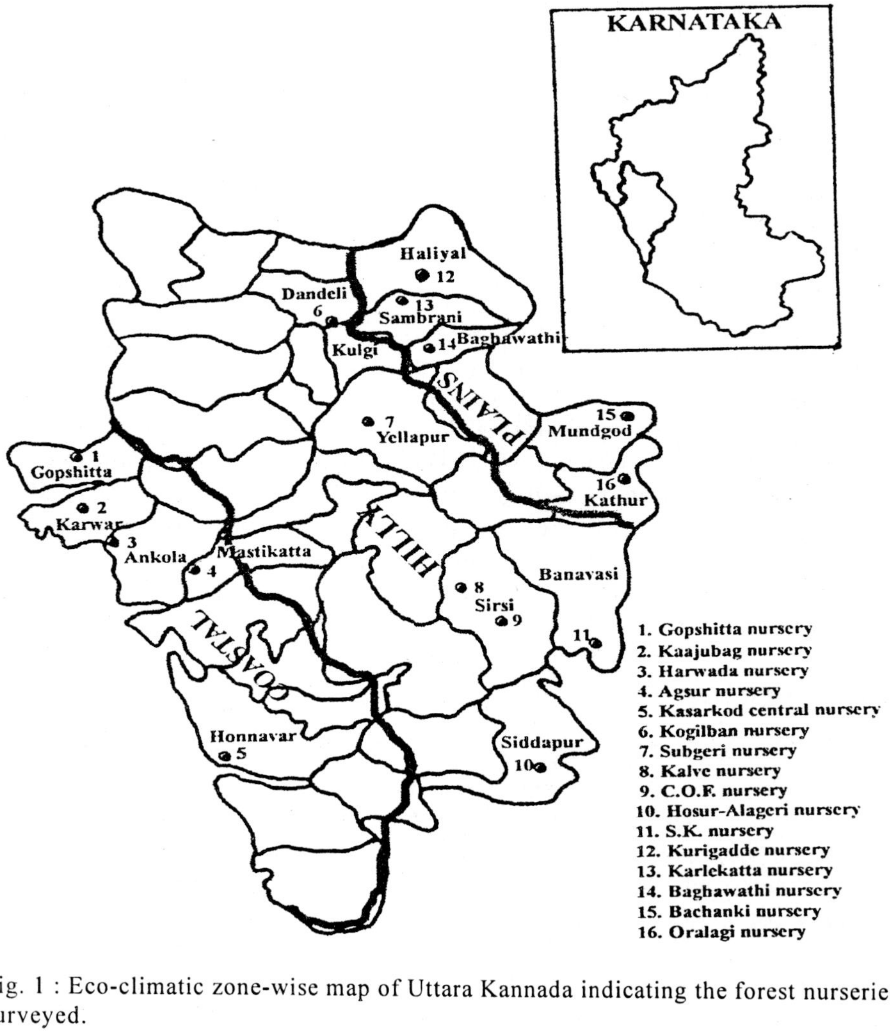

Fig. 1 : Eco-climatic zone-wise map of Uttara Kannada indicating the forest nurseries surveyed.

incidence was calculated by counting number of seedlings affected out of total number of seedlings in a lot and was expressed in percent scale in a tabular form.

Disease severity estimates

The severity of each disease was recorded by using the scale developed by Sharma and Mohanan (1984) with some modifications. The scale was as follows :

Description	Disease severity scale
No symptom	0
1 % leaf area affected	1
10 % leaf area affected, 1% defoliation	2
11 to 26 % leaf area affected, 5 % defoliation	3
26 to 50 % leaf area affected, 6-10 % defoliation	4
> 50% leaf area affected, 11-25 % defoliation	5

The disease severity levels of different seedling diseases were estimated by Percent Disease Index by using the following formula given by Wheeler (1969).

$$\text{PDI (\%)} = \frac{\text{Sum of numerical ratings of all the leaves}}{\text{Total number of leaves observed} \times \text{Maximum grade used}} \times 100$$

The estimated disease severity levels of different seedling diseases based on PDI were presented in a tabulated form.

Symptomatology

During the survey work a keen observation was made on different parts of the affected plants of different species in each of the nurseries and a detailed descriptions of the symptoms were noted down along with the photo documentation.

RESULTS AND DISCUSSION

Eco-climatic zone wise disease scenario

The results of eco-climatic zone-wise nursery diseases documentary on tree species of medicinal importance have been presented in Table-1. In plains, diseases of five species were found with a maximum disease incidence of leaf spot of *Sapindus emarginatus* (62.94%) fallowed by powdery mildew of *Tamerindus indica* (41.3%). Among the nurseries visited, Bachanki nursery was observed with maximum number of seedling diseases.

Table - 1 : Eco-climatic zone wise seedling disease estimates of medicinal plants in forest nurseries falling under Uttara Kannada District

Name of the nursery		Bachanki		Oralagi		Baghawathi		Kurigadde		Karlekatta		Average	
Disease	Host	Incid (%)	PDI (%)	Incid (%)	PDI (%)	Incid (%)	PDI (%)	Incid (%)	PDI (%)	Incid (%)	PDI (%)	Incid (%)	PDI (%)
PLAINS													
Leaf spot	*Sapindus emarginatus*	89.00	61.85	100.00	64.04	-	-	-	-	-	-	94.50	62.94
Powder mildew	*Tamarindus indica*	★	★	-	-	27.47	17.41	91.40	47.01	76.60	59.40	65.20	41.30
Leaf rust	*Emblica officinalis*	1.60	★	-	-	★	★	-	-	★	★	1.60	

Name of the nursery		Kogilbana		S K Nursery		Kalve		Hosur-Alageri		Subgeri		Average	
Disease	Host	Incid (%)	PDI (%)	Incid (%)	PDI (%)	Incid (%)	PDI (%)	Incid (%)	PDI (%)	Incid (%)	PDI (%)	Incid (%)	PDI (%)
HILLY													
Leaf spot	*Garcinia indica*	100.00	36.51	★	★	★	★	-	-	-	-	100.00	36.51
Leaf spot	*Vateria indica*	-	-	★	★	46.33	33.45	-	-	-	-	46.33	33.45
Leaf spot	*Terminalia bellirica*	★	★	52.60	29.44	★	★	★	★	-	-	52.60	29.44
Leaf spot	*Syzigium cuminii*	-	-	42.12	31.30	45.11	24.44		-	38.00	22.55	41.74	26.10

Leaf spot	*Mangifera indica*	★	★	★	★	45.65	24.68	★	★	42.60	23.60	44.12	24.14
Leaf spot	*Artocarpush-eterophyllus*	-	-	-	-	-	-	-	-	1.00	★	1.00	

Name of the nursery		Kasarkod		Agsur		Harwada		Gopshitta		Kaajubag		Average	
Disease	**Host**	Incid (%)	PDI (%)	Incid (%)	PDI (%)	Incid (%)	PDI (%)	Incid (%)	PDI (%)	Incid (%)	PDI (%)	Incid (%)	PDI (%)
COASTAL													
Leaf spot	*Mangifera indica*	★	★	★	★	52.20	31.03	68.40	33.95	2.00	★	40.87	32.50
Leaf spot	*Syzigium cuminii*	★	★	★	★	40.76	22.84	67.25	34.59	41.87	35.48	49.96	30.97
Leaf spot	*Artocarpus heterophyllus*	★	★	45.20	33.16	50.00	28.35	1.00	★	27.20	22.80	30.85	28.10
Leaf spot	*Garcinia gummigutta*	-	-	1.37	★	-	-	-	-	-	-	1.37	

★Species was raised in the nursery, Species not raised in the nursery.

Recent Advances in Medicinal Plant Research : Vision 21st Century, 2005
Editors : S.K. Prabhuji, G.P. Rao and S.K. Patil
Satish Publication House, New Delhi

Chapter 16

ETHNOMEDICINAL PLANTS FOR WOMENFOLK'S HEALTH CARE IN CENTRAL WESTERN GHAT REGION OF KARNATAKA

P. Ramana[1★], Siddappa Kannur[2] and S.K. Patil[3]
College of Forestry, Sirsi - 581 401, Karnataka
[1]Assistant Professor, Department of Forest Utilization, College of Forestry, Sirsi - 581401, Karnataka
[2] Junior Merit Associate, Department of Silviculture, College of Forestry, Sirsi - 581401, Karnataka
[3] Director of Instruction(For), College of Forestry, Sirsi - 581401, Karnataka
★For correspondance

During ethnopharmacobotanical investigations in Uttara Kannada district, Karnataka, 33 species of 23 families useful in womenfolk's health care were recorded along with their vernacular names, plant parts used and uses. The plants reported in this study are claimed to be the most effective remedies for conditions such as leucorrhoea, gonorrhea, disorders during delivery, pregnancy, menstruation and other fertility related complaints in women.

INTRODUCTION

Indians have one of the world's richest medicinal plants heritage. Around 8000 species of plants referred to by over 2,00,000 vernacular names, are used by the people of India in local health cultures for human, veterinary and agriculture (bio pesticides, seed treatment and bio-fertilizers) related applications, across her ten bio-geographic zones and 25 biotic provinces and 4635 ethnic communities. The knowledge of these plants is undocumented and transmitted through an 'oral' tradition. Around 1800 species are systematically documented in the codified Indian System of Medicine like Ayurveda, Unani, Sidha and Gso-rig-pa, which are expressed in thousands of medical manuscripts. These medical manuscripts contain details of the biological properties of plants as well as the products that can be made from the plants. Around 1,00,000 herbal formulations alongside their indications of use are described in local cultures and in the traditional medicine literature. They are extensively used as anti-cancer, antidiabetics, purgatives, hepatoprotectives, anti-inflammatory, anti-oxidants and antidotes (Anonymous, 2003).

Most of these plants are still in the main stay of our population for primary health care because of their easy accessibility, better compatibility with human being and lesser side effects. It is assured that search for potential medicinal plants through ethno botany gives clues for further research and development. Indigenous knowledge on medicinal plants serves as an important pointer to the species, which might contain significant phytochemicals (Sequeria, 1994). Earlier investigations on ethnomedicinal plants for womenfolk's health care have been carried out in India and abroad (Tarafder, 1983a, 1983b, 1984; Siddque *et al*, 1988; Lakshmanan and Sankara Narayanan, 1990; Singh and Zaheer Anvar Ali, 1996; Balik *et al*, 2000; Sashikumar and Janardhanan, 2002). The present communication highlights the scope of safe green ethnomedicine for womenfolk's health care in Uttara Kannada district of Central Western Ghats. Even though modern health care centers are in their reach, tribals of this region have got faith in local health cures and are still depending on the traditional medicines. The present study envisages the revival of the traditional knowledge of the tribes on flora and identifying the plant wealth in tribal habitations, which is under serious threat of extinction.

STUDY AREA

Uttara Kannada district is a unique rich biogeographic region in the Central Western Ghat region of Karnataka, holding multifarious endemic floral and faunal wealth. It is located between 13 ^{0} 55' and 15^{0} 31' N latitude and 74^{0} 9' and 75^{0} 10' E longitude. It has an area of 10,007 km^2. It is about 176 km from north to south and from 16 to 96 km from east to west. The Central Kanara includes three belts, a coast tract with broad winding lagoons, rich plains and wooded hills running to the sea; a central belt of lofty Sahyadris covered with magnificent forest; and a eastern upland, which is wild waving and thickly wooded in the west and in the east passes into a bare level and thickly populated plains.

Dry deciduous, moist deciduous, semi-evergreen and evergreen forests primarily represent the vegetation of the study area. The climate is primarily monsoonal and rains affected by Southwest monsoon during June to September. The rainfall is also variable from year to year. The region received an annual rainfall of 1933 mm with a minimum of 692 mm and a maximum of 3922 mm. The length of dry season varied from a mean of three months at higher altitudes in South to eight months on eastern side of the Ghat in the north. The relative humidity is highest exceeding 90 per cent during July-August and it is as low as 40 per cent in March-April. There is substantial seasonally in temperature. The mean annual range temperature varies from 18^0 C to 31^0 C under extreme condition drops down to 6.7^0 C and rises up to 39^0 C. In general, the soils are shallow to medium deep, reddish brown, dark brown and dark yellow brown in colour. The soils are acidic to slightly alkaline in nature, high in nitrogen, low in phosphorus and low to medium in potassium content (Anonymous, 1991).

METHODOLOGY

The data presented in this paper are the outcome of intensive studies conducted among Halakki Okkaligas, the local traditional medicine practioners. Local herbal informants were selected and interviewed extensively. The women chieftains are accorded a significant role in discussions since they possess more cognizance about the utility of local herbal products in curing various diseases. Data on the ethno herbal medicines along with their application were gathered from experienced and knowledgeable medical men and women. Information on plants used in women folk care was gathered by quantitative and qualitative elicitation methods (Martin, 1995; Jain, 1989). The collected data were verified by cross verification or triangulation (Cochran and Cornfield, 1951; Martin, 1995).

RESULTS AND DISCUSSION

An analysis of data in the present communication reveals the uses of 33 species belonging to 23 families for womenfolk's health care among local tribals of Uttara Kannada district. Correct botanical nomenclature with family names, vernacular names, habitat, plant parts used and uses is reported (Table-1). The reported data were compared with those gathered in other ethnobotanical studies carried out in Uttara Kannada earlier (Thakru Sukru Gouda, 1997).

Table - 1 : Ethnomedicinal plants used by Womenfolk's of Uttara Kannada District, Karnataka

Plant Name	Family	Vernacular Name	Habitat	Parts used	Uses
Abelmoschus moschatus Medic	Malvaceae	Kasturibende	Herb	Seeds	As antiseptic after delivery
Abrus precatorius L.	Fabaceae	Biligulaganji	Climber	Roots	As antiseptic after delivery
Abutilon indicum (L.)	Malvaceae	Mudregida	Shrub	Leaves	As emmenagogue
Acalypha indica L.	Euphorbiaceae	Kuppigida	Herb	Leaves	To check excessive pain during menstruation
Aegle marmelos (L.) Correa	Ruteaceae	Patregida	Tree	Seeds	To treat gonorrhea
Aloe vera L.	Malvaceae	Loolisara	Herb	Roots	To treat gonorrhea
Alternanthera sessilis (L.) DC.	Amaranthaceae	Honaganne	Creeper	Roots, Whole Plant	To cure menorrhagia As emmenagogue
Apama siliquosa Lam.	Aristolachiaceae	Chakranigida	Herb	Roots	To treat anemia during pregnancy
Breynia retusa (Dennst.) Alst.	Euphorbiaceae	Bettadanalli	Herb	Leaves	As emmenagogue
Calotropis gigantean (L.) Dryand	Asclepidaceae	Biliyekkegida	Shrub	Leaves	To cure malnutrition during pregnancy
Celastrus paniculatus Willd.	Celastraceae	Gangamman-aballi	Creeper	Roots	To cure excessive pain during menstruation and to induce fertility

Table - 1 : Continued .....

Plant Name	Family	Vernacular Name	Habitat	Parts used	Uses
Centella asiatica (L.)	Apiaceae	Omdelaga	Creeper	Stem and Root, Roots	During pregnancy for easy delivery As emmenagogue
Clerodendrum viscosum Vent.	Verbenaceae	Taggigida	Shrub	Leaves, Roots	To cure menorrhagia To treat stomachache during pregnancy
Cythula prostrata (L.) Bl.	Acnthaceae	Nelauttarani	Creeper	Leaves, Stem	To cure urinary disorders To cure leucorrhoea
Cynodon dactylon (L.) Peers	Poaceae	Garikehullu	Grass	Whole plant	As emmenagogue
Evolvulus alsinoides L.	Convulaceae	Vishnukranti	Creeper	Whole plant	As emmenagogue
Hibiscus rosa sinensis L.	Malvaceae	Bilidaasavala	Shrub	Leaves	To cure leucorrhoea
Kalanchoe pinnata (Lam.) Persoon	Crasullaceae	Patapategida	Herb	Leaves	To cure malnutrition during pregnancy
Lawsonia inermis L.	Lythraceae	Madarangi	Shrub	Leaves	To treat anemia during pregnancy
Mirabalis jalapa L.	Nyctaginaceae	Bili madhy-anhamallige	Creeper	Tuber	To cure leucorrhoea and induce fertility

Table - 1 : Continued .....

Plant Name	Family	Vernacular Name	Habitat	Parts used	Uses
Musa superba L.	Musaceae	Kallu baale	Shrub	Seeds	To cure excessive pain during menstruation
Ocimum sanctum L.	Lamiaceae	Tulasigida	Herb	Seeds	As emmenagogue
Phragmites karka (retz.) Trin-exstendel	Poaceae	Jeevaladakaddi	Herb	Tuber	To cure urinary disorders
Phyllanthus fraternus Webester	Euphorbiaceae	Nelanalli	Herb	Whole plant, Roots	As emmenagogue To cure urinary disorders
Piper langum L.	Piperaceae	Hippaliballi	Creeper	Fruits	During pregnancy for easy delivery and to cure sprue
Putranjiva rox burghii Wall.	Euphorbiaceae	Putranjeevi	Tree	Seeds	To cure leucorrhoea
Ricinus communis L.	Euphorbiaceae	Oudalugida	Shrub	Roots	During pregnancy for easy delivery
Securinega leucopyrus (Willd.) Mueller	Euphorbiaceae	Tondajjigida	Herb	Seeds	To cure leucorrhoea
Sida rhombifolia L.	Malvaceae	Kallamgadale	Shrub	Leaves	To cure malnutrition during pregnancy
Solanum nigrum L.	Solanaceae	Kaakigida	Herb	Whole plant	To cure menorrhagia and for piles

Table - 1 : Continued .....

Plant Name	Family	Vernacular Name	Habitat	Parts used	Uses
Tabernaemontana divaricata (L.) R Br.	Apocynaceae	Nandibattalu	Shrub	Roots	As antiseptic after delivery
Tenospora cordifolia (Willd.) Miers	Menispermaceae	Amruthaballi	Climber	Flower	To cure menorrhagia
Vetiveria Zizanoides (L.) Nash	Poaceae	Madiwaalada baeru	Grass	Roots	To treat typhoid during pregnancy

The tribals prefer traditional remedial practices for womenfolk's health care. The tribals of the study area have deep faith in traditional medicine. Mostly women are more conscious about the choice of herbal therapy for their ailments and have got a better knowledge about medicinal herbals than men. Most of the herbal drugs are authoritatively claimed to be the most effective remedies for conditions such as leucorrhoea, gonorrhea, disorders during delivery, pregnancy, menstruation and other fertility related complaints.

This ethno botanical information on the tribals of Uttara Kannada provides clues for new leads for systematic pharmaco-therapeutics and bio-prospection. The medicinal plants recorded in the present study needed extensive scientific research work in future to exploit their full potential as herbal drugs.

ACKNOWLEDGMENT

The authors extend their sincere thanks to the tribal herbalists who revealed their knowledge on medicinal herbs used in Womenfolk's health care.

REFERENCES

Anonymous (1991). *Gazeteer of the Bombay Presidency, Kanara District.* Volume XV, Part I, Sagardeep Publications, Karwar, p. 1-18.

Anonymous (2003). The treasure house of medicinal plants. *Amruth,* **7(3)** : 2.

Balick, M.J., Fresi, Kronenberg, Andreana, L., Ososki, Marin Rieff, Adriane Fugh-Berman, Bonnie O' Conner, Maria Rubia, Patrica Lohr and Daniel Atha (2000). Medicinal plants used by Latino healers for women's health care conditions in New York City. *Eco. Bot.*, **54(3)** : 344-357.

Cochran, N.C. and Cornfield, J. (1951). Modern methods in sampling human population. *Amer. J. Public Health*, **41** : 647-661.

Jain, S.K. (1989). *Methods and Approaches in Ethnobotany.* Deep Publications. New Delhi, India.

Lakshmanan, K.K. and Sankara, Narayanan (1990). Antifertlity herbals used in Analkatty Hills. Coimbatore District, Tamil Nadu. *J. Econ. Tax. Bot.*, **14(1)** : 171-173.

Martin, G.J. (1995). Ethnobotany: *A Methods Manual.* Chapman and Hall, London.

Sashikumar, J.M. and Janardhanan, K. (2002). Ethnomedicinal plants for Womenfolk's Health care in Nilgiri Biosphere Reserve, Western Ghats. *Indian Jour. For.*, **9(3/4)** : 138-143.

Sequiera, V. (1994). Medicinal plants and conservation in Soa Tome. *Biodiv. Conser.*, **3** : 910-926.

Siddique, M.B., Alam, M.M., Hussain, W. and Sharma, G.K. (1988). Ethnomedicinal studies of plants used for terminating pregnancy. *Fitoterpia*, **59** : 250-252.

Singh, V.K. and Zaheer, Anvar Ali (1996). *Folk medicinal plants used for family planning in India. In Ethnobiology in Human Welfare.* Ed. S.K. Jain. Deep Publications, New Delhi, pp. 184-186.

Tarafder, C.R. (1983a). Ethnogynecology in relation to plants-I. Plants used in antifertility and conception. *J. Econ. Tax. Bot.*, **4** : 483-490.

Tarafder, C.R. (1983b). Ethnogynecology in relation to plants-II. Plants used in abortion. *J. Econ. Tax. Bot.*, **4** : 507-516.

Tarafder, C.R. (1984). Ethnogynecology in relation to plants-III. Plants used to accelerate pre and post natal complaints. *J. Econ Tax. Bot.*, **5**: 572-576.

Thakru Sukru Gouda (1997). *Karnataka Oushadi Sasyagala Parampare –1. Uttara Kannadadha Halakki Okkaligaru.* INCERT, Bangalore.

Recent Advances in Medicinal Plant Research : Vision 21st Century, 2005
Editors : S.K. Prabhuji, G.P. Rao and S.K. Patil
Satish Publication House, New Delhi

Chapter 17

ETHNO-MEDICINAL PLANTS USED IN RHEUMATISM, GOUT AND ARTHRITIS AMONG THE TRIBALS IN INDIA

Anand Prakash and K.K. Singh
National Botanical Research Institute, Lucknow - 226 001, Uttar Pradesh

In recent years, large number of diseases and disorders are increasing day by day. Some of the chronic diseases and disorders like : Asthma, Jaundice, Diabetes, Rheumatism, Gout, Arthritis, Paralysis, Epilepsy, Filaria and Cancer etc. are still incurable in traditional systems of medicine as well as modern allopathic system of medicine. Therefore, the resurgence of interest in green medicine (medicines derived from plant sources) in search for safe, economical, sustainable and effective healthcare is increasing throughout the world. The paper highlights some potential ethno-medicinal plants used in different tribal communities for the treatment of Rheumatism, Gout and Arthritis, like: *Aloe vera, Ampelocissus latifolia, Boerhavia diffusa, Calotropis gigantea, C. procera, Cassia tora, Cassine glauca, Celastrus paniculatus, Crinum defixum, Datura innoxia, Euphorbia fusiformis, Gloriosa superba, Hymenodiction orixense, Plumbago zeylanica, Pygmaeopremna herbacea, Pterocarpus marsupium, Rivea cuneata,*

Sapindus mukorossi, Solanum viarum, Urtica parviflora. The highlighted species needs conservation and further systemic investigation for the isolation and development of potential, safe, economic drugs for sustainable utilization in healthcare and human welfare.

INTRODUCTION

Rheumatism, Arthritis and Gout are the serious diseases among the large population all over the world. The most common kind of arthritis is Rheumatoid arthritis, affects an estimated 1-2% of the world's population. Mostly 90% male patients are suffering from this disease. It is autoimmune disease, affects smaller joints like fingers of legs and hands. It does not occur in younger men and pre-menopause in women. Rheumatic arthritis is another kind of arthritis, which affects the larger joints like Knee and elbow accompanied by fever. It occurs generally in children having bacteria as the causal organism. These diseases are still incurable in allopathic system of medicine as well as other indigenous systems of medicine like unani, ayurveda and siddha. In homoeopathic system of medicines, patients are treated symptomatically. The medical scientists and physicians are seriously searching for safe, effective and economical medicine against the cure. Gout occurs in men and women due to excess amount of uric acid deposition around the joints in advanced age. The uric acid formation takes place as a result of protein molecule breakdown and abnormal metabolism. In some cases, it is hereditary also and may be due to abnormalities in some chromosomes up to a certain limits. The Philippine and Tawainian people living in America and Chinese in Malaysia have maximum probability of this disease due to hereditary reasons.

India has a very rich biodiversity, unique physical and ethnic diversity, traditional culture, and indigenous knowledge or tribal wisdom (Rao, 1994). There are 500 tribal and aboriginal communities in India living in close proximity to forests since time immemorial. Due to the close and long association with forests they have acquired unique knowledge of plants, plant produce and their uses in their daily needs and health care (Ambasta, 1986; Maheshwari *et al.,* 1981, 1986; Dhawan *et al.,* 1977; Singh *et al.,* 1984; Singh *et al.*, 1994; Singh and Prakash, 1994, 1996; Jain, 1975, 1991; Jain and Puri, 1984; Singh, 1997; Prakash and Singh, 2000).

The paper highlights several species, which are used and practiced by the tribal and aboriginal population for the treatment of rheumatism, gout and arthritis in Uttar Pradesh and Uttaranchal and provides baseline data for further validation and isolation of promising safe, economical and effective polyherbal medicines for human welfare.

METHODOLOGY

The tribals and aboriginals like Kols, Kharwar, Gonds,Tharu in Uttar Pradesh and Bhoxa, Bhotiya and Jaunsari in Uttaranchal were surveyed in different seasons and data on plants used for the treatment of rheumatism, gout and arthritis were documented. The informations were gathered with the help of knowledgeable local people, old and experienced persons, traditional healers and practitioners of herbal medicine after discussions and repeated verification. The names of the plants are arranged alphabetically, giving information on local names (Ln.), localities (Loc.), preparation of medicinal recipes, dosages and mode of administration. The voucher specimens have been deposited in herbarium of National Botanical Research Institute, Lucknow (LWG) for future reference and study.

RESULTS

***Achyranthes aspera* L. (Amaranthaceae)**

Ln. : Chircheria, Chichida; Loc. Varanasi, Gorakhpur; Tribe : Kol, Tharu

Root powder along with black pepper-one teaspoonful thrice a day for 15 days is given for the treatment of rheumatism and gout. Leaf paste together with mustard oil and camphor is applied locally in rheumatism and gout.

***Ainslaiaea aptera* DC. (Asteraceae)**

Ln. : Karu; Loc. Jaunsar; Tribe : Jaunsari

Root decoction is given orally in body inflammation.

***Aloe vera* (L.) Burm.f. (Liliaceae)**

Syn. *A. barbadensis* Mill.

Ln. : Ghekwar; Loc. : Nainital; Tribe : Tharu

The pulp of the plant, with salt and fermented sugarcane juice, is mixed together and taken to treat pain and inflammation of the body.

***Ampelocissus latifolia* (Roxb.) Planch. (Vitaceae)**

Ln. : Gulbas; Loc. : Jaunsar; Tribe : Jaunsari

Root decoction is useful in rheumatic affections.

***Boerhavia diffusa* L. (Nyctaginaceae)**

Ln. : Gadapurna; Loc. Gorakhpur; Tribe : Tharu

Cooked leaves are taken orally to check body inflammation.

***Calotropis gigantea* (L.) Dryand (Asclepiadaceae)**

Ln. : Akahua; Loc. : Kheri; Tribe : Tharu

The leaves warmed in Lahi (*Brassica campestris*) oil are applied in inflamed parts of the body.

C. procera **(Ait.) R. Br. (Asclepiadaceae)**

Ln. : Madar; Loc. : Gorakhpur; Tribe : Tharu

Latex of the plant together with mustard oil applied locally in rheumatism and gout.

Carduus nutans **L. (Asteraceae)**

Ln. : Bushiand; Loc. Jaunsar; Tribe : Jaunsari

Plant paste is warmed and applied externally in sprain and inflammation.

Cassia tora **L. (Caesalpiniaceae)**

Ln. : Pamar; Loc. : Bijnor; Tribe : Bhoxa

The young leaves are eaten as vegetable to cure rheumatic pain.

Cassine glauca **(Roxb.) Kuntze (Celastraceae)**

Ln. : Mamar; Loc. : Sonbhadra; Tribe : Gond

Aqueous extract of root bark is given orally once a day for one week for the treatment of gout.

Celastrus paniculatus **Willd. (Celastraceae)**

Ln. : Malkangani, Kakundan; Loc. : Nainital, Kheri; Tribe : Tharu

The paste of the fruit alongwith warm mustard oil is applied externally two times a day for one month for treating rheumatism and gout. Seed oil is massaged for joint pain.

Crinum defixum **Ker-Gawl. (Amaryllidaceae)**

Ln. : Biskanara; Loc. : Varanasi; Tribe : Kol

The tubers are cut from the base and are warmed in mustard oil. These are applied on injured and inflammed portion of the body to relieve pain.

Cuscuta reflexa **Roxb. (Cuscutaceae)**

Ln. : Saragbaburi; Loc. : Nainital; Tribe : Tharu

The warm paste of the plant is applied in rheumatism and gout and other affected parts of the body.

Dalbergia sissoo **Roxb. (Fabaceae)**

Ln. : Shisham; Loc. : Nainital; Tribe : Tharu

The seed oil and decoction of the wood is used in the treatment of rheumatic pain.

***Datura inoxia* Mill. (Solanaceae)**

Ln. : Datura; Loc. Varanasi; Tribe : Kol

Seed paste along with koina oil (*Madhuca longifolia*) is warmed and applied three times a day for seven days for the treatment of arthritis.

***Dioscorea bulbifera* L. (Dioscoreaceae)**

Ln. : Genthi; Loc. Varanasi; Tribe : Kol

Paste of the corm along with warm mustard oil is applied on the affected parts in rheumatism and gout.

***Dioscorea hispida* Dennst. (Dioscoreaceae)**

Ln. : Bank; Loc. Varanasi; Tribe : Kol

Paste of the corm along with warm mustard oil is applied on the affected parts in rheumatism and gout.

***Euphorbia fusiformis* Buch.-Ham. (Euphorbiaceae)**

Syn. *E. acaulis* Roxb.

Ln. : Ban muli; Loc. : Kheri, Bahraich; Tribe : Tharu

The paste of the root stock alongwith warm mustard oil is applied externally three times a day for a long time for the treatment of rheumatism and gout.

It has shown marked anti-inflammatory and anti-arthritic activities in rats. Its anti-arthritic activity was found to be superior to phenylbutazon.

***Gloriosa superba* L. (Liliaceae)**

Ln. : Kali-hari; Loc. : Gorakhpur, Varanasi; Tribe : Tharu, Kol

Corm paste together with kerosene and goat's milk is applied in for rheumatism and gout for a long period of time.

***Holoptelea integrifolia* (Roxb.) Planch. (Ulmaceae)**

Ln. : Chilbil, Papri; Loc. : Sonbhadra; Tribe : Gond

The leaf paste is warmed and applied externally to relive body inflammation.

***Hymenodictyon orixense* (Roxb.) Mabberley (Rubiaceae)**

Syn. *H. excelsum* Wall.

Ln. : Bhurkur; Loc. : Sonbhadra; Tribe : Gond

The powdered root mixed with cow's milk and given in the morning to relieve body inflammation.

Ipomea carnea **Jacq. subsp.** ***fistulosa*** **Austin (Convolvulaceae)**

Ln. : Behaya; Loc. : Nainital, Varanasi; Tribe : Tharu, Kol

The leaves are warmed in mustard oil and are applied in rheumatism and gout for one month.

Jatropha curcas **L. (Euphorbiaceae)**

Ln. : Ban Rendi, Bakend; Loc. : Sonbhadra; Tribe : Gond

The fomentation with leaves is practiced to relive body inflammation.

Litsea glutinosa **C B. Robin. (Lauraceae)**

Ln. : Maida; Loc. : Kheri; Tribe : Tharu

The paste of the stem bark mixed with common salt is applied externally in inflammation and sprains.

Ludwigia adscendens **(L.) Hara. (Onagraceae)**

Ln. : Kuand; Loc. : Nainital; Tribe : Bhoxa

The plant paste mixed with urine of cow is used in the treatment of rheumatism and gout.

Melia azedarach **L. (Meliaceae)**

Ln. : Bakain; Loc. : Varanasi; Tribe : Kol

The paste of the leaves mixed with Koyena oil (*Madhuca longifolia*) and common salt is applied externally in rheumatism and gout.

Ocimum sanctum **L. (Lamiaceae)**

Ln. : Tulasi; Loc. : Sonbhadra; Tribe : Gond

Powdered leaves along with powder of seeds of *Strychnos nuxvomica*, opium,black pepper and clove are made into pills. Two pills a day for 10 days is given for the treatment of arthritis.

Olax scandens **Roxb. (Olacaceae)**

Ln. : Batas; Loc. : Varanasi; Tribe : Kol

Leaf is made into pills. Two pills twice a day is given for the treatment of rheumatoid arthritis.

Perilla frutescens **(L.) Britt. (Lamiaceae)**

Ln. : Bhangiri; Loc. : Jaunsar; Tribe : Jaunsari

Seed oil is massaged twice a day in arthritis for a long time.

Peristrophe paniculata **(Fork.) Burm.f. (Acanthaceae)**

Syn. *P. bicalyculata* Nees

Ln. : Chitchitiya; Loc. : Sonbhadra; Tribe : Tharu, Gond

The paste of the root is applied externally two times a day for 15 days for the treatment of rheumatism and gout.

Plumbago zeylanica **L. (Plumbaginaceae)**

Ln. : Chitra; Loc. Jaunsar; Tribe : Jaunsari

Root paste along with buttermilk is applied locally in rheumatism and gout twice a day for two weeks. Extract of the roots and leaves are useful in rheumatism.

Pygmaeopremna herbacea **(Roxb.) Mold. (Verbenaceae)**

Ln. : Gahtia bad, Bhant; Loc. : Varanasi, Kheri; Tribe : Kol, Tharu.

The paste of the whole plant is made into pills. Two pills are given three times a day for fifteen days for the treatment of rheumatism and gout. Warm root paste is applied externally two times a day for one month for the treatment of rheumatoid arthritis and gout.

Pterocarpus marsupium **Roxb. (Fabaceae)**

Ln. : Bijasar; Loc. : Nainital; Tribe : Tharu

The powdered gum, powdered alum and hot milk, is given for rheumatic pain. The powdered stem bark is given orally in inflammation of the body for ten days.

Rivea cuneata **Wight (Convolvulaceae)**

Ln. : Barwat; Loc. : Sonbhadra; Tribe : Gond

Aqueous extract of root alongwith root extract of *Marsdenia tenacissima*, 2-table spoonful is given once a day in the morning for 21 days for the treatment of arthritis.

Rubus niveus **Thunb. (Rosaceae)**

Ln. : Aghoi; Loc. Jaunsar; Tribe : Jaunsari

Root paste is applied on swellings.

Sapindus mukorossi **Gaertn. (Sapindaceae)**

Ln. : Athu; Loc. : Jaunsar; Tribe : Jaunsari

Seed paste is applied on the affected parts in rheumatism.

Schleichera oleosa **(Lour.) Oken. (Sapindaceae)**

Ln. : Kusum; Loc. : Sonbhadra; Tribe : Gond

The stem bark is warmed and used as a poultice to cure rheumatic pains.

Solanum viarum **Dunal (Solanaceae)**

Ln. : Khalar Kawaoe, Khem; Loc. : Jaunsar; Tribe : Jaunsari

Fruit paste is applied to the affected parts in rheumatism and gout.

S. xanthocarpum **Schrad (Solanaceae)**

Ln. : Katli, Uwain; Loc. : Jaunsar; Tribe : Jaunsari

Fruit paste is rubbed on swellings of body.

Urtica parviflora Roxb. (Urticaceae)

Ln. : Kali, Kishki; Loc. : Jaunsar; Tribe : Jaunsari

Leaves rubbed on the affected parts in rheumatism.

CONCLUSIONS

The study provides some new indigenous herbal technology on the traditional uses of some potential plants for the treatment of rheumatism, gout and arthritis among the tribal and rural population in India. The emperical indigenous knowledge of these plants may serve as useful source of information to the chemist, pharmacologist and practioners of herbal medicines for isolation, validation and formulation of polyherbal products, useful for its sustained utilization in healthcare for human welfare.

ACKNOWLEDGEMENTS

The authors are thankful to Dr. P. Pushpangadan, Director, NBRI, for providing facilities, guidance and encouragements; authorities and officers of Forest Department, Block Development Officers for their helps and suggestions. We also express our thanks to various tribal chiefs, medicine-men, knowledgeable informers and traditional healers, who have provided the valuable information and various technical and field staff who have helped in various ways during the field work and studies.

REFERENCES

Ambasta, S.P. (1986). The useful plants of India, CSIR, New Delhi, and pp. 1-918.

Anonymous India (2001). Publication Division, Ministry of information and Broadcasting, Government of India, New Delhi - 110 001, pp. 1-873.

Anand, Prakash and Singh, K.K. (2000). Observations on some high valued ethnomedicinal plants among the tribals of Uttar Pradesh. *J. Med. Arom. Pl.,* : 519-522.

Jain, S.K. (1975). Medicinal Plants. National Book Trust, New Delhi, pp. 1-180.

Jain, S.K. (1991). Dictionary of Indian Folk medicine and Ethnobotany, Deep publications, New Delhi - 110 063, pp. 1-311.

Maheshwari, J.K., Singh, K.K. and Saha, S. (1981). The ethnobotany of the Tharus of Kheri district, Uttar Pradesh. Economic Botany Information Service, NBRI, Lucknow, pp. 1-48.

Maheshwari, J.K., Singh, K.K. and Saha, S. (1986). Ethnobotany of tribals of Mirzapur district, Uttar Pradesh. Economic Botany Information Service, NBRI, Lucknow, pp. 1- 38.

Rao, R.R. (1994). Biodiversity in India, Bishen Singh Mahendra Pal Singh , Dehra Dun, pp. 1-315.

Singh, K.K., Kalakoti, B.S. and Anand, Prakash (1994). Traditional Phytotherapy in the health care of Gond tribals of Sonnbhadra district Uttar Pradesh. *J. Bombay Nat. Hist. Soc.,* **91(3)** : 386-390.

Singh, K.K. and Anand, Prakash (1994). Indigenous Phytotherapy among the Gond tribe of Uttar Pradesh, India. *Ethnobotany,* **6(1&2)** : 37-41.

Singh, K.K. and Anand, Prakash (1996). Observations on ethnobotany of Kol tribe of Varanasi district Uttar Pradesh, India. *J. Econ. Tax. Bot.* (Add. series), **12** : 133-137.

Singh, K.K. and Anand, Prakash (1996). Traditional medicinal plant therapy used for skincare by the tribals of Uttar Pradesh, India. *J. non-timber forest product,* **3(1/2)** : 51-55.

Singh, K.K. (1997). Studies on Native Medicine of Jaunsari tribe of DehraDun District, Uttar Pradesh, India. *Int. J. Pharmacog.,* **35(2)** : 105-110.

Singh, K.K. (2000). Role of ethnoherbals in bioprospecting of safe phytomedicines and neutraceuticals for 21st century. *J. Med. Arom. Pl.,* **22/4A 23/1A** : 579-596.

Recent Advances in Medicinal Plant Research : Vision 21st Century, 2005
Editors : S.K. Prabhuji, G.P. Rao and S.K. Patil
Satish Publication House, New Delhi

Chapter 18

ETHNOMEDICINE OF HALAKKI OKKALIGAS FOR SNAKE BITE AND PEDIATRIC USE

P. Ramana*, A. Krishna and S.K. Patil
College of Forestry, Sirsi - 581 401, Karnataka, India
*e-mail : ramtarunsirsi@yahoo.com

The expensive medicines for snakebites and pediatric use are beyond the reach of the poor tribals. So, the indigenous knowledge and technologies are important natural resources that can facilitate the medicare in remote places. Hence, in the present chapter some medicinal plants identified by tribals of Uttara Kannada district, Karnataka which are being used for pediatric and also as antidote to snake bite purpose are discussed.

INTRODUCTION

Plants have been used in various folk systems for the treatment of human ailments. All ancient cultures of the world have evolved their own medicinal methods and practices to take care of health problems. Such health care practices were all culturally, socially and environmentally closure to the masses and were systematically recorded and incorporated into pharmacopoeias of many eastern cultures like India, China and become the Materia Medica of the traditional medicine (Tripathi, 1998).

Recording of information from traditional heelers in indigenous methods of treatment will go a long way in finding out locally available technologies for health care. Many times health care is not available to peoples in remote areas, because of the location of hospitals far away. Moreover, the expensive medicines for snakebites and pediatric use are beyond the reach of the poor tribals. So the indigenous technologies are important natural resources that can facilitate the development process, if moulded on scientific line. Hence, there is an urgent need for collection of indigenous knowledge on health care and use of herbal medicine in their disease control.

HISTORICAL REVIEW

Poisonous snakes have been creating fear and anxiety in the minds of the people since the dawn of the civilization. The annual mortality due to poisonous snake bites continues to be as high as 15-20 thousand in the world (Singh and Pandey, 1980). It has been estimated that every year over three lakh incidents of snake bites occur in India of which about 10,000 lead to the death of victims (Jena and Sarangi, 1993). Indian traditional medicinal system are rich in information pertaining to the effect of various types of poison and poisonous bites on the body and methods for their treatment. Ayurveda has a special branch called 'agada tantra' or 'visha tantra' dealing only with the poison of both animal and plant origin (Singh and Pandey, 1980). Only few reports concerned with scientific documentation of the indigenous knowledge on treatment snakes bites or available from India (Murthy *et al.,* 1986 and Joshi, 1993).

Use of plants as food and medicine is well known among all the ethnic groups in India. Mother and Child health acre is a priority area in Ayurveda. Hence, paediatrics is one among the main branch of ayurveda, which will help the house wife to manage the common health problems with locally available materials (Venugopal, 1992).

The present paper is an attempt in this direction to identify medicinal plants used for pediatric and also plants as antidote to snake bite purpose by the tribals of Uttara Kannada district, Karnataka.

METHODOLOGY

The data presented here is based on information gathered from medicine man, local healer, village heads, old experienced, knowledgeable man and women and from literature.

Study Area

Uttara Kannada district is a unique rich biogeographic region in the Central Western Ghat region of Karnataka, holding multifarious endemic floral and faunal wealth. It is located between 13° 55' and 15° 31' N latitude and 74° 9' and 75° 10' E longitude. It has an area of 10,007 km^2. It is about 176 km from north to south and from 16 to 96 km from east to west. The Central Kanara includes three belts, a coast tract with broad winding lagoons, rich plains and wooded hills running to the sea; a central belt of lofty Sahyadris covered with magnificent forest; and a eastern upland, which is wild waving and thickly wooded in the west and in the east passes into a bare level and thickly populated plains.

Dry deciduous, moist deciduous, semi-evergreen and evergreen forests primarily represent the vegetation of the study area. The climate is primarily monsoonal and rains affected by Southwest monsoon during June to September. The rainfall is also variable from year to year. The region receives an annual rainfall of 1933 mm with a minimum of 692 mm and a maximum of 3922 mm. The relative humidity is highest exceeding 90 per cent during July-August and it is as low as 40 per cent in March-April. There is substantial seasonally in temperature. The mean annual range temperature varies from 18° C to 31° C under, extreme condition drops

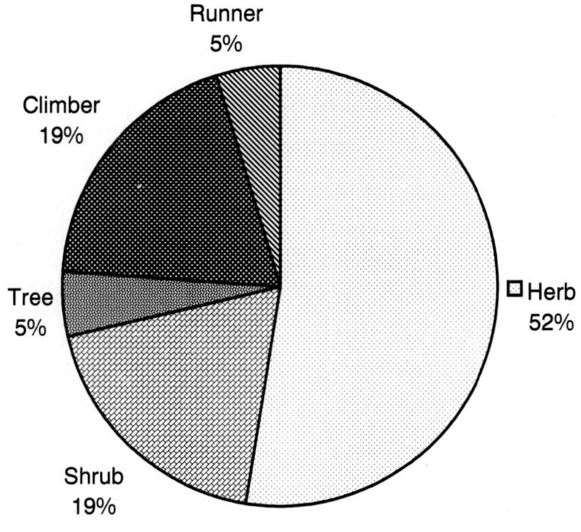

Fig. 1 : Plant types used for pediatric purpose

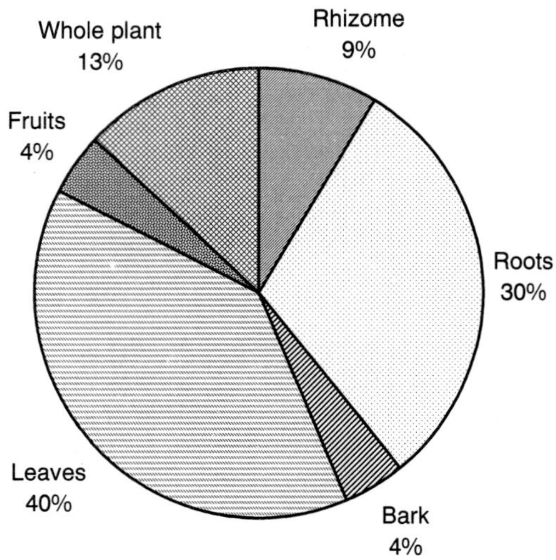

Fig. 2 : Plant parts used for pediatric purpose

down to 6.7° C and rises up to 39° C. In general, the soils are shallow to medium deep, reddish brown, dark brown and dark yellow brown in colour. The soils are acidic to slightly alkaline in nature, high in nitrogen, low in phosphorus and low to medium in potassium content (Anon., 1991).

The Tribal Population

Haalakki Okkaligas is one of the dominant tribal community of Uttara Kannada district. These communities are distributed mainly in the hilly areas of Sirsi, Yellapur, Honnavar and Ankola talukas of the district. The

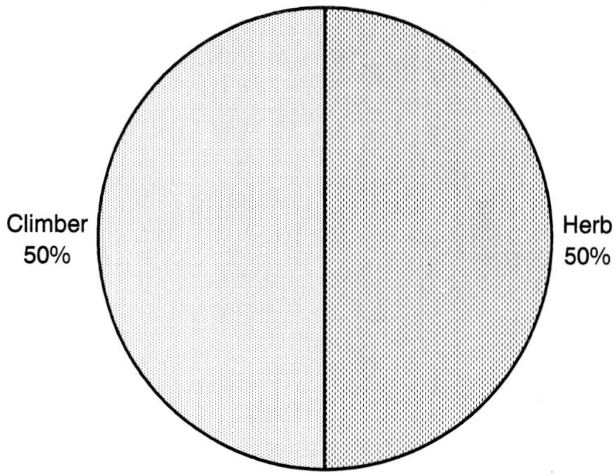

Fig. 3 : Plant type used for snake bite

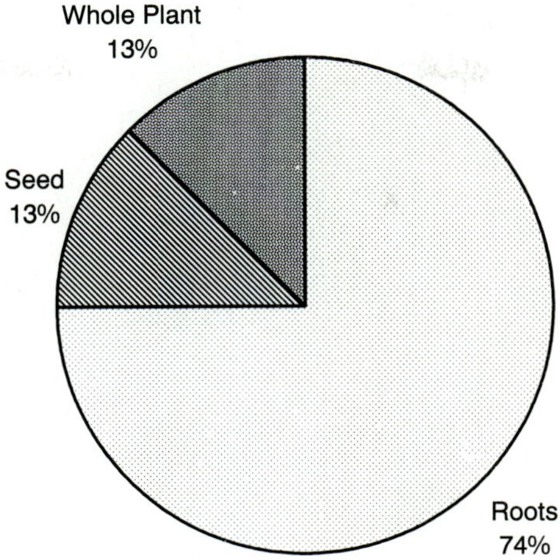

Fig. 4 : Plant parts used for snake bite

main occupation of this community is farming and livestock rearing. Most of these tribal with primitive living conditions have a direct and daily dependence on the natural resources for food, medicine, shelter, household goods, furniture and manure etc. A number of valuable medicinal plants used since time immoral and mentioned in ancient medical literature and endemic to this area. These people have accumulated considerable knowledge about plants and their medicinal utility owing to their constant intimacy with their immediate floral environment and through long experience from their treatments in the course of centuries. They are highly secretive and unwilling to reveal their medical practices due to suspicion and exploitation. The knowledge about plants is passed on from generation to generation by tribal medicine men. Many uses of the plant are not known outside their restricted community and locality. Interactions with villagers, tribal and local medicine men of the respective area revealed some important and interesting knowledge about plant drugs used by them to treat different ailments and their mode of administration.

Enumeration of Plants

In the following Table-1 and 2, botanical name is followed by the name of the family, common name, habitat, parts used, therapeutic use and prescription.

RESULTS AND DISCUSSION

An analysis of data in the present communication reveals local tribals are using the uses of twenty-one plants belonging to seventeen families for pediatric purpose and eight plants belonging to six families as antidote to

Table - 1 : Ethno medicinal plants for Pediatric Use

Sl. No.	Botanical name	Family	Common name	Habitat	Parts used	Therapeutic use	Treatment
1.	*Salvinia spp.*	Salvinaceae	Antharagange	Herb	Leaves	Bronchitis	5 fried leaves are crushed to get powder. Half spoon of the powder is mixed with one spoon honey, given thrice a day for 5 days.
2.	*Adhatoda vasica* Nees	Acanthaceae	Aadusooge	Shrub	Leaves	Cough and Cold	Leaves are macerated with water. 1 spoon of aqueous extract given once in a day for 2 to 3 days.
3.	*Costus specious* (Koen.) Sem.	Costaceae	Aarathikundige	Shrub	Rhizome /bulb	Bronchitis	Rhizome is burnt in medium fire, and then crushed to extract juice. The extracted juice is mixed with sugar. Given twice a day for 2 days.
4.	*Anthocephalus cadamba* Miq.	Rubiaceae	Aapathyadha mara	Tree	Bark	ChikenPox	Bark is macerated with water. 2 spoons of macerate are given twice a day for 9 days.
5.	*Caesalpinia crista* L.	Caesalpiniaceae	Gajagadha balli	Climber	Fruits	Hornia	4 to 5 fruits are tied to the abdomen.

Table - 1 : Continued .....

Sl. No.	Botanical name	Family	Common name	Habitat	Parts used	Therapeutic use	Treatment
6.	*Solanum indicum* L.	Solanaceae	Gullabadane	Herb	Root	Diarrhoea, dysentery	The root powder is mixed with milk and given half spoon twice a day for 4 days.
7.	*Swertia corymbosa* (Griseb.) Clarke	Gentianaceae	Chiraayata	Herb	Leaves	Abdominal complaints	The leaf decoction mixed with sugar and given twice a day for 2 days.
8.	*Pharagmites karka* (Retz.) Trin. Exstendel	Poaceae	Jeevaaladha kaddi	Herb	Rhizome	Nervous disorders	Rhizome is macerated with limejuice. Half spoon of macerated paste is given twice a day for 3 days.
9.	*Leucas aspera* (Wild.) Spreng.	Lamiaceae	Tumbegida	Shrub	Leaves	Vermifuge	Leaves are mixed with *Ocimum* and early tender coconut; the mixture is macerated with water. The macerate paste is applied on the affected part twice a day for 3 days.
10.	*Securinega leucopyrus* (Wild.) Mueller	Euphorbiaceae	Thondajjigida	Herb	Roots	Diarrhoea, dysentery	Roots are mixed with *Cyclea burmanii* in equal proportion and macerated with limejuice. 2 spoons of macerate are given twice a day for 3 days.

Table - 1 : Continued .....

Sl. No.	Botanical name	Family	Common name	Habitat	Parts used	Therapeutic use	Treatment
11.	*Elephanopus scaber* L.	Asteraceae	Nelaganagalu	Shrub	Roots	Bronchitis	Roots are mixed with *Coriander* and Jeera seeds, mixture is grinded and boiled with half-liter water to get 10 ml decoction. Then decoction is mixed with sugar, given once a day for 4 to 5 days.
12.	*Pogostemun heyneanus* Bent.	Lamiaceae	Pachhithene	Herb	Leaves	S .mach ache	Juice extracted from 4 to 5 leaves is given twice a day for 2 days.
						Apitite enhancer	Juice extracted from 4 to 5 leaves is given 4 spoons per day for 3 days.
13.	*Oxalis corniculata* L.	Oxalidaceae	Pullampurache	Runner	Whole plant	Bronchitis	Extract from whole plant is mixed with honey. 2 spoons of extract are given twice a day for 5 days.
14.	*Ixora nigricans* R.Br.	Rubiaceae	Bilikusumaale	Herb	Root	Urinary disorders	Roots are macerated with cow milk. Half spoon of macerate is given twice a day for 7 days.

Table - 1 : Continued .....

Sl. No.	Botanical name	Family	Common name	Habitat	Parts used	Therapeutic use	Treatment
15.	*Plumbago zeylanica* L.	Plumbaginaceae	Bilichitramulaa	Herb	Roots	Eczema	Crushed rocts are boiled in cow ghee. The boiled ghee is applied on affected part.
16.	*Artemisia pallens* (Wall.) et. DC	Asteraceae	Manjipathre	Herb	Leaves	Constipation	10 leaves are macerated in water. One spoon of macerate is given once in a day.
						Abdominal complaints	Half spoon of juice extracted from 5 leaves is given twice a day for 2 days.
						Blood Purifier	Half spoon of juice extracted from 5 leaves is given twice a day for 2 days.
17.	*Catharanthus roseus* (L.) Don.	Apocynaceae	Sadaapushpa	Herb	Leaves	Stomach ache	100 grams of leaves are boiled with 1 liter of water to get 250 ml. decoctions. 2 spoons of decoction are given once in a day for 8 days.
18.	*Coleus ambonicus* Lour.	Lamiaceae	Saavirasambhaara gida	Herb	Leaves	Bronchitis	Fresh juice is mixed with sugar. Juice is given twicc a day for 3 days.

Table - 1 : Continued .....

Sl. No.	Botanical name	Family	Common name	Habitat	Parts used	Therapeutic use	Treatment
						Fatigue	2 leaves are crushed with 4 leaves of rose and 4 leaves of *Crysanthemum* to get juice. The juice is taken before food.
19.	*Cyclea burmanii* Hk. & Th.	Menispermaceae	Haade balli	Climber	Whole plant	Abdominal complaints	The plant is tied to the abdomen. Used for one-year-old children only.
					Roots	Diarrhoea, dysentery	Roots are mixed with roots of *Hemidesmus indicus* in milk. The extract is given thrice a day for 2 days. Used for two-year children only.
20.	*Hemidesmus indicus* (L.) R.Br.	Asclepiadaceae	Haalu balli	Climber	Root	Cough	Roots are macerated with cow milk. 5 spoons of macerate are given thrice a day for 5 days.
21.	*Piper longum* L.	Piperaceae	Hippali balli	Climber	Whole plant	Cough	Whole plant is macerated with honey. Half spoon of macerate is given twice a day for 3 days.

Table - 2 : Ethno medicinal Plants for Snake Bite

Sl. No.	Botanical name	Family	Common name	Habitat	Parts used	Treatment
1.	*Atalantio monophylla* (Roxb.) DC	Rutaceae	Kaakamchi balli	Climber	Root	Roots are macerated with Mandarin juice. Luke warm macerate is applied on the bitten part.
2.	*Dracena ternifolia* Roxb.	Liliaceae	Kempibaeru	Herb	Root	Roots are macerated with limejuice. 1 spoon of macerate is taken orally and also applied on the affected part once in a day for 7 days.
3.	*Rawolfia serpentina* Benk ex. Kurtz	Apocyan-aceae	Garudapaat-haala	Herb	Root	Roots are mixed with roots of *Cyclea burmanii*, *Tabernamontana divaricata* and *Cyathula prostrata,* macerated with lime juice to get paste.2 spoons of paste is taken orally twice a day.
						Roots are mixed with roots of *Sapindus trifoliatus* the mixture is macerated with lime juice. The paste is applied on the bitten part.
4.	*Tabernamontana divaricata* (L.) R.Br.	Apocyn-aceae	Nandibattalu	Herb	Roots	Roots are mixed with roots of *Crotolaria calycina;* the mixture is macerated with limejuice. The paste is applied on the bitten part thrice a day for 7 days.

Table - 2 : Continued .....

Sl. No.	Botanical name	Family	Common name	Habitat	Parts used	Treatment
5.	*Abrus precatorius* L.	Fabaceae	Biligula-ganjji	Climber	Seed	Three seeds powder is mixed with 500ml water and orally taken twice a day for 2 days.
6.	*Cissus discolor* Bl.	Vitaceae	Hacchambh-arave balli	Climber	Roots	Roots are mixed with roots of *Cyathula prostrata;* the mixture is macerated with limejuice to get paste. 1 spoon of paste taken orally once in a day.
7.	*Cyclea burmanii* Hk. & Th.	Menisper-maceae	Haade balli	Climber	Root	Roots are macerated with limejuice. 2 spoons of paste is taken orally once in a day for 3 days.
8.	*Crotolaria calycina* Schrank	Fabaceaea	Hullusaa-ranga gida	Herb	Whole plant	Whole plant is macerated with 1 liter water. 1 glass of aqueous extract is orally taken twice a day for 3 days.

snake bite. The present studies yielded interesting information on the plants used in ethno-medicinal practices. The age old culture and traditions associated with flora and fauna certainly illuminate the prosperity of our ancient heritage. Alam *et al* (1996), Houghton *et al* (1993), Lewis and Lewis (1991), Seetharam *et al* (1998) and Pereira *et al* (1994) have discussed about the plants used in ethno-medicinal system.

Most of these plants are still in the mainstay of tribal population for primary health care, because of their easy accessibility and lesser side effects. It is assured that search for potential medicinal plants through ethno-botany gives clues for further research and development.

Plant drugs used in tribal and other classical system of medicine deserves detailed studies. The efficacy and safety of most of the traditional remedies as well as those of lesser-known remedies of the tribes are required to be subjected to scientific verification particularly chemical investigations (Sequeira, 1994). The traditional knowledge is valuable not merely as an indicator of plant to be screened but also equally important to evolve new methodological approach and strategies towards treatment and research.

There is an urgent need for the detailed scientific studies on some unexploited plants and also necessary steps should be taken to conserve the threatened species. So, that the traditional knowledge can be preserved for the future generations.

ACKNOWLEDGMENT

Authors are grateful to the people who have revealed the precious information on medicinal plants

REFERENCES

Alam, M.I., Auddy, B. and Gomes, A. (1996). Viper venom neutralization by Indian Medicinal Plants. *Phytotherapy Research,* **10(1)** : 58-61.

Anonymous (1991). *Gazeteer of the Bombay Presidency, Kanara District.* Volume XV, Part I, Sagardeep Publications, Karwar, p. 1-18.

Houghton, P.J., Skeri, K.P. and Tabesses, S. (1993). *Medicines and Food: The Ethno pharmacological Approach,* 2nd European colloquium on Ethno pharmacology, 24-27 March, Heidelburg, Germany.

Jain, S.K. (1991). Dictionary of Indian Ethno botany and folk medicine. Deep Publishers, New Delhi.

Jena, I. and Sarangi, A. (1993). Snakes of medicinal importance and snake-bite treatment. Ashish Publishing House, New Delhi.

Lewis, W.H. and Elvin-Lewis, M.P. (1991). *Pentagonia gigantofolia* as a snakebite remedy, Empirical methodology functioning in Amazonian traditional medicine. *Economic Botany,* **45(1)** : 137-138.

Murthy, K.S., Sharma, P.C. and Premkishore (1986). Tribal remedies for snake-bite from Orissa. *Anc Sci Life*, **6** : 122-123.

Pereira, N.A., Pereira, B.M.R., do Nascimento, M.C., Parente, J.P. and Mors, W.B. (1994). Pharmacological screening of plants recommended by folk medicine as snake venom antidotes; IV. Protection against Jararaca venom by isolated constituents. *Plant Medica*, **60(2)** : 99-100.

Seetharam, Y. N., Haleshi, C. and Vijay (1998). Medicinal Plants of North–Eastern Karnataka and their status, *My Forest*, **34(2)** : 767-772.

Sequeira, V. (1994). Medicinal plants and conservation in Soa Tome. *Biodive. Conser.*, **3** : 910-926.

Singh, V.P. and Pandey, R.P. (1980). Medicinal Plant lore of the tribal of Eastern Rajasthan. *J. Econ. Taxon. Bot.*, **1** : 137-147.

Tripathi, Y.C (1998). Advances in Phytotherapeutic Research. *J. Non-Timber For. Prod.*, **5** : 37-43.

Venugopal, S.N. (1992). Medicinal Plants useful in Mother and Child Health. Department of Family Welfare, Minitry of Health & Family Welfare, Govt. of India, New Delhi.

Recent Advances in Medicinal Plant Research : Vision 21st Century, 2005
Editors : S.K. Prabhuji, G.P. Rao and S.K. Patil
Satish Publication House, New Delhi

Chapter 19

STUDY OF LITHONTRIPTIC PLANTS USED BY THE TRIBALS OF KHARGONE DISTRICT OF MADHYA PRADESH

S.K. Mahajan and Pushpa Patel
Post graduate Department of Botany, Govt. College, Khargone - 451 001, (M.P.) India

The present paper deals with 24 taxa belonging to 22 genera and 18 families that are used as antilithic plants by the tribals of Khargone district of M.P. Important suggestions are given to protect and propogate these valuable plants.

INTRODUCTION

Lithontriptic (Lithotriptic or antilithic) plants are used to dissolve or remove caliculi or stones present in gall bladder or urinary system of human beings. The formation of caliculi or stone is known as Lithiasis. In India, about 47 taxa are known to possess antilithic properties. Earlier, a considerable ethnobotanical work has been done in India in general and M.P. in particular (Jain, 1963; Jain and Tarafdar, 1968; Maheshwari *et al.*, 1981; Oomachan and Masih, 1991; Kumar and Yadav, 2004). In M.P. such type of work has been limited only to few districts (Balaghat, Bastar, Chhindwara, Mandla and Sidhi). So far as Khargone district in concerned, a little work has been done (Mahajan, 1990, 1995; Mahajan and Patel, 2002, 2003). Hence this work was undertaken which deals with the antilithic plants used by the tribals (Barela, Mankar, Bhilala and Korku) of Khargone district.

METHODOLOGY

The present work was carried out during 2002-2003 and for this purpose four tribal villages were surveyed (Pipaljhopa, Sirwel, Bhagwanpura and Vijaygarh). Old experienced and tribal medicinemen (Barwa) were contacted to get the firsthand information regarding such plants which are used by the tribals to dissolve caliculi or stones present in gall bladder or urinary system of human body. Herbarium sheets of such plants were prepared and

Plate I : Mr. Shiv Narayan Nhal Tribal medicineman - "Barwa" (Bhagawanpura Tehsil).

identification of plants was done following standard literature (Cooke, 1957; Kirtikar and Basu, 1935; and Solanki, 1984). These plants are enumerated below in alphabetical order with their scientific names, synonyms (if any), local names, name of family and plant parts used for this purpose. The herbarium sheets were deposited in Botany Deptt. of Govt. P.G. College, Khargone for future reference (Plate I - II).

ENUMERATATION OF PLANTS

1. *Achyranthes aspera* L. Local name - Hathijhara, Family - Amaranthaceae. Use - Plant ash is used.
2. *Ageratum conyzoides* L. Local name - Gandhari, Family - Asteraceae. Use - Plant extract is used to dissolve gall bladder stone.
3. *Argemone mexicana* L. Local name - Pili Katari, Family - Papaveraceae. Use - Root decoction is given in vesicular caliculus.
4. *Asteracantha longifolia* N. Local name - Ikshura, Family - Acanthaceae. Use - Root is used to dissolve stone or gravel in the kidney.
5. *Asparagus adscendens* R. Local name - Safed musli, Family - Liliaceae. Use - Fleshy roots are used in stone trouble.

Plate II : *Ocimum basilicum* L. plant growing in its natural habitat, with flowers and fruits.

6. *Azadirachta indica* L. Local name - Neem, Family - Meliaceae. Use - Leaves are used to remove caliculus or stone.

7. *Bryophyllum pinnatum* S. Local name - Ghamari, Family - Crassulaceae. Use - Leaf juice is used in caliculus affections.

8. *Butea monosperma* L. Local name - Palas, Family - Papilionaceae (Fabaceae). Use - Decoction of seeds is given with nitre in gravel complaints.

9. *Coriandrum sativum* L. Local name - Dhania, Family - Umbelliferae (Apiaceae). Use - Plant extract is used in caliculus complaints and urinary disorder.

10. *Cuminum cyminum* L. Local name - Jira, Family - Umbelliferae (Apiaceae). Use - Fruits are used with Jaggery for stone trouble.

11. *Cynodon dactylon* P. Local name - Doob, Family - Poaceae. Use - Root decoction is used in vesicular caliculus.

12. *Dolichos biflorus* L. Local name - Kulthi, Family - Papilionaceae (Fabaceae). Use - Seeds are lithontriptic.

13. *Lawsonia inermis* L. Local name - Mhendi, Family - Lythraceae. Use - Bark is used in urinary caliculi.

14. *Moringa olefera* L. Local name - Sigru, Family - Moringaceae. Use - Flower juice is antilithic.

15. *Ocimum basilicum* L. var. *thysiflora* Local name - Ban tulsi, Family - Lamiaceae. Use - Plant extract is taken.

16. *Solanum surattense* B. (Syn. *S. xanthocarpum*) Local name - Bhatkatai, Family - Solanaceae. Use - Fruits or roots are used in stone trouble of bladder.

17. *Saraca indica* L. Local name - Ashok, Family - Caesalpiniaceae. Use - Seed powder with water is used to remove kidney stone.

18. *Syzygium cuminii* S. Local name - Bara jamun, Family - Myrtaceae. Use - Fruit juice with 'Sendhia namak' is used in urinary stone.

19. *Tephrosea purpurea* P. Local name - Sharponkha, Family - Papilionaceae (Fabaceae). Use - One and a half ounce of leaf juice is taken with sugar in caliculus affections.

20. *Terminalia arjuna* W & A. Local name - Arjun, Family - Combretaceae. Use - Plant juice is used in kidney stone trouble.

21. *Terminalia balerica* R. Local name - Bahera, Family - Combretaceae. Use - Fruit is given in caliculus trouble.

22. *Terminalia chebula* W & A. Local name - Harad, Family - Combretaceae. Use - Fruit is given in caliculus trouble.

23. *Tribulus terrestris* L. Local name - Chhota gokharu, Family - Zygophyllaceae. Use - Fruit extract is used in urinary caliculus.

24. *Tinospora cordifolia* M. Local name - Gurvel, Family - Menispermaceae. Use - Plant juice is used in kidney stone trouble.

RESULTS AND DISCUSSION

The study revealed that in all 24 taxa belonging to 22 genera and 18 families of angiosperms are used as antilithics by the tribals of Khargone district. Important genera that are used as antilithics are *Asparagus, Bryophyllum, Solanum, Terminalia* and *Tinospora.* These valuable taxa need protection and conservation for future use. Forest department and local administration may take action to protect these plants. Further work in this direction is in progress.

ACKNOWLEDGEMENTS

Authors are thankful to the Principal, Govt. P.G. College, Khargone for facilities and encouragement.

REFERENCES

Cooke, T. (1957). *Flora of the Presidency of Bombay.* Vol. I-II.

Jain, S.K. (1963). Plants used in medicine by tribals of Madhya Pradesh. *Bull. Reg. Res. Lab.,* Jammu-Tawi, **1** : 126-128.

Jain, S.K. and Tarafdar, C.R. (1968). Native plant remedies for snakebite among adivasis of Central India. *Ind. Med. J.,* **57(12)** : 307-309

Kirtikar, K.R. and Basu, B.D. (1935). *Indian Medicinal Plants,* Calcutta.

Kumar, A. and Yadav, D.K. (2004). *Role of traditional beliefs in conservation of sacred groves in Magadh region.* In "Recent Advances in Environmental Biology and Ethnobotany" (ed.) R.K. Mishra, System and Software, C-Sector, Itanagar, A.P., p. 1-5.

Mahajan, S.K. (1990). A preliminary Ethnobotanical survey in the Sirwel hills of Western M.P. Natl. Seminar on Ethnobotany of tribes of M.P., organised by S.G.S. Govt. P.G. College, Sidhi, M.P. and sponsored by MAPCOST, Bhopal. Abst. 14.

Mahajan, S.K. (1995).. A preliminary ethonobotanical survey in the Bijagarh hills, M.P. A seminar on recent researches on Science and Technology, organised by Karimganj under the ageis of Assam University, Silchar. Abst. p. 4.

Mahajan, S.K. and Patel, P. (2002a). Antiseptic plants of Khargone and its neighbourhood areas of Madhya Pradesh; A seminar on Plant Taxonomy in National Developement. Deptt. of Botany, Goa Univ. Abst. p. 73.

Mahajan, S.K. and Patel, P. (2002b). A preliminary survey of anthelmintic plants used by the tribes of West Nimar district, Madhya Pradesh. Ibid Abst. p. 74.

Mahajan, S.K. and Patel, P. (2003). A preliminary survey of diuretic plants of west nimar district M.P. An International symposium on plant systematics, Advances and Relavances Abst. p.

Maheshwari, J.K., Singh, K.K. and Saha, S. (1981). The ethnobotany of Tharus of Kheri district, U.P. Economic Botany Information Service, N.B.R.I., Lucknow.

Oomachan and Masih, S.K. (1991). Ethnobotanical and conservational aspects of medicianal plants of Madhya Pradesh. *Indian J. Applied and Pure Biol.*, **6(1)** : 39-44.

Solanki, C.M. (1984). *Flora of Indore.* Ph.D. Thesis, Devi Ahilya Vishwavidyalaya.

Recent Advances in Medicinal Plant Research : Vision 21st Century, 2005
Editors : S.K. Prabhuji, G.P. Rao and S.K. Patil
Satish Publication House, New Delhi

Chapter 20

ETHNO-VETERINARY PLANTS USED BY HALAKKI OKKALIGAS OF CENTRAL WESTERN GHATS REGION OF KARNATAKA

P. Ramana*, **A. Krishna and Siddappa Kannur**
College of Forestry, Sirsi - 581 401, Karnataka, India
*e-mail : ramtarunsirsi@yahoo.com

This chapter enlists thirty-six plants of twenty-five families used as veterinary medicine along with description for curing various aliments in cattle. Besides this, stress has also been laid for the need to mould the indigenous technologies for veterinary care on scientific lines for the benefit of farmers living in remote area where modern veterinary facilities are not available.

INTRODUCTION

Recording of information from traditional heelers in indigenous methods of treatment will go a long way in finding out locally available technologies for veterinary care. Many times' veterinary care is not available to farmers in remote areas, because of the large population of the animals in a particular block and the location of hospitals far away. Moreover, the expensive medicines for live stock are behind the reach of the poor farmers. So the indigenous technologies are an important natural resources that can facilitate the development process, if molded on scientific line. Hence there is an urgent need for collection of indigenous knowledge on live stock and use of herbal medicine in their disease control.

HISTORICAL REVIEWS

Though a large amount of information has been documented with regard to herbs for using human medicine, search documentation is lacking in veterinary medicine. The age old culture and traditions associated with flora and fauna certainly illuminate the prosperity of our ancient heritage. Girach *et al* (1998), Saiprasad and Pullaih (1996), Seetharam and Vijay (1998), Seetharam *et al.* (1998), Sumer Chandra (2000) and Sharma and Singh (2001) have discussed about the plants used in folk and ethno-veterinary system. The present paper is an attempt in this direction and enlists thirty six such plants commonly used in veterinary medicine by tribals of Uttara Kannada district, Karnataka.

METHODOLOGY

The data presented here is based on information gathered from medicine man, local healer, village heads, old experienced, knowledgeable man and women and from literature.

Study Area

Uttara Kannada district is a unique rich biogeographic region in the Central Western Ghat region of Karnataka, holding multifarious endemic floral and faunal wealth. It is located between 13 ^{0} 55' and 15^{0} 31' N latitude and 74^{0} 9' and 75^{0} 10' E longitude. It has an area of 10,007 km^{2}. It is about 176 km from north to south and from 16 to 96 km from east to west. The Central Kanara includes three belts, a coast tract with broad winding lagoons, rich plains and wooded hills running to the sea; a central belt of lofty Sahyadris covered with magnificent forest; and a eastern upland, which is wild waving and thickly wooded in the west and in the east passes into a bare level and thickly populated plains.

Dry deciduous, moist deciduous, semi-evergreen and evergreen forests primarily represent the vegetation of the study area. The climate is primarily monsoonal and rains affected by Southwest monsoon during June to September. The rainfall is also variable from year to year. The region received an annual rainfall of 1933 mm with a minimum of 692 mm and a maximum of 3922 mm. The length of dry season varied from a mean of three months at higher altitudes in South to eight months on eastern side of the Ghat in the north. The relative humidity is highest exceeding 90 per cent during July-August and it is as low as 40 per cent in March-April. There is substantial seasonally in temperature. The mean annual range temperature varies from 18^0 C to 31^0 C under extreme condition drops down to 6.7^0 C and rises up to 39^0 C. In general, the soils are shallow to medium deep, reddish brown, dark brown and dark yellow brown in colour. The soils are acidic to slightly alkaline in nature, high in nitrogen, low in phosphorus and low to medium in potassium content (Anonymous, 1991)

The Tribal Population

The data presented in this paper are the outcome of intensive studies conducted among Halakki Okkaligas, the local traditional medicine practioners. Local herbal informants were selected and interviewed extensively. The women chieftains are accorded a significant role in discussions since they possess more cognizance about the utility of local herbal products in curing various diseases. Data on the ethno herbal medicines along with their application were gathered from experienced and knowledgeable medical men and women.

Enumeration of plants

In the following Table-1, botanical name is followed by the name of the family, common name, habitat, flowering/fruiting season, parts used and veterinary uses.

RESULTS AND DISCUSSION

An analysis of data in the present communication reveals the uses of thirty six species belonging to twenty-five families for veterinary practices among local tribes. The present studies yielded interesting information on the plants used in ethno-veterinary practices. The reported data were compared with those gathered in other ethnobotanical studies carried out in Uttara Kannada earlier (Thakru Sukru Gouda, 1997) The age old culture and traditions associated with flora and fauna certainly illuminate the prosperity of our ancient heritage. Girach *et al.* (1998), Saiprasad and Pullaih (1996), Seetharam and Vijay (1998), Seetharam *et al.* (1998), Sumer Chandra (2000) and Sharma and Singh (2001) have discussed about the plants used in folk and ethno-veterinary system.

Most of these plants are still in the mainstay of our population for primary health care of cattle, because of their easy accessibility and lesser side

Table - 1 : Ethno-veterinary medicinal plants of Uttara Kannada District, Karnataka

Botanical name	Family	Common name	Habitat	Flowering/ fruiting season	Parts used	Veterinary Uses	Treatment
Justica simplex Don	Acanthaceae	Elavusandhigida	Herb	Dec	Leaves	For broken bones	100 g of whole plant macerated with small amount of water. Macerate is mixed with ½ l curd. Given orally once a day for 3 days.
Holarrhena antidysentrica (Roth.) D.C	Apocynaceae	Kurchi, Karchi, Kura	Shrub	Sept.-Nov.	Root	To expel intestinal worms in colfs	Roots are macerated with 20 g of sulphur. Given orally once in a week.
Ervatania heyneana (Wall). Cooke	Apocynaceae	Nagkunda	Shrub	Jan./April	Root and bark	For diarrhea	150 g of bark powder is mixed with ½ l water. Given orally
Lagenandra ovata (L.) Thw	Araceae	Rukh alu	Herb	Feb.-Mar.	Roots	For foot and mouth disease	Macerate 100 g roots, *Sapandus, Elaegnus*

Table - 1 : Continued .....

Botanical name	Family	Common name	Habitat	Flowering/ fruiting season	Parts used	Veterinary Uses	Treatment
							conferta bark and 10 pepper seeds with ½ l water. Filtrate is given thrice a day
Remusatia vivipera	Araceae	Rukalu	Herb	June-Sept.	Rhizome	For curing collection of pus and for broken bones	100 g of small pieces of rhizome are mixed with ghee. Given orally
Aristolochia indica L.	Aristolo-chiacae	Ishwari mul	Climber	Aug.-Oct.	Root	For dental disorders	Roots are macerated with lime juice. Macerate is given orally twice a day for 5 days
Calotropis gigantea (L.) Dryand	Asclipiada-cea	Ak	Shrub	Annual	Leaves, milk and root	For diarrhea and for foot and mouth disease	100 g of leaves are macerated in 1 l of water. Boil to get ½ l

Table - 1 : Continued .....

Botanical name	Family	Common name	Habitat	Flowering/ fruiting season	Parts used	Veterinary Uses	Treatment
							decoction, decoction given thrice a day
Gymnema sylvestre (Retz.) Schultes	Asclipiada-cea	Gur-mara, Mersingi	Creaper	Oct.-Nov.	Leaves and bark	For fever	10 g bark, 50 bark of *Ervatamia heyneana*, 5 g *Anacardium oxidentale*, 20 g of pepper seeds are powdered and mixed with honey. Given twice a day
Wattakaka valubilis (l.f) Stapf	Asclipiada-cea	Nakchhikni	Creaper	April-May	Leaves and root	For curing collection of pus	Macerate 100 g of leaves and 8 pepper seeds. Paste is applied on affected parts.

Table - 1 : Continued .....

Botanical name	Family	Common name	Habitat	Flowering/ fruiting season	Parts used	Veterinary Uses	Treatment
Elephantopus scaber L.	Asteraceae	Gobhi, Samudulan	Creeper	Sept.-Nov.	Root	To stimulate milk production	1 kg small pieces of whole plant is mixed with rice cake. Given for 20 days
Artemisia pallens (Wall) et. DC	Asteraceae	Davana	Herb	-	Leaves	To expel intestinal worms	100 g of leaves are macerated in 1 l of water. Boil to get ½ l decoction, decoction given thrice a day
Celastrus paniculatus Willd	Celastraceae	Malkangni	Creeper	Aug.-Sept.	Root and seeds	For gas trouble and to prevent premature delivary	Macerate 250 g of bark, 1 ginger with 4 l water and boil to get 1 l decoction. Decoction given twice a day for 5 days

Table - 1 : Continued .....

Botanical name	Family	Common name	Habitat	Flowering/ fruiting season	Parts used	Veterinary Uses	Treatment
Costus specious (Koen) Sm.	Costaceae	Keu	Shrub	Aug.-Sept.	Stem and rhizome	To prevent premature delivary	Bark & rhizome are macerated with 10 g of rice. The macerate is given once in a week
Momordica dioica	Cucurbitac-eae	Kaksa, Golkandra	Creeper	Aug.-Sept.	Leaves, fruit and rhizome	For insect bite	Macerate the rhizome to get juice. Juice is applied on the body
Breynia retusa (Dennst.) Alst	Euphorbia-ceae	Kandoyi	Herb	July-Aug.	Leaves	To prevent pus formation	½ kg leaves and *Clerodendrum viscosum* leaves are fried separately. Powdered and is applied on affected part
Antidesma diandrum Schl.	Euphorbia-ceae	Aamatie	Shrub	Sept.-Oct.	Bark	As nutrition	½ kg bark, ¼ kgUrd dal, ¼

Table - 1 : Continued .....

Botanical name	Family	Common name	Habitat	Flowering/ fruiting season	Parts used	Veterinary Uses	Treatment
							kg menthe and 1 coconut are macerated. Macerate is mixed with ¼ kg ground nut oil and egg, the paste is given orally once in a year
Bridelia scandens (Roxb.) Willd	Euphorbia-ceae	Chikana	Shrub	Jan.-Feb.	Bark and leaves	For diarrhea	1 kg small pieces of leaves given along with fodder, twice a day for three days
Putranjiva roxburghii Wall.	Euphorbia-ceae	Putranjiva, Juti	Tree	Mar.-May	Leaves, bark and seeds	To prevent premature delivary	1 kg leaves are macerated with starch and given once in a day for 15 days

Table - 1 : Continued .....

Botanical name	Family	Common name	Habitat	Flowering/ fruiting season	Parts used	Veterinary Uses	Treatment
Abrus precatorius L.	Fabaceae	Ghungchi, Rati	Creeper	Aug.-Nov.	Leaves and seeds	As vasoconstrictor	4 seeds, 1 spoon jeera, 4 clove, 1 almond, 1 piece of camphor fried and powdered. Powder is mixed with 100 g of sugar, given orally
Leucas aspera (Willd) Spreng	Lamiaceae	Chota halkusa	Herb	Aug.-Sept.	Leaves	To prevent swelling in bottom part of the neck	Macerate 100 g of leaves,100 g of *Vitex nigundo* leaves with ½ l water. Filtrate given through nasal passage
Asparagus racemosus Willd	Liliaceae	Satawar	Creeper	Aug.-Sept.	Rhizome	To stimulate milk production	100 g small pieces of rhizome is mixed with cotton seed cake. Given orally thrice a week

Table - 1 : Continued .....

Botanical name	Family	Common name	Habitat	Flowering/ fruiting season	Parts used	Veterinary Uses	Treatment
Leea indica Merr.	Leeaceae	Kurkur, -jihwa	Shrub	July-Oct.	Leaves, bark and root	For amiboic dysentry	100 g of tender leaves macerated in butter. Macerate is given orally
Tinospora cordifolia (Willd) Miers	Menisper-maceae	Amrita, Jivantica	Creeper	Dec.-Jan.	Bark and flowers	For muscle pains	5 spoons of bark extract from 40 g of bark , given twice a day for 2 days
Coscinium fenestratum maceae	Menisper-	Jhar-I-huldi	Creeper	Sept.-Oct.	Bark	For respiratory problems	50 g bark, 50 g of *Elaegnus conferta*, 50 g of *Emblia ribes*, 50 g of *Ervatamia heyneana* , 50 g of *Casunerina equisitifolia*, 1 garlic are macerated. Macerate is given orally.

Table - 1 : Continued .....

Botanical name	**Family**	**Common name**	**Habitat**	**Flowering/ fruiting season**	**Parts used**	**Veterinary Uses**	**Treatment**
Musa superba	Musaceae	Jangalekelu	Shrub		Seeds	To prevent premature delivary	Mix 50 g seed powder in boiled water. Given orally once a day for 3 days
Mimosa pudica (L.)	Mimosaceae	Lajwanti, Chui-mui	Creeper	Sept.-Oct.	Roots and leaves	To stimulate milk production and to facilitate delivary	½ g of small pieces from whole plant is given along with fodder. Given once in a day for 7 days
Syzygium jambos (L.)	Myrtaceae	Gulabjamba	Shrub	April-May	Root and bark	For diarrhea	Macerate 100 g of bark with 4 l of water, boil it to get 1 l decoction. Decoction is mixed with cotton seed cake and given once in a day for 3 days

Table - 1 : Continued .....

Botanical name	Family	Common name	Habitat	Flowering/ fruiting season	Parts used	Veterinary Uses	Treatment
Vetiveria zizanoids (L.) Nash	Poaceae	Khas-khas, Bena	Shrub	Annual	Root	For physical wounds	Powdered root is mixed with coconut oil and boiled. Paste is applied on affected parts
Pholidota Spp.	Orchidaceae	Marabaalekai	Epyphite	Aug.-Oct.	Fruit	To stimulate milk production and leg wounds	10 –15 fruits are macerated and mixed with 2 l water. Macerate is mixed with sugar and given orally
Anthocephalus cadamba Mig	Rubiaceae	Kadamba	Tree	Aug.-Sept.	Bark	For leg disorder and for fever	Bark is macerated with 10 g of rice. Macerate is mixed with ½ l water, given once a day for 3 days

Table - 1 : Continued .....

Botanical name	Family	Common name	Habitat	Flowering/ fruiting season	Parts used	Veterinary Uses	Treatment
Gardenia resinifera Roth.	Rubiaceae	Mar	Shrub	Mar.-June	Leaves, gum and root	For amiboic dysentry	Macerate 1 kg of leaves with 2 spoons of edible oil and sugar. Macerate is given orally twice a day
Aegle marmelos (L.) Correa	Rutaceae	Bel	Tree	April-May	Leaves, bark and fruit	To prevent premature delivary	100 g of leaves, 100 g of *Celastrus paniculatus* and 100 g of *Putrangiva roxburghii* are cut into small pieces. Mixed with cotton seed cake. Given once in a day for 7 days
Naravelia zeylanica (L.) DC	Ranuncul-aceae	Chagal-bati, Murcha	Creeper	Oct.-Nov.	Root	For inflam-mation of the joints	Macerateleaves with 1 white onion and 4

Table - 1 : Continued .....

Botanical name	Family	Common name	Habitat	Flowering/ fruiting season	Parts used	Veterinary Uses	Treatment
							pepper seeds. Macerate is given through nasal passage
Vitex negundo L.	Verbenaceae	Sambalu, Shivari	Shrub	Aug.-Sept.	Root, leaves and bark	For prevention of puss formation	100 g of leaves are macerated. Macerate is applied on affected part
Cissus discolor Bl.	Vitaceae	Thelichaavaela	Creeper	Sept.-Nov.	Leaves, stem and root	For amiboic dysentry and dysentry associated with blood	Macerate 100 g of leaves in water, fitrate is given orally
Cissus rependa	Vitaceae	Diboria	Creeper	Sept.-Nov.	Leaves	For foot and mouth disease	Bark free stem is cut into small pieces are fried later macerate. Macerate is applied on the affccted parts

effects. It is assured that search for potential medicinal plants through ethnobotany gives clues for further research and development.

Plant drugs used in tribal and other classical system of medicine deserves detailed studies. The efficacy and safety of most of the traditional remedies as well as those of lesser known remedies of the tribes are required to be subjected to scientific verification particularly chemical investigations (Sequeira, 1994). The traditional knowledge is valuable not merely as an indicator of pant to be screened but also equally important to evolve new methodological approach and strategies towards treatment and research.

There is an urgent need for the detailed scientific studies on some unexploited plants and also necessary steps should be taken to conserve the threatened species. So, that the traditional knowledge can be preserved for the future generations.

ACKNOWLEDGMENTS

Authors are grateful to the people who have revealed the precious information on medicinal plants

REFERENCES

Anonymous (1991). *Gazeteer of the Bombay Presidency, Kanara District.* Volume XV, Part I, Sagardeep Publications, Karwar, p. 1-18.

Girach, R.D., Bhamhan, M. and Misra, M.K. (1998). Folk Veterinary herbal medicines of Bhaadrak District, Orissa, India. *Ethnobotany*, **10** : 85-88.

Saiprasad, P. and Pullaiah, T. (1996). Folk-Veterinary medicines of Kurnool District, A.P., *Ethnobotany*, **8** : 71-74.

Seetharam, Y.N. and Vijay (1998). Folk and Ethno-Veterinary medicinal plants of Bidar District. *Myforest*, **34(4)** : 917-919.

Seetharam, Y.N., Haleshi, C. and Vijay (1998). Medicinal plants of North–eastern Karnataka and their status, *Myforest*, **34(2)** : 767-772.

Sequeira, V. (1994). Medicinal plants and conservation in Soa Tome. *Biodive. Conser.*, **3** : 910-926.

Sharma, P.P. and Singh, N.P. (2001). Plants in Ethno-veterinary practces in Dadra and Nagar Haveli (VT). *Journal of Non-Timber Forest Products*, **8(3/4)** : 150-153.

Sumer, Chandra (2000). Common Ethno-Veterinary plants. *Journal of Non-Timber Forest Product*, **7(1/2)** : 116-122.

Thakru Sukru Gouda (1997). Karnataka Oushadi Sasyagala Prampare-1, Uttara Kannadha Halakki Okkaligaru INCERT, Bangalore.

Recent Advances in Medicinal Plant Research : Vision 21st Century, 2005
Editors : S.K. Prabhuji, G.P. Rao and S.K. Patil
Satish Publication House, New Delhi

Chapter 21

BIRTH CONTROL PRACTICE AMONG RURAL AND TRIBAL WOMEN OF CHHATARPUR DISTRICT (M.P.)

Amita Arjariya[1] and Anjali Rawat[2]
[1]*Department of Botany, Govt. Autonomous Maharaja College, Chhatarpur Madhya Pradesh*
[2]*University Teaching Department, U.T.D. Sagar, Madhya Pradesh*

The present communication reports first hand information gathered on 20 medicinal plants traditionally used by rural and tribal woman of Chhatarpur district (M.P.) for the birth control and various disorders. The methods adopted by the rural and tribal woman for birth control have been discussed in the present paper. The study provides information on local names, locality ethanomedicinal recipes, dosages and made of administrations as practised by rural and tribal woman.

INTRODUCTION

The present paper provides information on 20 plants that are commonly used for bith control by many tribal communities and rural populations in different parts of India.

Data on falk medicinal uses of plants were obtained through interviews of local Vaidya, Hakeems, Bharat Medicine-men and old ladies among tribal population in the course of ethanobotinal survey.

PLANTS WITH ANTIFERTILITY AND ABORTIFICANT PROPERTIES

In the following listing, the plants have been grouped under two catagories one as antifertility agents and plants as abortificants. Data are presented in following sequence botanical name, family, local name and uses.

Antifertility Character Possessing Plants

1. ***Azardirachta indica* (A.); Fam. : Meliaceae 'Neem'**

   5-10 ml oil of neem seeds insert into vegina during intercourse; induces sterility.

2. ***Buchanania lanjan* (Sprong); Fam. : Anacardiaceae 'Gumchi'**

   3-5 seeds powder with milk after mensuration take orally in the morning; induces sterility.

3. ***Bombyx ceiba* (L.); Fam. : Bombacaceae 'Semal'**

   5-6 dried capsules epicarp of semal and equal amount of *Pistacia-chinensis* after grinding take two times daily with banana pulp upto ten days; induces sterility in man.

4. ***Bute amonosperma* (L.); Fam. : Fabaceae 'Palash'**

   3-5 g seed powder of Palash with the help of water at the night upto 2-5 days after mensuration; induces sterility in woman.

5. ***Cuscuta reflexa* (R); Fam. : Cuscutaceae 'Amarbel'**

   Juice of whole plant given to woman once only after menses claims to make the woman barren for ever.

6. ***Curcuma longa* (L.); Fam. : Zingiberaceae 'Haldi'**

   10-15 g of turmeric powder taken with water two times a day for 5 days after menses; induces sterility in woman.

7. ***Euphorbia orbiculata* (L.); Fam. : Euphorbiaceae 'Choti dudhi'**

   50-100 g of whole plant ground and mixed in 250 ml of cow's milk, and taken once daily from 5th day after menses to 15th day; induce sterility in woman.

8. ***Ficus religiosa*** **(L.); Fam. : Moraceae 'Peepal'**

10-15 receptacles taken with local liquor just before one week of menses is used to check fertility.

9. ***Foeniculum vulgare*** **(Mill); Fam. : Apiaceae 'Saunf'**

25-50 g powder of fruits given with hot water once daily in the morning for 7-10 days after menses, induces sterility.

10. ***Lounea coromondelica*** **(H); Fam. : Anacardiaceae 'Gunjan'**

Stem bark juice of Gunjan and flower's liquor of *Madhuca longifolia* mixed in equal quantity; 25 ml of this taken twice daily upto 10-25 days after menses, induces sterility.

11. ***Mentha arvensis*** **(L.); Fam. : Lamiaceae 'Podina'**

About 20 ml juice of mentha plant given to woman just before intercourse is induced sterility.

12. ***Mentha longifolia*** **(L.); Fam. : Lamiaceae 'Podina'**

5 g leaves paste mixed with candy sugar (Misri) and given once daily for 4 days after menses. Repeat after every menses.

13. ***Mimosa pudica*** **(L.); Fam. : Minosoideae 'Lajwanti'**

12 to 15 g powdered roots mixed with candy sugar and are given twice daily for 3 days after menses.

14. ***Ricinus communis*** **(L.); Fam. : Euphorbiaceae 'Arand'**

One seed of *Ricinus communis* taken daily from the first day of menses upto seven to ten days make the woman sterile.

15. ***Soymida febrituga*** **(A.); Fam. : Meliaceae "Rohan'**

Decoction of 500 g stem bark mixed with 1 g sulphur powder and coal mixture drunk once daily for three days the same recipe repeated at each menses produces sterility.

16. ***Vitex negunda*** **(L.); Fam. : Verbenaceae 'Meodi'**

12-15 g seed powder given three times a day upto 15 days makes the man sterile.

Abortificant Character Possessing Plants

1. ***Achyranthus aspera*** **(L.); Fam. : Amaranthaceae 'Chidchida'**

Paste of roots inserted into vagina of pregnant lady and left 2-3 hrs. causes bleeding and subsequent abortion.

2. *Cassia tora* (L.); **Fam. : Casalpiniaceae 'Chikoda'**

4-5 seeds of *Cassia tora* with hot water taken twice a day upto 5 days, bring about abortion of two/three months.

3. *Carica papaya* (L.); **Fam. : Caricaceae 'Papita'**

Latex of unripe fruit is dried, and given in the dose of 5 g with cow's milk twice daily 10 days. The dried milky juice is also used as suppository inside vagina to bring about abortion of two-three month's pregnancy.

4. *Ruellia tuberosa* (L.); **Fam. : Acanthaceae "Dabori'**

Paste of 2-3 roots inserted into the vagina and left overnight, causes bleeding and subsequent abortion.

DISCUSSION

The data were compared with important ancient and recently published papers, reports and magazines (Agrawal, 1986; Basu, 1987; Choudhari, 1993; Day, 1896; Kirtikar and Basu, 1935; Nadkarni, 1954; Garg, 2004; Singh and Ali, 1994). Most of the uses reported here are contempory and do not seem to have been reported earlier.

REFERENCES

Agrawal, V.S. (1986). Economic plants of India. Kailas Prakashan, Calcutta.

Basu, A.K. (1987). Folk beliefs and practices in producing sterlity and abortion (accidental) and for the treatment of sterility herbs and drugs (herbal) which are to be axoided. First Int. Sem. Unani. Med. New Delhi Abst. 61.

Choudhari, R.R. (1993). Quest for a herbal contraceptive. *National Med. J. India,* 6 : 199-201.

Day, K.L. (1896). The Indigenous Drugs of India. Pama Primlane of Chronica, New Delhi, India.

Garg, G.S. (2004). Sudhanidhi Magazine 2004. Sudhanidhi Office Vijaygar, Aligarh (U.P.).

Kirtikar, K.R. and Basu, B.D. (1935). Indian medicinal plants. Vol. I-IV Periodical Exports Delhi, India.

Nadkarni, A.K. (1954). Indian Materia-Medica, Vol. I & II. 3rd Edition, Popular Book Depo. Bombay.

Singh, V.K. and Ali, Zaheer Anwar (1996). Folk Medicinal plants used for family planning in India Ethanobiology in human welfare, Ed. S.K. Jain 1996. Deep Publication New Delhi, pp. 184-186.

Recent Advances in Medicinal Plant Research : Vision 21st Century, 2005
Editors : S.K. Prabhuji, G.P. Rao and S.K. Patil
Satish Publication House, New Delhi

Chapter 22

DYE YIELDING TREE SPECIES OF UTTARA KANNADA DISTRICT OF KARNATAKA

P. Ramana★, A. Krishna and S.K. Patil
College of Forestry, Sirsi - 581 401, Karnataka, India
★e-mail : ramtarunsirsi@yahoo.com

In India, the use of natural dyes dates back to time immemorial. Due to easy solubility of the natural colouring material, it is readily absorbed by the fabrics. The colouring matters may be obtained from the leaves, flowers, fruits, seed, bark and root of the plants. *Acacia arabica, Adenanthera pavonina, Bauhinia variegata, Bixa orellana, Cedrela toona, Mallotus philipinensis, Memecylon umbellatum, Morinda citrifolia, Mammea longifolia, Tamarindus indica, Tectona grandis, Terminalia bellirica, Terminalia chebula, Thespesia populnea* and *Wrightia tinctoria* have been identified as dye-yielding tree species of Uttara Kannada District of Karnataka.

INTRODUCTION

The use of natural dyestuffs dates back to the remote past. It is to be presumed that the first steps in the art of dyeing resulted from the accidental staining of the skin or fabric by a vegetable material and this was followed by the use of their extracts for the purpose of dyeing. Mordanting was known in India at a very early date but these mordents consisted of naturally occurring sulphates of aluminium and iron. A knowledge of the use of these salts gradually spread from India to other parts (Anon., 1991).

HISTORICAL REVIEWS

Owing to the easy solubility of the natural colouring matters the fabric readily absorbs them during the dyeing process. Natural dyestuffs are less permanent than those obtained by other methods. The colouring matter can be found in the leaf, flower, fruit, seed, bark or root. There was considerable demand for many of the natural organic colouring materials like madder *Rubia tinctoria* L., Manjeet *R. cordifolia* L., Kamala, *Mallotus philippinensis* Muell, Arg. and Indigo *Indigofera tinctoria* L. in the dyeing industry. They were used in textile industry and also for colouring paints, varnishes, leather, inks, paper, wool, medicines and foods. Owing to non-toxic effects some are employed for colouring edible oils, fats, cheese, margarine, chocolate, butter, ghee and the like products (Beutner *et al.*, 1996; Martin, 1995 and Martin *et al.*, 1999).

METHODOLOGY

The data presented here is based on information gathered from medicine man, village heads, old experienced, knowledgeable men and women and from literature.

Study area

Uttara Kannada district is an unique, rich biogeographic area in the Central Western Ghat region of Karnataka, holding multifarious endemic floral and faunal wealth. It is located between 13 ^{0} 55' and 15^{0} 31' N latitude and 74^{0} 9' and 75^{0} 10' E longitude. It has an area of 10,007 km^{2}. It is about 176 km from north to south and 16 to 96 km from east to west. The Central Kanara includes three belts, a coast tract with broad winding lagoons, rich plains and wooded hills running to the sea; a central belt of lofty Sahyadris covered with magnificent forest; and an eastern upland, which is wild waving and thickly wooded in the west and in the east passes into a bare level and thickly populated plains.

Dry deciduous, moist deciduous, semi-evergreen and evergreen forests primarily represent the vegetation of the study area. The climate is primarily monsoonal and rains affected by Southwest monsoon during June to

Table - 1 : Dye yielding tree species of Uttara Kannada District, Karnataka

S.No	Plant Name	Family	Vernacular Name	Parts used	Uses
1.	*Acacia arabica* Willd.	Mimosaceae	Jali mara	Bark	Dye
2.	*Adenanthera pavonina* Linn.	Mimosaceae	Manjuti mara	Wood	Red dye
3.	*Bauhinia variegata* Linn.	Caesalpiniaceae	Mandar mara	Bark	Dye
4.	*Bixa orellana* Linn.	Bixaceae	Arnattu mara	Seeds	Orange dye
5.	*Cedrela toona* Roxb.	Meliaceae	Devdari mara	Flowers	Red or Yellowish dye
6.	*Mallotus philippinensis* Muell. Arg.	Euphorbiaceae	Kunkuma mara	Bark	Orange dye
7.	*Memecylon umbellatum* Burm.	Melastomaceae	Limbtoli mara	Leaves	Yellow dye
8.	*Morinda citrifolia* Linn.	Rubiaceae	Akk mara	Root	Yellow and Red dye
9.	*Mammea longifolia* Planch & Triana	Clusiaceae	Suragi mara	Flower buds	Red dye
10.	*Tamarindus indica* Linn.	Caesalpiniaceae	Hunase mara	Leaves	Reddish yellow dye
11.	*Tectona grandis* Linn.	Verbenaceae	Tegina mara	Leaves	Dark red dye
12	*Terminalia bellirica* Roxb.	Combretaceae	Tare mara	Fruit	Red dye
13	*Terminalia chebula* Retz.	Combretaceae	Alale mara	Fruit	Red dye
14	*Thespesia populnea* Soland ex correa	Malvaceae	Adavi bende mara	Flowers and Capsules	Yellow Dye
15	*Wrightia tinctoria* R. Br.	Apocynaceae	Kodmurki mara	Leaves	Inferior indigo and blue dye

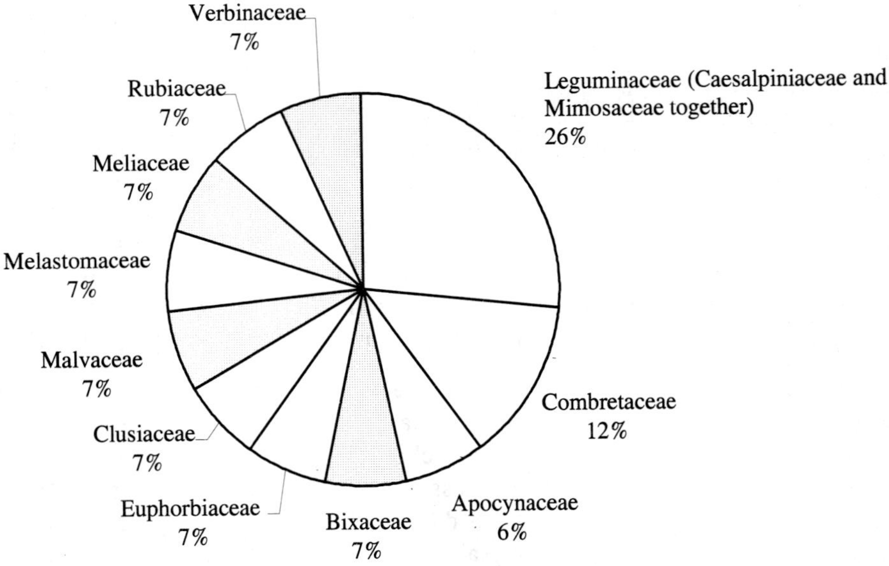

Fig. 1 : Dye yielding tree families (%)

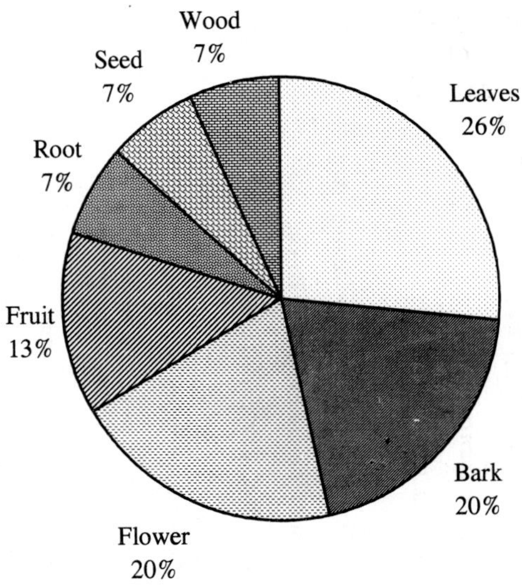

Fig. 2 : Plant parts used for dyeing (%)

September. The rainfall is also variable from year to year. The region received an annual rainfall of 1933 mm with a minimum of 692 mm and a maximum of 3922 mm. The length of dry season varied from a mean of three months at higher altitudes in South to eight months on eastern side of the Ghat in the north. The relative humidity is highest exceeding 90 per cent during July-August and it is as low as 40 per cent in March-April. There is substantial change in temperature. The mean annual range temperature varies from 18 °C to 31 °C, under extreme conditions it drops down to 6.7 °C and rises up to 39 °C. In general, the soils are shallow to medium deep, reddish brown, dark brown and dark yellow brown in colour. The soils are acidic to slightly alkaline in nature, high in nitrogen, low in phosphorus and low to medium in potassium content (Anonymous, 1991).

The data presented in this paper are the outcome of intensive studies conducted among the local people. Local informers were selected and were interviewed extensively. The women chieftains were accorded a significant role in discussions since they possess more cognizance about the utility of local herbal products. Data on the dye yielding tree species were gathered from experienced and knowledgeable men and women.

Table-1 shows botanical name followed by the name of the family, common name, parts used and colour shade obtained from the dye yielding tree species.

RESULTS AND DISCUSSION

An analysis of data in the present communication reveals the uses of fifteen species belonging to eleven families for obtaining dyes by local people (Fig. 1). The present studies yielded interesting information on the tree species and its parts used by local people in fulfilling their day to day needs (Fig. 2). Most of these tree species are still in the mainstay of our population for dyeing purpose because of their easy accessibility. It is assured that search for potential natural dye yielding tree species through ethno-botany gives clues for further research and development.

The traditional knowledge is valuable not merely as an indicator of plant to be screened but also equally important to evolve new methodological approach and strategies towards research. There is an urgent need for the detailed scientific studies on some unexploited dye yielding tree species and also necessary steps should be taken to conserve the threatened species so that the traditional knowledge may be preserved for the future generations.

ACKNOWLEDGMENTS

Authors are grateful to the local people who have revealed the precious information on dye yielding trees.

REFERENCES

Anonymous (1991). *Gazeteer of the Bombay Presidency, Kanara District.* Volume XV, Part I, Sagardeep Publications, Karwar, p. 1-18.

Beutner, S., Broszeit, G., Diepenbrock, F., Kersting, D., Klesczewski, B., Köpsel, C., Peters, M., Schaper, K., Stenhorst, F., Martin, H.D., Bettermann, H., Heinze, J., Sies, H. and Strehblow, H.H. (1996). Carotenoids and Related Dyes: Synthesis, Properties, Chemical and Biological Functions. *Colorchem.*, 14-17.

Martin, H.D. (1995). The Function of Natural Colorants : The Biochromes. *Chimia,* **49** : 45-68

Martin, H.D., Jäger, C., Ruck, C., Schmidt, M., Walsh, R. and Paust, J. (1999). Anti– and Prooxidant Properties of Carotenoids. *J. Prakt. Chem.*, **341** : 302-308.

Recent Advances in Medicinal Plant Research : Vision 21st Century, 2005
Editors : S.K. Prabhuji, G.P. Rao and S.K. Patil
Satish Publication House, New Delhi

Chapter 23

TANNIN YIELDING TREE SPECIES OF CENTRAL WESTERN GHATS REGION OF KARNATAKA

P. Ramana★, A. Krishna and S.K. Patil
College of Forestry, Sirsi - 581 401, Karnataka, India
★e-mail : ramtarunsirsi@yahoo.com

Tannins are complex uncrystalizable colloidal substances having astringent properties. Tannin yielding plants occur in temperate, sub-tropical and tropical climates. Being phenolic in nature, tannins are used for the development of adhesives. They are mostly extracted from the bark, fruits or pods and to some extent from the leaves of trees. *Acacia arabica, Cassia auriculata, Garuga pinnata, Mallotus philippinesis, Tamarindus indica, Tectona grandis, Tarminalia alata, T. arjuna, T. bellirica, T. chebula, Wagatea spicata, Zizyphus jujuba* and *Z. xylopyra* have been identified as tannin-yielding tree species of Central Western Ghats region of Karnataka.

INTRODUCTION

Tannins are complex organic compounds often of very large and high molecular weight. They are mostly uncrystalizable colloidal substances with astringent properties and with the ability to precipitate gelatin from the solution and to form insoluble compounds with gelatin yielding tissues. It is this property, which enables them to convert hides and skins into leather. In the cycle of organic degradation the tannins get converted into sugars and gums. The tannins are more prevalent among the angiosperms or higher plants especially in certain dicotyledons than the lower plants. Among the dicotyledons there are many families in which tannin occurs very frequently, the important ones being Leguminosae (Wattle), Anacardiaceae (*Rhus* spp.), Combretaceae (myrobalan), Rhizophoraceae (mangroves) and Myrtaceae (Eucalyptus). Plants, rich in tannin, occur both in temperate and in tropical or sub-tropical climates but the frequency of such species is more in the tropics. Tannins are commercially important plant products and are also known to protect plant cell protoplast against desiccation, decay or injury by animals (Lakshmanan, 1985). In the entire world there is a tendency towards traditional and more natural ways of life. Now-a-days, people prefer natural curing practices, herbal remedies, natural foods and even biological farming without using synthetic pesticides and fertilizers. Much awareness has been created towards the use of natural agro-based products and ecofriendly standards have become a part of consumer demand. All this has happened due to excessive use of synthetic chemicals and their derived substances in the last one and half century. Moreover, their production causes pollution and disturbs global environment. As a part of this trend, there is a lobby of environmentalists, scientists, educationalists and social workers advocating the use of natural products.

HISTORICAL REVIEWS

Tannins consist of highly complex mixture of poly phenolic compounds. Tannins being phenolic in nature have been used for the development of adhesives for various wood based panels (Pizzi and Scharafetter, 1978). Guptal *et al.* (1978) have developed adhesives from tannins of Wattle bark, mangrove bark, myrobalan nuts and a mixture of tannins from Wattle bark (85%) + myrobalan nuts (10%) + cashew testa (5%) separately and used these adhesives for making tea chest grade plywood. The results showed that plywood pass the requirements of IS specification for tea chest grade. However, the performance of tannin-based adhesives has not been uniform in all cases due to the variation in the chemical composition and structure of tannins from different sources and even in different portions of the same tree. The chemical analysis and physio-chemical properties of tannins, extracted from different tree species, have been reported (Gupta *et al.*, 1981, 1982).

Kathiresan and Veera Ravi (1990) have determined the tannin content of mangrove bark from Pichavaram forest. However, no scientific attempt has been made so far to document tannin yielding tree species of Uttara Kannada District, Karnataka. The present work is, therefore, aimed at generating the information on tannin rich tree species of this region.

METHODOLOGY

The data presented here is based on information gathered from medicine men, local healer, village heads, old experienced, knowledgeable men and women and from literature.

Study area

Uttara Kannada district is an unique, rich biogeographic area in the Central Western Ghat region of Karnataka, holding multifarious endemic floral and faunal wealth. It is located between 13° 55' and 15° 31' N latitude and 74° 9' and 75° 10' E longitude. It has an area of 10,007 km^2. It is about 176 km from north to south and 16 to 96 km from east to west. The Central Kanara includes three belts, a coast tract with broad winding lagoons, rich plains and wooded hills running to the sea; a central belt of lofty Sahyadris covered with magnificent forest; and an eastern upland, which is wild, waving and thickly wooded in the west and in the east passes into a bare level and thickly populated plains.

Dry deciduous, moist deciduous, semi-evergreen and evergreen forests primarily represent the vegetation of the study area. The climate is primarily monsoonal and rains affected by Southwest monsoon during June to September. The rainfall is also variable from year to year. The region receives an annual rainfall of 1933 mm with a minimum of 692 mm and a maximum of 3922 mm. The length of dry season varies from a mean of three months at higher altitudes in South to eight months on eastern side of the Ghat in the north. The relative humidity is highest exceeding 90 per cent during July-August and it is as low as 40 percent in March-April. There is substantial seasonal change in temperature. The mean annual temperature varies from 18 °C to 31 °C, under extreme conditions it drops down to 6.7 °C and rises up to 39 °C. In general, the soils are shallow to medium deep, reddish brown, dark brown and dark yellow brown in colour. The soils are acidic to slightly alkaline in nature, high in nitrogen, low in phosphorus and low to medium in potassium content.

The data presented in this paper are the outcome of intensive studies conducted among the local people. Local informers have been selected and interviewed extensively. The women chieftains have been accorded a significant role in discussions since they possess more cognizance about the utility of local herbal products. Data on the tannin yielding tree species were gathered from experienced and knowledgeable men and women. Table-

Table - 1 : Tannin yielding tree species of Uttara Kannada District, Karnataka

S. No.	Plant Name	Family	Vernacular Name	Parts used	Use (% Tannin)
1.	*Acacia arabica* Willd.	Leguminosae	Jali mara	Bark gum	Tanning (12-20%)
2.	*Cassia auriculata* Linn.	Leguminosae	Talvad mara	Bark	Tanning (15-19%)
3	*Garuga pinnata* Roxb.	Burseraceae	Halabalage mara	Bark	Tanning (1-3%)
4	*Mallotus philippinensis* Muell. Arg.	Euphorbiaceae	Kunkuma mara	Leaves and bark	Tanning (6-10%)
5	*Tamarindus indica* Linn.	Leguminaceae	Hunshe mara	Bark	Tanning (7%)
6	*Tectona grandis* Linn.	Verbenaceae	Tegina mara	Leaves	Tanning (6%)
7	*Terminalia alata* Hyne ex. Roth.	Combretaceae	Matti mara	Bark	Tanning (16-19%)
8	*Terminalia arjuna* (Roxb.) Wight & Arn.	Combretaceae	Holematti mara	Bark	Tanning (15-24%)
9	*Terminalia bellirica* Roxb.	Combretaceae	Tare mara	Bark and fruits	Tanning (1.4 –7%)
10	*Terminalia chebula* Retz.	Combretaceae	Alale mara	Fruits	Tanning (30-32%)
11	*Wagatea spicata* Dalz.	Leguminaceae	Gajaja mara	Pod	Tanning (1-4%)
12	*Zizyphus jujuba* Mill.	Rhamnaceae	Ilanji mara	Bark	Tanning (7%)
13	*Zizyphus xylopyra* Willd.	Rhamnaceae	Mullukare mara	Fruit	Tanning (7.2%)

1 shows the botanical name and common name of the trees, family, parts used and the tannin content.

RESULTS AND DISCUSSION

An analysis of data in the present communication reveals the uses of thirteen species belonging to six families for obtaining tannin by local people (Fig. 1). The present studies have yielded interesting information on the trees species used by local people in fulfilling their day-to-day needs. Most of these tree species are still in the mainstay of our population for tanning purpose because of their easy accessibility. It is assured that search for potential tannin yielding tree species through ethno-botany provides clues for further research and development.

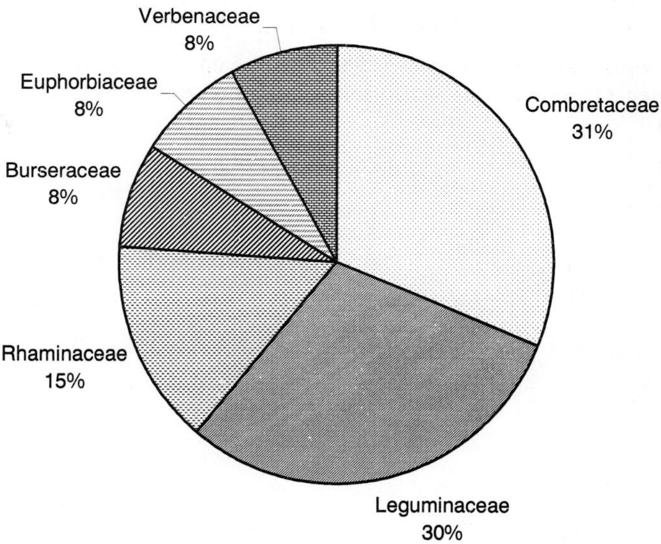

Fig. 1 : Tannin yielding tree families (%)

The traditional knowledge is valuable not merely as an indicator of plant to be screened but also equally important to evolve new methodological approach and strategies towards research. There is an urgent need for the detailed scientific studies on some unexploited tannin yielding tree species and also necessary steps should be taken to conserve the threatened species, so that the traditional knowledge may be preserved for the future generations.

ACKNOWLEDGMENTS

Authors are grateful to the local people who have revealed the precious information on tannin yielding trees.

REFERENCES

Gupta, R.C., Singh, S.P. and Singh, J. (1978). Tannin adhesive for tea chest plywood. *IPIPI Journal,* **8(1)** : Jan-June.

Gupta, Sushma, Singh, S.P. and Gupta, R.C. (1981). Studies on tannin from Indian Wattle (*Acacia mearnsii*)bark. *Indian J. Forestry,* **4(1)** : 18.

Gupta, Sushma, Singh, S.P. and Gupta, R.C. (1982). Studies on physio-chemical properties on Wattle (*Acacia mearnsii*) bark extract. *Indian J. Forestry,* **5(3)** : 185.

Kathiresan, K. and Veera Ravi, A. (1990). Seasonal changes in tannin content of mangrove leaves. *Indian Forester,* May : 390-392.

Laksmanan, K.K. (1985). Mangrove as the source of tannin- past, present and prospectus. The Mangroves proc. *Nat. Symp. Biol. Util. Cons. Mangroves,* 99-156.

Pizzi, A. and Scharafetter, H.O. (1978). The chemistry and development of tannin based adhesives for exterior plywood. *J. Appl. Poly. Soc.,* 22 : 7.

Sushma, Prakash and Singh, S.P. (198). Studies on tannins from mangrove bark and myrobalan (*Terminalia chebula*) nuts. *Indian J. For.,* **15(3)** : 250-253.

Recent Advances in Medicinal Plant Research : Vision 21st Century, 2005
Editors : S.K. Prabhuji, G.P. Rao and S.K. Patil
Satish Publication House, New Delhi

Chapter 24

IMPACT OF SELECTED HERBALS ON RESPIRATORY AILMENTS IN HUMAN BEINGS

R. Brindha and Parvathy
Department of Botany, Avinashilingam Deemed University
Coimbatore - 641 043, Tamil Nadu

The present work was made to elucidate and evaluate the response of drug originated from selected herbals, viz., *Adhatoda vasica, Ocimum sanctum* and *Solanum trilobatum* on respiratory ailments in human beings. It also deals with the studies on some important pharmacognostical and phytochemical characteristics of the aforesaid three plants.

Macroscopic and Microscopic characters of the sputum were studied from the collection of Asthma patient's sputum. The sputum was cultured to analyse the sensitivity of newly formulated capsule. The treatment with the capsule (1 g) and the triple compound extract (10 ml) completely suppressed the growth of gram negative bacillus of the sputum under *in vitro* conditions.

INTRODUCTION

Medicinal plants, containing active principles, are known to modern and ancient civilizations for their healing properties. Until the development of chemistry and, particularly, of the synthesis of organic compounds in the Nineteenth century, medicinal plants and herbs have been found to be the sole source of active principles capable of curing human ailments. They continue to be important to people who do not have access to modern medicines and, moreover, modern pharmaceutical industry rely heavily on the same active principles whether they are natural or synthetic. The active principles differ from plant to plant due to their biodiversity. The World Health Organisation (WHO) has estimated that about 90 per cent of the people in the world have faith on traditional medicines for their primary healthcare needs. Chemical investigation and biological screening of about 300 wild tribal medicinal plants have been carried out by many researchers (AICRPE, 1993).

In the present investigation, an effort has been made to find out the effect of the selected three herbal plants, viz., *Adhatoda vasica*, *Ocimum sanctum* and *Solanum trilobatum* in curing respiratory ailments. The study is aimed at :

1) To make a pharmacognostic study of three plants namely (i) *Adhatoda vasica* (ii) *Ocimum sanctum* (iii) *Solanum trilobatum*.
2) To analyse the biochemical constituents of the entire plants of *O. sanctum* and *S. trilobatum* whereas in A. *vasica*, only the leaves and stems are used.
3) To find out the effect of the above medicinal plants on human beings.
    a) Collection of sputum from the asthma patients.
    b) Culture of the sputum to know the sensitivity of newly formulated herbal capsule.

METHODOLOGY

About 10 kg shade dried *Adhatoda vasica, Ocimum sanctum* and *Solanum trilobatum* plants were powdered. The powdered plants and the triple compound extract were used to study the pharmacognostic characters. The sputum was collected from the asthma patients of Government Hospital, Vettaikaranpudur and analysed by Gram staining process. The macroscopic and the microscopic examination was carried out with the sputum of Asthma patients to report the presence and absence of the elasic fibres, Curschmann's spirals, Charcot-Leyden crystals, pigmented fungi and parasities. The cultured sputum was analysed to know the sensitivity of Gram negative and Gram positive bacteria to the newly formulated herbal capsule. The effect of the *Adhatoda vasica*, *Ocimum sanctum*, *Solanum trilobatum* powders was studied on the respiratory ailments especially for asthma in human beings. The powder study, phytochemical and the microbial study of the sputum of asthma patients were done in the laboratory.

RESULTS

Physical Characteristics

The study on powder of three experimental plants, viz., *A. vasica*, *O. sanctum* and *S. trilobatum* have shown the organolepic characters like colour, odour and taste (Table-1).

Table - 1 : Physical characteristics of the powder.

Name of the Plant	Colour	Taste
Adhatoda vasica	Dark Green	Pungent, Bitter, Acrid
Ocimum sanctum	Pale Green	Pungent, Bitter
Solanum trilobatum	Pale Green	Bitter

Fluorescence analysis

The plant powder was treated with certain chemicals and the colour changes were observed under visible light (Table-2).

The mixture of the powders of three experimental plants was dissolved (treated) in water as well as in some organic solvents. These solvent-powder mixtures exhibited fluorescence under UV-irradiation having dominance of green colour (Table-3).

Table - 2 : Behaviour of Powder with Different Chemicals

Treatment	Colour Under Visible Light
Powder	Pale Green
Powder + NaOH	Brown
Powder + NaOH + Methanol	Dark Green
Powder + HCl	Green
Powder + NaOH + Water	Light Brown
Powder + Ethanol	Dark Green
Powder + 50% HNO_3	Red
Powder + 50% H_2SO_4	Pale Green

Table - 3 : Fluorescence Analysis

Treatments with Powder	Long UV	Short UV
Petroleum Ether	Dark Green	Green
Solvent Ether	Pale Green	Green
Chloroform	Green	Pale Green
Acetone	Dark Green	Green
Alcohol	Dark Green	Pale Green
Water	Pale Brown	Brown

Phytochemical Screening

The active principles in a drug are most important ingradients for curing the diseases. During the present study, it has been found that the plant extract contains cellulose, protein, starch, flavanoid, suberin, sugar, steroid, phenol, fat and saponin. All the photochemical tests showed the positive result except the Quinone test (Table-4).

In the present investigation the combined extract exhibited the positive test for presence of alkaloids (Table-5).

Sputum study

In the macroscopic study the sputum of asthmatic patients was greenish white in appearance with bad odour and the consistency was frothy (Table-6).

The observation of the sputum of asthmatic patients under microscopic examination showed the presence of curschmann's spirals, Charcot-Leyden crystals, Actinomycetes and gram negative bacteria. The absence of elastic fibres, pigmented cells and parasites were also noticed (Table-7). The presence of above features has indicated that the patients were suffering from asthma.

The treatment with the capsule (1 g) and the triple compound extract (10 ml) completely suppressed the growth of gram negative bacteria in the sputum of Sumathy, Anaimalai within a week under *in vivo* condition (Table-8).

DISCUSSION

The results of the present study prove that the newly formulated herbal capsule was effective to cure the bronchial asthma under *in vitro* and *in vivo* condition.

Table - 4 : Phytochemical Test for the Plant Extract

Test for	Reagents	Nature of Colour Change	Phytochemical Change
Cellulose	Extract + Iodine followed by H_2SO_4	Black Colour	Cellulose present
Starch	Extract + Iodine solution	Blue Colour	Starch present
Protein	Extract + Millons reagent	Yellow Colour	Protein present
Protein	Picric Acid solution was added to the substance	Yellow Colour	Protein present
Fixed Oil and Fat	The powder was mixed with Sudan-III	Shiny Orange Colour	Fat and Oil present
Flavanoid	The substance was mixed with 10% NaOH	Yellow Colour	Flavanoid present
Quinone	The substance was added to Con. HCl	No Reaction	Quinone absent
Saponin	Substance shaken in water	Frothing Absent	Saponin present
Saponin	Substance was added to Lead Acetate	White Precipitate	Saponin present
Suberin	The powder was heated with Con. H_2SO_4	Black Colour Develops	Suberin present
Sugar	The Powder was mixed with Anthrone reagent followed by Con. H_2SO_4 and heated	Blue Colour	Sugar present
Steroid	Substance was mixed with Chloroform and a drop of Acetic acid and heated, a drop of Con. H_2SO_4 was added along the side of the Tube	Red Colour	Steroid present
Steroid	Chloroform and Con. H_2SO_4 added to Alcoholic Plant Extract	Light Red Colour (Pink)	Steroid present
Phenolic Content	Plant Extract + Ferric Chloride solution	Blackish Green	Phenolic present

Table - 5 : Test for Alkaloids

Test for	Reagents	Nature of Colour Change	Phytochemical Change (Result)
Alkaloid	Extract + Mayer's Reagent	Black Colour	Alkaloid present
Alkaloid	Extract + Maner's Reagent	Black Colour	Alkaloid present
Alkaloid	Extract + Suberin	Black Colour	Alkaloid present
Alkaloid	**Salakowski Test :** Extract + Chloroform + Con. H_2SO_4	Black Colour	Alkaloid present
Alkaloid	Dragendroff's Reagent + Extract	Black Colour	Alkaloid present

Table - 6 : Macroscopic Examination of Sputum

Name of the Patient : Mrs. Sumathy Anaimalai	
Macroscopic Examination of Sputum	**Results**
Quantity or Volume	5 ml
Consistency : A. Serous B. Frothy C. Mucoid D. Purulent E. Seropurulent F. Haemorrhagic	Frothy
Colour : A. Whitish Yellow B. Greenish C. Blackish D. Dark Reddish Brown	Greenish with White
Odour	Bad Odour

According to Purushothaman *et al.* (1979) B-solamarine, the major glycoalkaloid present in the plant, has antibacterial, antifungal, antimeiotic and antitumour activities. It shows promising result in two cancer tests (System – K.B. cells and sarcoma 180 in mice).

In *Solanum trilobatum* the fruits contain solamarine and the flowers and fruits are used to cure cough. Solanum glycoalkaloids are usually more

Table - 7 : Macroscopic Examination of Sputum

Name of the Patient : Mrs. Sumathy Anaimalai	
Macroscopic Examination of Sputum	**Results**
Elastic Fibres	Absent
Curschmann's Spirals	Present (Asthma)
Charcot-Leyden Crystals	Present (Asthma)
Pigmented Cells	Absent
Fungi	Actinomycetes
Parasite	Absent
Bacteria	Gram Negative Bacillus

Table - 8 : Study of Microbial Factor in Sputum

***Invitro* Clinical Trial to Analyse the Sensitivity of the Capsule**			
Name of the Patient	**Sputum - I**	**Sputum - II**	**Result**
Sumathi	Gram Stain shows **Negative** stain	-	High Sensitive
Sutha	-	Gram Stain shows **Positive** stain	*Moderate* Sensitive

concentrated in fruits particularly in unripe fruits. Siddha physicians consider this drug to be very specific and prepare a ghee from this for its use in tuberculosis and is recommended as food in all kind of lung diseases.

Several plants are used in traditional medicine for the treatment of bronchial asthma (Dorsch and Wagner, 1991). Children's hospital of the Johannes Guternberg University also identified that the unknown alkaloid from *A. vasica* showed pronounced protection against allergy-induced bronchial obstruction in guinea pigs. Pharmacological effects of the plant extracts and pure compounds in man are under investigation.

Pandey *et al.* (1984) and Gupta (1985) treated plant powders with different chemical reagents and observed their behaviours. In the present research similar pharmacognostic study has been made.

The findings of the present investigation may facilitate to formulate a new herbal capsule to cure asthma without any side-effects.

higher activity than the crude drug at 3 and 6 h, but there was no anticonvulsant activity at 1 h.
The crude drug, aqueous and chloroform extracts of *B. monnieri* (500 mg/kg) showed mild to moderate activity from 1 h to 6 h but there was no activity at 24 h. The methanolic extract of *B. monnieri* showed a lesser degree of activity only at 3 h and 6 h. The solubilised extract of *B. monnieri* (500 mg/kg and 1000 mg/kg)) showed mild to moderate activity at 3 h and 6 h with minimal activity at 24 h.
In the Pentylenetetrazol (PTZ) chemoshock seizure test, no activity was detected for both plants (500 mg/kg).
C. asiatica lowered the anticonvulsant activity of PB at 1, 3 and 6 hours, of CBZ at 6 hours and of PHT at 6 h. On the other hand *B. monnieri* did not lower the peak anticonvulsant activity of any of the AEDs given. In fact, when combined with PB, CBZ and PHT, the peak activity was increased at 3, 6 and 6 h respectively. Both *C. asiatica* and *B. monnieri* shifted the T max of PHT from 4 to 6 h. Overall, *B. monnieri* has a faster onset of action while *C. asiatica* had quantitatively lesser activity but had a longer duration of action. The herbal plant products interact pharmacologically with standard AEDs and hence caution should be exercised to avoid any possible adverse interactions.

INTRODUCTION

The recent worldwide interest in traditional forms of healthcare has given an impetus to "bio-prospecting" of plant sources for diverse therapeutic applications and currently, well planned studies for detecting their potential pharmacological activity are being actively pursued (Jain, 1994). In the traditional system of Indian medicine, Ayurveda, *Centella asiatica Linn.* and *Bacopa monnieri Linn.* have been used in the treatment of insanity, epilepsy and hysteria (Chopra *et al.*, 1986). Ashtanga Hridaya states that *B. monnieri*, Brahmi, is the best remedy for epilepsy (Chopra *et al.*, 1986; Ashtange Hridaya). *C. asiatica* commonly known in India as Mandookaparni, has considerable reputation in indigenous medicine and is frequently confused for *B. monnieri* (Patnaik and Dhawan, 1986). The plants often get substituted for each other in the market as both are commonly sold under the same vernacular name 'Brahmi". But Ayurvedic texts and related literature are clear in mentioning the name Brahmi to *B. monnieri*. Comparative anatomical features can differentiate the two species (Indian Herbal Pharmacopoeia, 1998). Chemically both the species are rich in saponins. Madecassoside and asiaticoside are the important saponins of *C. asiatica* whereas *B. monnieri* contains bacosides A & B.

B. monnieri has been shown to have sedative, cardiotonic, vasoconstrictor and neuromuscular blocking actions (Malhotra and Das, 1959). A tranquilizing and hypotensive effect has also been reported (Aithal and Sirsi,

1961). In earlier acute pharmacological studies on *B. monnieri* (Martis *et al.*, 1992), 0.6 g/kg of the aqueous extract and 7 g/kg of the alcoholic extract, administered i.p. to rats, did not alter the flexor or clonic phases of maximal electroshock (MES) induced seizures, but reduced the duration of the extensor phase, though not to a significant extent. Similar results were obtained when the same doses were administered orally to rats for 15 days. Thus, at these doses, there was no abolition of the extensor phase of MES seizures, which is the criterion that has been established to denote evidence of anticonvulsant activity in the MES test (Swinyard *et al.*, 1989). In the pentylenetetrazol (PTZ) test, acute and chronic administration of the aqueous and alcoholic extracts (at the same doses mentioned above), significantly increased the time to onset of convulsions for 4 hours post drug, while mortality was marginally reduced in comparison to controls. These studies showed that the aqueous and alcoholic extracts of *B. monnieri* at the doses employed, did not provide complete seizure protection in either the MES or PTZ tests (Martis *et al.*, 1992).

C. asiatica has been used as brain tonic, as an adjuvant in the treatment of leprosy ulcers, as a diuretic, for the treatment of skin disorders and tried in the treatment of obesity (DeLucia *et al.*, 1990; Reynolds, 1996) C.*asiatica*, (aqueous extract, 50 mg/kg)given orally to rats did not influence flexion, extension or clonus but significantly reduced the duration of post seizure recovery in the MES induced seizures (Sakina and Dandiya, 1991). In the PTZ test, *C. asiatica* at 50 mg/kg i.p. administered to mice, prolonged the time to onset of convulsions and caused 60% mortality, while control mice had 100% mortality (Diwan *et al.*, 1991). In Amerindian and African traditional medicine, anticonvulsant activity of *C. asiatica* has yielded differing results. The hydroalcoholic extract has been shown to have anticonvulsant activity against PTZ (DeLucia *et al.*, 1997) while the alcoholic extract has been reported to have no anti PTZ activity (Adesina, 1982).

In view of the lack of significant consistent activity in classical anticonvulsant tests in rodents and the lack of agreement with proven use of the above two herbal remedies in epilepsy in Ayurveda, the first study was undertaken to systematically evaluate the anticonvulsant activity in different extracts of the whole plants of *C. asiatica* and *B. monnieri.*

Classical pharmacological screening tests such as the MES test and the chemoshock or PTZ test were employed and standard drugs were used and standard antiepileptic drugs such as Phenytoin and Sodium Valproate were used for comparison (Swinyard *et al.*, 1989). The MES test has predictive value for efficacy in clinical Generalized Tonic Clonic Seizures (GTCS) and Phenytoin is the prototype drug for these seizures, while the PTZ test predicts activity in Absence Seizures and Sodium Valproate is the standard anti-absence agent of choice (Swinyard *et al.*, 1989).

Established guidelines for the clinical trials of new antiepileptic agents make it obligatory for the new agent to be administered as "add-on" therapy to the existing antiepileptic medication (Guidelines, 1989). Concurrent use of herbs may mimic, magnify, or oppose the use of drugs. A previous study has shown that *Shankapushpi*, when added to Phenytoin treatment, caused breakthrough seizures in patients, suggesting that herbal remedies when co-administered with phenytoin may result in lowering blood levels and loss of seizure control (Dandekar *et al.*, 1991). Previous reports of herb-drug interactions are sketchy and lack systematic scientific investigations. Hence the objective of the second study was to determine if the concurrent administration of single, effective anticonvulsant doses of either *C. asiatica* or *B. monnieri* would alter the anticonvulsant efficacy of standard antiepileptic drugs such as phenytoin, phenobarbitone and carbamazepine in rats.

METHODOLOGY

Study 1

Plant material

Whole plants of *Centella asiatica* (Linn.) Urban (Syn. : *Hydrocotyle asiatica* Linn; Family : Apiaceae) and *Bacopa monnieri* (L.) Pennell (Syn. : *Bramia monnieri Pennell, Moniera cuneifolia* Michx., *Herpestis monniera* (L.) H.B.&K. of family : Scrophullariaceae); were collected, in the fresh condition, from the electronic city, Bangalore, India in February 1999 and taxonomically identified. The plants were sun dried for 3-4 days, herbaria were prepared and voucher specimens (CA/E/01) and (BM/E/01) were deposited at the Pharmacognosy Department of Natural Remedies Pvt. Ltd. Voucher specimens were also sent to the National Institute of Science Communication (NISCOM), New Delhi for authentication. The dried crude drug of *C. asiatica* was found to have asiaticoside-0.6% w/w, madecassoside-0.9%w/w, estimated by HPLC technique using chromatographic reference standards obtained from M/s Extrasynthese, France (Lot No 99091005 and 99091006 respectively). The crude drug of *C. asiatica* was found to contain 10.5%w/w of total saponins, estimated according to the procedure of Marston *et al.* (1995). The dried crude drug of *B. monnieri* was found to have bacoside A- 4.2%w/w estimated by HPLC and HPTLC techniques using chromatographic reference standard provided by Phytochemistry Department of Natural Remedies Pvt. Ltd. The crude drug of *B. monnieri* was found to contain 11% w/w of total saponins (Marston *et al.*, 1995).

Preparation of extracts

Methanolic extracts

The crude drugs, thus obtained, were extracted thoroughly with methanol.

The dried methanolic extract of *C. asiatica* (yield 25%w/w) had 5.2% w/w asiaticoside, 7.8% madecassoside and 55%w/w of total saponins.

Whole plant of *B. monnieri* was thoroughly extracted with methanol (yield 14% w/w). The methanolic extract was, then, partitioned between butanol and water. Extraction with butanol was carried out to separate the organic components from inorganic and high molecular weight components like carbohydrates, etc. The dried butanolic extract (yield 7% w/w) had a content of 40% w/w bacoside-A.

Solubilised extracts

The dried methanolic extract of *C. asiatica* and the dried butanolic extract of *B. monnieri* were solubilised using 4%w/w glycerol polyethylene glycol hydroxy stearate (Cresmer RH 40) and propylene glycol 6%w/w. The pH of 10% aqueous solution of Cresmer RH 40 was found to be in the range of 6 to 7.

Aqueous extracts

The aqueous extracts of *C. asiatica* and *B. monnieri* were prepared by placing 1 Kg of powdered material in percolator and adding 8-10 *l* of double distilled water at room temperature. The water extract from the percolator was drained and filtered through syrup pads and concentrated to dryness on a water bath.

Chloroform extracts

The chloroform extract of *C. asiatica* and *B. monnieri* were prepared by refluxing 1 Kg of powdered material with 5 *l* of chloroform in Soxhlet apparatus for 4 hours. The boiling solvent was cooled and filtered through Whatmann No. 1 filter paper and resultant solvent extract concentrated to dryness on a rotatory evaporator at 45°-50°C under reduced pressure.

Leaf of *C. asiatica* (Teka Brahmi capsules) was obtained from Juggat Pharma, Bangalore.

The standard reference agents - Phenytoin/Dilantin suspension, 30 mg/kg (Parke Davis, India, Ltd.) in the MES test and Sodium Valproate (Reckitt Colman of India Ltd.) 200 mg/kg in the PTZ test were administered orally.

Crude drug of both plants as well as all extracts were triturated with distilled water and administered orally to rats in a volume of 0.4ml/100g. The slurries thus obtained were thick and difficulty was encountered during gavaging.

Animals

Adult male Wistar rats (220-250g) and male albino mice (20-25g), were housed in standard environmental conditions of temperature of 27 ± 1°C. The dark/light cycle was daylight driven. They were fed with standard rat pellet diet (Hindustan Lever India Ltd.) and water *ad libitum*, except during experimental sessions. The Institutional Ethics Committee approved the protocol of the animal experiments that were conducted in humane conditions, in accordance with NIH guidelines (NIH publication, No.80-23; revised 1978).

Anticonvulsant activity against Maximal Electroshock Seizures (MES) in rats

Animals were randomly divided into groups of eight (n>=8). Independent groups of animals were used for each time point tested. Groups received the crude drug, methanolic extracts of *C. asiatica* or *B. monnieri* at 500 mg/kg (po), solubilised extracts of *C. asiatica* or *B. monnieri* at doses of 500 and 1000 mg/kg (po), aqueous extract of *C. asiatica* or *B. monnieri* (5 g/kg), chloroform extract of *C. asiatica* or *B. monnieri* (500 mg/kg), or Phenytoin 30 mg/kg (po). Control animals have received equivalent amounts of the respective vehicle.

Experiments were conducted at the same time each day and the rats were subjected to MES at 150 mA, 60 Hz for 0.2 sec (Swinyard *et al.*, 1989) through pinnal electrodes at 1, 3, 6 and 24 h after vehicle/drug administration. In all electrically induced convulsions, the rats are manually restrained and released immediately after stimulation to permit observation of the seizure during its entire course. MES results in Hind Limb Tonic Extension (HLTE), and the duration was being measured in seconds. Rats were pre-tested 24 h prior to drugging and those failing to give HLTE were rejected. The criterion for anticonvulsant activity and protection against MES induced seizures is abolition of HLTE, which is taken as the end point of the test. The tonic extensor component is considered abolished if the HLTE does not exceed a 90° angle with the plane of the body. The MES test was done according to the criteria followed in NIH, USA and followed in the Anticonvulsant Drug Development Programme (Krall *et al.*, 1978).

The percentage protection was calculated for each group using the following formula :

$$\frac{\text{Number of rats in each group that do not exhibit HLTE}}{\text{Total number of rats in the group}} \times 100$$

Anticonvulsant activity against Chemoshock (PTZ) induced seizures in mice and rats

Seizures were induced in mice or rats with PTZ at 70 mg/kg ip, which is the convulsive dose in 97% of the animals (Swinyard *et al.*, 1989). PTZ was dissolved in 0,9% saline and injected ip in mice at 0.1 ml/10g and rats at 0.2 ml/100 g. Animals were placed in isolation cages and observed for the next 30 min for the presence or absence of clonic spasms, persisting for

Table - 3 : Levels of Phenytoin in mg/ml (Mean + SD) in the plasma of rats at different times after drug administration and pharmacokinetic parameters of different groups of rats (n=9).

Time after drug administration	PHT	*C. asiatica* + PHT	*B. monnieri* + PHT
0.5 h (mg/ml)	3.16 + 0.90	0.91 + 0.08*	1.71 + 0.65*
1 h (mg/ml)	3.49 + 0.80	1.22 + 0.62*	2.37 + 1.09*
3 h (mg/ml)	4.30 + 1.02	2.12 + 0.91*	2.77 + 1.09*
6 h (mg/ml)	3.48 + 1.39	2.91 + 1.39	4.55 + 1.50
8 h (mg/ml)	1.82 + 0.65	1.59 + 0.68	2.70 + 0.65
24 h (mg/ml)	0.44 + 0.08	0.38 + 0.10	0.71 + 0.16
Pharmacokinetic parameters			
C max (mg/ml)	4.30 + 1.02	2.91 + 1.39	4.55 + 1.50
T max (h)	4.00 + 1.50	6.00 + 0.00*	6.00 + 0.00*
t 1/2 abs (h)	1.30 + 0.61	1.09 + 0.48	2.45 + 1.17
t ½ elm (h)	7.85 + 3.23	5.88 + 2.38	6.62 + 1.97
AUC 0 – tn	33.01 + 15.53	20.89 + 15.4	33.44 + 20.41
Vd (l)	648.6 + 244.5	1094.3 + 996.2	570.2 + 306.6
Cl (l/min)	995 + 637	2065 + 1260	1108 + 736

$p < 0.05$ when compared to phenytoin group.

with PHT, the seizure protective effect of PHT was increased at 6 h and persisted for 24 h.

Pharmacokinetic aspects: Plasma levels of PHT per se and in combination with *C. asiatica* and *B. monnieri* are shown in Table 3. Both *C. asiatica* and *B. monnieri* significantly lowered plasma levels of PHT at 0.5, 1 and 3 h after drug administration. PHT, when given alone, had a mean peak plasma concentration (C max) of 4.3 mg/ml at 3 h which correlates with its maximum pharmacodynamic activity; when *C. asiatica* is combined with PHT C max is reduced from 4.30 mg/ml to 2.91 mg/ml. On the other hand, when *B. monnieri* is combined with PHT, C max is not reduced but actually increased to 4.55 mg/ml at the 6 h. When *C. asiatica* and *B. monnieri* are combined with PHT, the time to peak plasma concentration (T max) is shifted from

4 to 6 h. The AUC of PHT and *B. monnieri* + PHT is the same but that of *C. asiatica* + PHT is reduced.

The results obtained in the MES test in rats showed that crude drug as well as different extracts of *C. asiatica* and *B. monnieri*, protects against MES induced seizures. *C. asiatica* had a slower onset of action than *B. monnieri* and had a quantitatively lesser degree of anticonvulsant activity (approximately, one third the activity of Phenytoin) but the activity lasted for 24 h. This could suggest the possibility of active metabolite/s. In comparison, *B. monnieri* had a fast onset of action which was about half to three fourths the activity of Phenytoin, but like Phenytoin lasted only for 6 h. The anticonvulsant profile of activity of the crude drug of *B. monnieri* as well as BM-II at 1000 mg/kg was similar to that of Phenytoin, but of lesser intensity. The results obtained in these studies, demonstrate unequivocally that like Phenytoin, both *C. asiatica* and *B. monnieri*, possessed anticonvulsant activity in the MES test, predicting their potential to control clinical GTCS, and corresponding to the observation in Ayurvedic medicine (Sudha *et al.*, 2002). Abolition of HLTE, the decisive laboratory parameter for establishing anticonvulsant activity (Swinyard *et al*, 1989), was not evident in previous studies for *B. monnieri* following acute or chronic administration (Martis *et al.*, 1992). However, our results differ from previous studies (Martis *et al.*, 1992; DeLucia *et al.*, 1997) as neither plant product protected rats against PTZ induced seizures. This could be attributed to species difference between the plants or the method of extraction. Moreover, in the study of De Lucia *et al.* (1997), only the leaves of *C. asiatica* were used, whereas in our study the whole plant has been used.

The crude drug of both plants had moderate activity. In comparison the different extracts, in general, showed less activity suggesting that the glycosides, asiaticosides and bacosides, did not contribute to the anticonvulsant activity. Solubilization improved activity of both plants, were presumably due to better absorption, especially for *B. monnieri* (500 and 1000 mg/kg) which showed almost 39% protection at 6h post drug. Notably, at the higher dose of 1000 mg/kg, both plant products showed 50% reduction in HLTE duration suggesting the possibility that like Phenytoin, they could limit seizure spread.

The results have conclusively demonstrated that in accepted standard pharmacological screening tests (Swinyard *et al.*, 1989), single doses of both *C. asiatica* and *B. monnieri* possessed anticonvulsant activity in the MES test and it is likely that following chronic administration their efficacy may improve considerably.

In traditional medicine, if a plant is claimed to treat epilepsy in one or more countries, or potential anticonvulsant activity has been reported from different continents, as in the case of *C. asiatica* (Diwan *et al.*, 1991; DeLucia *et al.*, 1997), this would support the scientifically obtained evidence that the claims are authentic and worth investigating.

Laboratory studies are currently fortifying the tenets of ancient wisdom derived from herbal remedies. Researchers are gaining new insight into traditional medicines in assisting the body to maintain its own self-healing systems while preventing debilitating effects of chronic diseases, like epilepsy. This study addressed the question of defining the pharmacological activity of *C. asiatica* and *B. monnieri*, which are well known remedies in Ayurveda for the treatment of epilepsy, and confirmed the presence of their anticonvulsant activity in animals.

Concurrent administration of antiepileptic drugs and herbal medicines may result in interactive synergistic or antagonistic clinical effects and complicate the purpose of long term medication (Sudha *et al.*, 2003a; 2003b Chen *et al.*, 2000). Pharmacological or toxicological effects of either component may increase or decrease during combined therapy. Decreased concentrations of phenytoin (PHT), one of the most commonly used antiepileptic drugs, occurred when combined with a traditional Ayurvedic preparation, Shankapushpi, which contains Centella asiatica (*C.a.*), resulting in breakthrough seizures in epileptics (Dandekar *et al.*, 1991).

Salient pharmacodynamic features obtained in rats were that *C. asiatica* and *B. monnieri*, at equivalent doses of 500 mg/kg, showed quantitatively less anticonvulsant activity that PHT, 30 mg/kg, with *B. monnieri* being marginally more active than *C. asiatica.* The anticonvulsant profile of activity of *B. monnieri* had overall similarity to that of PHT with a fast onset of action which declined at 24 h. When *C. asiatica* or *B. monnieri* were combined with PHT, only the combination of *B. monnieri* with PT showed significant protection ($p < 0.05$) at 6 h and was almost twice as active as *B. monnieri* or PHT per se. One other feature of interest was that when *C. asiatica* and *B. monnieri* were combined together, the 6 h anticonvulsant activity of the two herbal medicines almost equaled that shown by the combination of *B. monnieri* and PHT.

The enhanced anticonvulsant activity of *B. monnieri* and PHT at the post drug 6 h was further supported by pharmacokinetic results, showing that the plasma level of *B. monnieri* and PHT was 4.55 + 1.50 and was greater as compared to PHT which was 3.48 + 1.39 mg/ml.

The results obtained with combination of either *C. asiatica* or *B. monnieri* with PHT show a significant increase in the T max from 4 h to 6 h without corresponding significant changes in the AUC 0 – tn or t ½ elm. The absorption of PHT is dependent on the pH of the medium in which it is dissolved and its concentration. Little absorption takes place in the stomach, because PHT is insoluble at the pH (approx. 2.0) of gastric juice. In the intestine, where the pH is 7.0 – 7.5, more of the PHT exists in the ionized form and is, therefore, more soluble, rendering PHT to be predominantly absorbed from the small intestine. The large surface area, vascularity and blood flow at this site also facilitate absorption. When *C. asiatica* or *B. monnieri* is added to PHT, they further slow gastric emptying and delay absorption. It is not known with certainity whether *C. asiatica* or *B. monnieri* possess anticholinergic activity, as this would also affect gastric emptying and peristaltic movement of the small intestine, further contributing to the delayed absorption of PHT. The delayed absorption could also account for the late onset of maximal anticonvulsant activity particularly in the case of *B. monnieri* co-administered with PHT.

PHT is metabolized in the liver primarily by the cytochrome P 450 enzyme system and eliminated almost exclusively by metabolism. Any agent that can induce/inhibit the P 450 enzyme system may potentially interact with PHT. In the above study, there were no significant differences obtained in the C max, t ½ and Cl between the groups. Hence, *C. asiatica* or *B. monnieri* may not significantly influence the extent of absorption, metabolism and elimination of PHT. The herbal extracts did not significantly reduce the duration of the righting reflex and did not affect spread of the discharge, confirming the results obtained in previous studies (Vaidya, 1997).

It has been reported that 16% of epileptic patients, being treated with PHT, were also using Chinese herbal medicines concomitantly (Ohnishi *et al.*, 1999). Other studies showed that PHT pharmacokinetics in the rat were not affected after a single dose with the plant *Paeoniae radix*, possessing anticonvulsant activity, except for significant differences in T max and Vd (Sudha *et al.*, 2003a). Depending on the pharmacological properties of herbal medicines, it is feasible to co-administer them along with PHT, without compromising anticonvulsant efficacy provided comprehensive preclinical profiling is done. The main concern is the possibility of unknown herbal drug/allopathic drug interaction that may adversely influence the therapy.

In conclusion, the moderate anticonvulsant activity of *C. asiatica* or *B. monnieri*, their favourable interaction with phenytoin and absence of significant adverse effects merit consideration for further investigations.

REFERENCES

Adesina (1982). Studies on some plants used as anticonvulsants in Amerindian and African traditional medicine. *Fitoterapia*, **53** : 147-162.

Aithal, H.N. and Sirsi, M. (1961). Pharmacological Investigations on *Herpestis monniera* H.B.&K. *Ind J Pharm.*, **23(1)** : 2-6.

Ashtanga Hridaya : Uttara Tantra, Chapter 40: verse 51.

Chen, L.C., Chen, Y.F., Yang, L.L., Chou, M.H. and Lin, M.F. (2000). Drug utilization patterns of antiepileptic drugs and traditional Chinese medicines in a general hospital in Taiwan – a pharmaco-epidemiologic study. *J. Clin. Pharm. Therap.*, **25(2)** : 125-129.

Chopra, R.N., Nayar, S.L. and Chopra, I.C. (1986). *Glossary of Indian Medicinal Plants*, CSIR, New Delhi. 32, 58.

Dandekar, U.P., Chandra, R.S., Dalvi, S.S., Joshi, M.V., Gokhale, P.C., Sharma, A.V., Shah, P.U. and Kshirsagar, N.A. (1991). Analysis of a clinically important interaction between phenytoin and Shankapushpi, an Ayurvedic preparation. *J. Ethnopharmacology,* **35(3)** : 285-288.

De Lucia, R., Pannizza, S., Oshiro, T.T., Silva, F.D., Basile, A.C., Sertie, J.A.A. (1990). *Annals of II Congresso Nacional da Sociedade Brasileira de Alimentacao.* 222.

De Lucia, R., Sertie, J.A.A., Carmargo, E.A. and Panniza, S. (1997). Pharmacological and toxicological studies on *Centella asiatica. Fitoterapia.*, **68** : 413-416.

Dill, W.A., Chucot. L., Chang, T. and Glazko, A.J. (1971). Simplified benzophenone procedure for determination of diphenylhydantoin in plasma. *Clinical Chemistry,* **17(12)**.

Diwan, P.V., Karwande, L. and Singh, A.K. (1991). *Fitoterapia*, **62** : 253-257.

Guidelines of ILAE Commission on Anti-epileptic Drugs (AEDs) (1989). *Epilepsia,* **30** : 400-408.

Indian Herbal Pharmacopoeia (1998). RRL, Jammu and IDMA, Mumbai, **1** : 53.

Jain, S.K. (1994). *Ethnobotany and the Search for New Drugs (Ciba Foundation Symposium 185)*, John Wiley & Sons, New York; 153-168.

Krall, R.L., Penry, J.K., White, B.G., Kupferberg, H.J. and Swinyard, E.A. (1978). *Epilepsia,* **19** : 409-428.

Malhotra, C.L. and Das, P.K. (1959). *Ind. J. Med. Res.*, **47** : 294-6.

Marston, A. and Hostettmann, K. (1995). In *Saponins*, Cambridge University Press: 144.

Martindale (1996). The extra pharmacopeia, 31st ed., JEF Reynolds, The Royal Pharmaceutical Society, London, p 1084.

Martis, G., Rao, A. and Karanth, K.S. (1992). *Fitoterapia*, **631** : 399-404.

Ohnishi, N., Yonekawa, Y., Nakasako, S., Nagasawa, K., Yokoyama, T., Yoshioka, M. and Kuroda, K. (1999). Studies on interactions between traditional herbal and Western medicines. *Biol. Pharm. Bull,* **22** : 527-531.

Patnaik, G.K. and Dhawan, B.N. (1986). Dhawan BN (Ed.) *Current Research on Indian Medicinal Plants in India*, INSA, New Delhi, India; 45-71.

Sakina, M.R. and Dandiya, P.C. (1991). *Fitoterapia,* **61** : 291-5.

Sudha, S., Bindu, R., David, J., Venkataraman, B.V. and Kulkarni, K.S. (2003a). Effect of Centella asiatica and Bacopa monnieri on the pharmacodynamics and pharmacokinetics of phenytoin. *Neurosciences Today,* **7** : 47-53.

Sudha, S., Bindu, R., Joyce, G., David, J., Venkataraman, B.V. and Kulkarni, K.S. (2003b). Pharmacological interaction of Centella asiatica and Bacopa monnieri with antiepileptic drugs – An experimental study in rats. *Ind. J. Pharmacol.,* (submitted).

Sudha, S., Kumaresan, S., Amit, A., David, J., Venkataraman, B.V. (2002). Anticonvulsant activity of different extracts of Centella asiatica and Bacopa monnieri in animals. *J. Nat. Rem.*, **2** : 33-41.

Swinyard, E.A., Woodhead, J.H., White, H.S. and Franklin, M.R. (1989). In Levy, R., Dreifuss, F.E., Mattson, R., Meldrum, B. and Penry, J.K. (Eds.) *Antiepileptic Drugs*, Raven Press, New York; 85-102.

Vaidya, A.B. (1997). The status and scope of Indian medicinal plants acting on central nervous system. *Indian J. Pharmacol.,* **29** : S340-343.

Recent Advances in Medicinal Plant Research : Vision 21st Century, 2005
Editors : S.K. Prabhuji, G.P. Rao and S.K. Patil
Satish Publication House, New Delhi

Chapter 26

ANTHELMINTIC ACTIVITY OF *SEMECARPUS KATHALEKANENSIS*

P. Ramana[1★], S.K Patil[1] and K.S. Rao[2]
[1]*College of Forestry, Sirsi - 581 401, Karnataka, India*
[2]*Director, Institute of Wood Science and Technology, Bangalore, India*
★e-mail : ramtarunsirsi@yahoo.com

Drug development from plant-based compounds could be useful in meeting the demand for newer drugs with minimal size effects. The vast biodiversity of Indian herbal heritage is under exploration for getting source of some novel drugs. In the present work, leaf and bark extracts of *Semecarpus kathalekanensis* were found to possess strong anthelmintic activity against *Eudrilus eugeniae*. The alcoholic extracts had always better anthelmintic activity than the extracts prepared in distilled water.

INTRODUCTION

Anthelmintics are therapeutic agents used to destroy parasitic worms or remove them from the infected host. The majority of helminthes infections are acquired by contact with infected animals, ground contaminated by human or animal excrement, water infected with cercaride and ingestion of infected meat. Although helminthes infections are usually associated with tropical regions. Diseases caused by helminthes have a serious economic problem to the animals, as animal is vulnerable to large number of parasitic worm infections. The ultimate test of anthelmintic activity is the ability of a chemical agent to eliminate worms from a specifically parasitized animal with a minimum toxic effect in the host (Gadaginmath, 1974). Drug development from plant-based compounds could be useful in meeting the demand for newer drugs with minimal side effects. The vast biodiversity of our country is presently under explored and many plants could be the source of novel drugs.

HISTORICAL REVIEWS

Six species of *Semecarpus* are reported in India. Most of the species yield a dark, vesicant, resinous juice from the pericarp of the drupes. Plants belonging to genus *Semecarpus* are known for anticancer (Prasad, 1987), antirheumatic (Patwardhan, 1989), anthelmintic (Chattopadhyay and Khare, 1969) and antibacterial (Patwardhan *et al.*, 1982) properties. *Semecarpus kathalekanensis* Dasappa belongs to the family Anacardiaceae. The *Semecarpus kathalekanensis* is a tree, 10-20 m tall, seen in myristica swamps and water logged localities of Western Ghats of North Canara. The bark is astringent; the tree exudes a gum, which turns black after exposure (Dasappa and Swaminath, 2000). There are several reports available on chemical and biological properties of Semecarpus species (Murthy, 1983; Premalatha and Sachdanandam 1999). The present study was undertaken to evaluate the anthelmintic efficacy of various extracts of *Semecarpus kathalekanensis*.

METHODOLOGY

Preparation of plant extracts

Fresh leaves and bark of *S. kathalekanensis* were collected from the myristica swamps of Western Ghats of North Canara. The plant material was shade dried and powdered. The powdered plant material (200 gm) was subjected to soxhlet extraction separately and successively with petroleum ether (60-80 °C), chloroform, ethanol and distilled water. The crude plant extracts were concentrated to dryness in a flash-evaporator under reduced pressure and controlled temperature. All the extracts were stored in a refrigerator.

Bioassay of Plant Extracts

Anthelmintic activity

The anthelmintic activity of the crude plant extracts was carried out against *Eudrilus eugeniae* (Kinb.) by following reported procedure (Giand,1965) using piperazine citrate (7.5 mg/ml and 3.75 mg/ml) suspensions as standard. Saline solution (0.9%) was used as control. The ethanol and distilled water extracts were used for testing, the test solutions of each extract was prepared by dissolving suitable quantity of crude plant extract to get a 1.0, 0.5 and 0.25 per cent concentrated solutions respectively.

RESULTS AND DISCUSSION

The results of anthelmintic activity of the extracts of *S. kathalekanensis* are summarized in Table-1. Both ethanol and distilled water extracts were found to be effective against the *Eudrilus eugeniae* (Kinb.). The ethanol extract has showed high anthelmintic activity against test organism at all the concentrations, the anthelmintic activity of ethanol extract was higher than distilled water extract at all the concentrations. The distilled water extract showed moderate anthelmintic activity against *Eudrilus eugeniae* (Kinb.).

Table - 1 : Anthelmintic Activity of Ethanol and Distilled Water Extracts of *Semecarpus kathalekanensis*

Crude Plant extracts and standards	Concentration of solutions	Observation period in hrs	Time in minutes for*	
			Paralysis	Death
Ethanol extract	1.0 %	10	40	98
	0.5 %	10	49	105
	0.25 %	10	50	179
Distilled water extract	1.0 %	10	145	438
	0.5 %	10	185	500
	0.25 %	10	202	559
Control Normal Saline	0.9 %	10	NP	ND
Standard (Piperazine citrate)	7.5 mg/ml	10	13	55
	3.75 mg/ml	10	21	113

*Mean values of three independent replications observed until 10 hrs.
ND = No death and NP = No paralysis

The anthelmintic activity of plant extracts was found to vary with the concentrations of the extracts, highest anthelmintic activity was observed at higher concentration, while low anthelmintic activity was observed at lower concentration.

The present investigation of *S. kathalekanensis* extracts strongly suggests that the ethanol extract has potent anthelmintic activity and the results are in agreement with the earlier reports on anthelmintic activity (Chattopadhyay and Khare, 1969). The present findings in this communication are an addition to the list of bioactivities exhibited by plants belonging to genus *Semecarpus*. Drug development from plant-based compounds could be useful in meeting the demand for newer drugs with minimal side effects. The vast biodiversity of our country is presently under exploration and many plants could be the source of novel drugs. Solvents play an important role in extracting active ingredients of specific nature. Further purification needs to be done before any such conclusions are drawn. Work is under way to identify compounds responsible for this bioefficacy.

REFERENCES

Chattopadhyay, M.K. and Khare, R.L. (1969). Isolation of anacardic acid from *Semecarpus anacardium* and study of its anthelmintic activity. *Indian J. Pharm.*, **31(4)** : 104.

Dasappa and Swaminath, M. (2000). A new species of *Semecarpus* L. (Anacardiaceae) from the myristica swamps of Western Ghats of North Canara, Karnataka, India. *Indian Forester,* **126(1)** : **78-82.**

Gadaginmath, G.S. (1974). Ph.D. Thesis, Karnataka University, Dharwad, Karnataka, India.

Giand, K.N., Dar, R.N., Chopra, B.M. and Kaul, R.N. (1965). *Indian J. Pharm.*, **27** : 141.

Murthy, S.S.N. (1983). Naturally occurring biflavonoid derivatives: Part III-a biflavanone from the nutshells of *Semecarpus anacardium. Indian J. Chem.*, **22B** : 1167.

Patwardhan, B. (1989). Studies on the mechanism of action of *Semecarpus anacardium* in rheumatoid arthritis. *J. Ethnopharmacol.*, **25(2)** : 159-164.

Patwardhan, B.K., Francis, R.P., Kapre, S.V. and Sharma, K.D. (1982). Antibacterial activity of *Semecarpus anacardium* extracts. *Bull Haffkine Inst.*, **10(2)** : 27-30.

Prasad, G.C (1987). Studies on Cancer in Ayurveda and its management. *J.Res Ayurveda Sidha.*, **8(3-4)** : 146-167.

Premalatha, B. and Sachdanandam, P (1999). Effect of *Semecarpus anacardium* nut milk extract on rat serum alpha-fetoprotein level in aflatoxin B1-mediated hepatocelluar carcinoma. *Fitoterapia,* **70(3)** : 279-283.

Recent Advances in Medicinal Plant Research : Vision 21st Century, 2005
Editors : S.K. Prabhuji, G.P. Rao and S.K. Patil
Satish Publication House, New Delhi

Chapter 27

LATEX OF *CARICA PAPAYA* FOR HEPATOPROTECTIVE, ACUTE AND SUB ACUTE TOXICITY STUDIES IN RATS

V.P. Rasal*, Mahesh Kumbhar, Hussain Jakir and S. Arulmozhi
Department of Pharmacology, K.L.E.S's College of Pharmacy, J.N.M.C Campus, Nehru Nagar, Belgaum - 590 010, Karnataka, India
*e-mail : rasalvp@rediffmail.com

Latex of *Carica papaya* has been found to protect the liver from CCl_4 induced damage. Mechanism of action is by the production of the structural integrity of hepatocyte cell membrane or regeneration of damaged liver cells by the latex as revealed by the histopathological and biochemical studies. Acute and sub acute toxicity studies on LCP was performed. Cage observation, changes in morphology, behaviour and body weight were noted weekly. At the end of the exposure period, biochemical, haematological and pathological changes were examined. Liver damage was induced in albino rats (150–220 g) by administering CCl_4 (0.5 ml/kg,ip) once daily for 7 days. LCP (130 mg/kg, p.o) was given for one week. Enzyme levels, viz., SGPT, SGOT, ALP and bilirubin were estimated in serum. Histopathological studies were also done to confirm the biochemical changes. The LD_{50}

of LCP was found to be 1300 mg/kg. LCP (130 mg/kg, p.o) for 90 days showed significant decrease in SGPT and SGOT levels. The mean ± SEM serum SGOT, SGPT and ACP levels in control animals were 18.33 ± 0.643, 134.16 ± 1.606, 27.61 ± 1.119 IU/l respectively whereas in CCl_4 treated rats, the level rose to 75.02 ± 1.719, 222.33 ± 1.525, 92.16 ± 1.14 IU/l respectively. LCP reduced the SGOT, SGPT, ALP levels to 21.83 ± 1.280, 131.33 ± 1.312, 28.16 ± 1.238 IU/l respectively. There was a significant increase in serum bilirubin levels after CCl_4 i.e, 1.29 ± 0.0233 mg/dl, which was reversed by LCP i.e, 0.56 ± 0.0478 mg/dl.

INTRODUCTION

Carica papaya (Caricaceae) is a herbaceous perennial with milky latex which attains height of 5 m and is grown in all tropical and sub tropical areas. The latex has been used to treat eczema, ringworms, warts, corns, calluses, infected wounds, malignant tumors, pain of burns, bleeding haemorrahoids (Bhattacharjee, 1998) and also as an anticoagulant (Ambasta, 1986). Small quantity of latex mixed with honey or sugar is given for the treatment of enlarged liver and spleen (Bhattacharjee, 1998), whooping cough and also in expelling out the round worms. Papain and chymopapain are the main active components of LCP. Papain is known for its digestive properties to relieve dyspepsia and also for antidiarrhoeal activity. Papain containing nasal spray is prescribed in some pollen causing allergies. Chymopapain is now-a-days used for neuro surgery and orthopaedics to overcome the pain of pinch nerves (Bhattacharjee, 1998).

There are reports that latex of *Carica papaya* helps in proper functioning of liver and is beneficial in treatment of enlarged liver. Presently, there is a resurgence of interest among healthcare professionals in treating hepatotoxicity with various drugs of plant origin (Bhattacharjee, 1998; Kurinan, 1995).

So, an investigation regarding hepatotoxicity caused by carbon tetrachloride is undertaken in the present study (Jayshekar *et al.*, 1997). The hepatoprotective activity is monitored biochemically by estimating serum transaminase, serum alkaline phosphatase and serum bilirubin. The histopathological changes in the liver were also examined during this study.

METHODOLOGY

The present study is carried out in the following steps:

a) Acute toxicity studies, i.e., determination of LD_{50} value (both oral and i.p.) in mice,
b) Sub-acute toxicity studies (45 and 90 days) in rats,
c) Assessment of liver protective activity.

a) Acute toxicity studies in mice

Albino mice (20-25 gm) of either sex (Vaccine Institute, Tilakwadi, Belgaum) were used. Latex of *Carica papaya* was collected from the College garden and prepared in the form of suspension by using 2% gum tragacanth.

Animals have been inducted on fasting overnight before the experiment. They were divided into five groups, each containing ten animals among which one was receiving 0.5 ml of 2% gum tragacanth while others received one each of test sample in the dose of 3000 mg/kg b.w. suspended in 2% gum tragacanth.

Then, subsequently the doses of test sample were decreased to bear a ratio of 1/v2 (3000, 2100, 1500, 1060, 750 mg/kg b.w.) to its antecedent dose till all the animals survived in treated groups. The test sample was administered both by oral and i.p. routes in different sub-groups.

b) Sub-acute toxicity studies in rats

Albino rats (150-200gm) of either sex were used . They were maintained on standard diet and water provided *ad libitum*. The suspension was administered orally once daily to the rats for the period of 45 and 90 days.

Twenty-four rats were randomly divided into four groups of six rats each. Group I served as control and received 0.5 ml of 2% gum tragacanth. The remaining groups (II, III, IV) were orally administered latex for 45 and 90 days at the dose rate of 26, 52 and 130 mg/kg/day respectively, which are $1/50^{th}$, $1/25^{th}$ and $1/10^{th}$ of the LD_{50} value (1300 mg/kg).

On day 46 and 91 all the rats were anesthetized and blood was collected by cutting carotid artery in test tube containing Disodium EDTA. Whole blood and plasma were used for haematological and biochemical studies respectively.

After collecting the blood from all the rats, their liver, kidney, spleen and adrenals were also excised.

c) Hepatoprotective studies

Albino rats (120-150 gm) of either sex (Venkateshwara Enterprises, Bangalore) were used. Rats were housed for one week in a group of four or six animals under natural day and night cycle for acclimatization with laboratory environment.

In acute liver damage experiment group I served as control and dose was given i.p. with 4 ml/kg vehicle, i.e., 2% gum tragacanth at 0, 8, 16 hours intervals. One hour after the first injection, coconut oil (4 ml/kg s.c.) was administered as control-the vehicle of the hepato-toxin. The group II was used to study the effect of test sample alone on liver and test sample was administered at 8 hr interval. Group III was treated with single dose of

CCl_4 (50% v/v CCl_4) in coconut oil (4 ml/kg s.c.). The group IV was given a single dose of CCl_4 injection as group III and three doses of test sample (first dose being 1hr before intoxication).

In chronic liver experiment, group I served as control which received 2% gum tragacanth (2 ml/kg i.p.) twice a week. Group II received the test sample twice a week (130 mg/kg b.w., i.p.), group III was given CCl_4 in coconut oil (50% v/v, 2ml/kg s.c.) twice a week. Group IV was injected with test substance twice weekly as group II and CCl_4 as in group III. All the animals were given treatment for 14 days.

RESULTS

a) Acute toxicity studies

Acute toxicity studies was carried out according to method of Miller and Tainter in albino mice, the results of the same are indicated in table and graphs.

The LD_{50} of latex of *Carica papaya* (Tables-1 and 2) suspended in 2% gum tragacanth was found to be

Oral - 1300 mg/kg b.w.

I.p. - 900 mg/kg b.w.

The corrected values in Tables-1 and 2 may be calculated as follows :

$$\text{Correction for 0\% death} = 100 \times \frac{0.25}{n} \qquad (n = 10)$$

$$= 100 \times \frac{0.25}{10} = 2.5\%$$

$$\text{Correction for 100\% death} = 100 \times \frac{(n - 0.25)}{n}$$

$$= 100 \times \frac{(10 - 0.25)}{10} = 97.5\%$$

b) Sub-acute toxicity studies on rats:

The data obtained after haematological and biochemical analysis (Table-3) indicate :

1) Oral administration of latex of *Carica papaya* for 45 days at the dose of 52 mg/kg/day produced significant ($P<0.02$) reduction in Hb values and also significant ($P<0.01$) reduction in TLC values.

Table - 1 : Mortality of animals receiving different oral doses of *Carica papaya* latex

Group	No. of animals	Dose in mg/kg B.W.	Log Dose	Dead/ Total	Death %	Corr-ected %	Probit	LD_{50} value from graph
I	10	370	2.568	0/10	0.00	2.5	3.04	1318 mg/kg body weight
II	10	530	2.724	2/10	20	20	4.16	
III	10	750	2.875	2/10	20	20	4.16	
IV	10	1060	3.027	3/10	30	30	4.48	
V	10	1300	3.1139	5/10	50	50	5.00	
VI	10	1500	3.1761	10/10	100	97.5	6.96	

Table - 2 : Mortality of animals receiving different I.P. doses of *Carica papaya* latex

Group	No. of animals	Dose in mg/kg B.W.	Log Dose	Dead/ Total	Death %	Corr-ected %	Probit	LD_{50} value from graph
I	10	270	2.431	0/10	0	2.5	3.04	891.25 mg/kg body weight
II	10	370	2.568	0/10	0	2.5	3.04	
III	10	530	2.724	2/10	20	20	4.16	
IV	10	750	2.875	3/10	30	30	4.48	
V	10	900	2.954	5/10	50	50	5.00	
VI	10	1060	3.0275	10/10	100	97.5	6.96	

Table - 3 : Effect of daily oral administration of latex of *Carica papaya* for 45 and 90 days on haematobiochemical profiles of rats (n=6)

Parameter	Control		Latex of *Carica papaya* (mg/kg/day)					
			26		52		130	
	Day 46	Day 91	Day 46	Day 91	Day 46	Day 91	Day 46	Day 91
Hb (g%)	11.93 ± 1.10	10.65 ± 0.73	9.51 ± 0.25	9.25 ± 0.34	8.6^{b} ± 0.21	10.15 ± 0.42	11.25 ± 0.55	10.48 ± 0.28
TEC ($10 \times 6/cm^2$)	3.82 ± 0.16	3.55 ± 0.13	3.53 ± 0.11	3.47 ± 0.13	3.40 ± 0.10	3.64 ± 0.06	3.89 ± 0.14	3.89 ± 0.02
TLC ($10 \times 3/cm^2$)	5.09 ± 0.80	5.30 ± 0.24	3.40 ± 0.13	5.68 ± 0.12	1.66^{c} ± 0.04	5.13 ± 0.04	4.32 ± 0.01	7.17^{d} ± 0.03
Glucose (mg %)	61.16 ± 2.61	5.16 ± 1.22	77.1^{d} ± 0.87	79.66^{d} ± 1.28	91.5^{d} ± 1.06	77^{d} ± 1.06	95.66^{d} ± 1.25	61.16^{d} ± 1.08
Cholesterol (mg %)	95.43 ± 1.28	86.83 ± 1.62	115.09^{a} ± 1.52	122.5^{a} ± 1.34	110.09^{a} ± 1.54	125.36^{a} ± 1.09	100.07^{a} ± 1.46	117.43^{a} ± 0.99
Total Proteins (g %)	7.1 ± 0.05	6.5 ± 0.16	7.28^{a} ± 0.08	7.16^{a} ± 0.03	7.05^{a} ± 0.11	6.70^{a} ± 0.1	7.30^{a} ± 0.07	6.38^{a} ± 0.1
BUN (mg %)	23.8 ± 1.25	19.3 ± 0.84	18.66^{a} ± 1.54	16.0^{a} ± 0.96	20.16^{a} ± 1.08	17.66^{a} ± 1.08	22.66^{a} ± 1.54	16.16^{a} ± 1.24
SGPT (i/ml)	26.1 ± 1.22	24.5 ± 1.09	20.40^{a} ± 1.35	18.66^{a} ± 0.92	22.66^{a} ± 1.31	25.83^{a} ± 0.88	18.83^{a} ± 0.98	21.66^{a} ± 1.41
SGOT (i/ml)	47.16 ± 1.01	45.06 ± 1.33	32.83^{a} ± 0.91	42.5^{a} ± 1.18	30.33^{a} ± 1.48	41.16^{a} ± 2.50	45.33^{a} ± 1.41	35.03^{a} ± 0.73

Values are expressed as mean ± SEM. $^{a}P < 0.5$, $^{b}P < 0.05$, $^{c}P < 0.01$, $^{d}P < 0.001$ as compared to control.

2) Oral administration of latex for 90 days at the dose of 130 mg/kg/day produced significant ($P<0.001$) increase in the TLC value.

3) All the three dose levels (26,52 and 130 mg/kg/day) the latex caused significant ($P<0.05$) reduction in BUN, SGOT and SGPT levels in day 46 and 91.

4) However, a significant ($P<0.05$) increase in total protein and cholesterol levels was produced after 45 and 90 days at all the three dose levels.

5) There was also a significant ($P<0.001$) increase in glucose levels after administration of latex at 26,52,130 mg/kg/day orally, however, after 45 and 90 days it did not produce any appreciable alterations in Hb,TEC and TLC values.

The treatment also did not cause any apparent signs of over toxicity or mortality throughout the period of investigation.

Relative organ weights

After oral administration of latex of *Carica papaya* for 45 and 90 days all the rats were sacrificed and their liver, kidney, spleen and adrenals were excised, mopped with filter paper and the relative weights were calculated.

It was found that relative liver and adrenal weights were significantly ($P<0.05$) increased by higher doses (52 and 130 mg/kg/day) of latex of *Carica papaya* without altering significantly the relative weights of kidneys and spleen, at all the three dose levels.

c) Hepatoprotective activity

SGPT, SGOT, alkaline phosphatase and total bilirubin were observed.

Histopathological study was carried out as per the method given by Jayshekar *et al.* (1997).

Enzyme level observations

Group II, treated with test substance in acute chronic test methods, exhibited approximately the same enzyme level in comparison with that of control. This clearly indicates that the test substance, as such, is not hepatotoxic.

Serum enzyme level was very high in rats treated with CCl_4 in group III (Table-4).

In group IV (test substance and CCl_4 treated) the enzyme activity is significantly lower when compared to that of group III and the values are closer to that of the control. There is marked reduction of increased serum bilirubin content of CCl_4 induced acute and chronic liver damage experiments (group IV).

Table - 4 : Effect of latex of *Carica papaya* on serum enzymatic activity and total bilirubin content in acute and chronic liver damage induced by CCl_4

Group	Treatment/ Exposure	SGOT (U/L)	SGPT (U/L)	Alkaline Phosphate (U/L)	Total Bilirubin (mg/dl)
I Control	ACUTE	16.66 ± 0.957	125.66 ± 1.313	32.16 ± 1.049	0.6 ± 0.057
	CHRONIC	18.33 ± 0.643	134.16 ± 1.606	27.66 ± 1.119	0.69 ± 0.574
II T.S. Treated	ACUTE	14.5 ± 0.888	116.8 ± 1.081	26.66 ± 1.057	0.45 ± 0.143
	CHRONIC	21.16 ± 0.749	145.33 ± 1.119	22.5 ± 1.122	0.52 ± 0.194
III CCl_4 Treated	ACUTE	61.83 ± 1.626	205.33 ± 1.458	95.5 ± 0.995	1.33 ± 0.0355
	CHRONIC	75.02 ± 1.719	222.33 ± 1.525	92.16 ± 1.141	1.29 ± 0.0233
IV T.S. and CCl_4 Treated	ACUTE	26.66 ± 1.119	134.50 ± 1.180	23 ± 0.777	0.58 ± 0.055
	CHRONIC	21.83 ± 1.280	131.33 ± 1.312	28.16 ± 1.238	0.56 ± 0.0478

n = six animals in each group. Values are mean ± SEM T.S. = Test sample, CCl_4 = Carbon tetra-chloride.

DISCUSSION

1) Acute toxicity studies

Animals receiving the test substance in doses (1500 mg/kg b.w,i.p) died within 30 minutes of drug administration and they showed CNS stimulant symptoms like hyperactivity, erection of tail and convulsions prior to their death.

The doses of the test substances were decreased subsequently and the LD_{50} values of the latex of *Carica papaya* were found to be :

Oral - 1300 mg/kg

I.P. - 900 mg/kg

2) Sub-acute toxicity studies

Administration of latex of *Carica papaya* (26, 52 or 130 mg/kg) once daily for 45 and 90 days did not exert any observable toxic manifestations. However, slight variation in the values of Hb and TLC on the day 46 and 91 suggest that the chemical, on prolonged exposure, exerted some deleterious effects on haemopoetic system.

The latex produced increase in plasma cholesterol concentration and this may due to hyperadrenal activity, as reflected by the enlargement of adrenals, noted in the present study. Hyperadrenal activity is associated with increased synthesis and release of adreno cortical and medullary hormones which are known to possess lipolytic effect and mobilize the fats from the adipose tissue (Goyal *et al.*, 1986).

The latex did not produce any increase in transaminase (SGPT or SGOT) activities on BUN, thereby indicating that it did not exert any hepatotoxic effect. Instead, it is decreasing the SGPT or SGOT levels, indicating a hepatoprotective effect.

An increase in relative weights of liver and adrenals could be due to chemically induced physical stress.

Thus, the results of the present investigation suggest that the latex of *Carica papaya,* on prolonged exposure to rats, exerts mild haemotoxic effect and may alter the lipid metabolism.

3) Hepatoprotective activity

An elevation in the enzyme levels of cytoplasmic and mitochondrial enzymes in group III reflects that the liver injury is induced by CCl_4.

The latex decresed the CCl_4 induced elevated enzyme levels in acute and chronic experimental models in group IV which is suggestive of production of the structural integrity of hepatocyte cell membrane or regeneration of damaged liver cells as shown by the histopathological and biochemical studies.

Decrease in the serum bilirubin level, after treatment with the latex in both acute and chronic liver damage, indicates the effectiveness of the latex in restoring the normal functional status of liver. Thus, the result of present investigation suggests that the latex of *Carica papya* possesses significant hepatoprotective activity.

ACKNOWLEDGEMENT

We are thankful to Dr. F.V. Manvi, Principal, K.L.E.S's College of Pharmacy, Belgaum for providing the necessary facilities and guidance.

REFERENCES

Ambsta, S.P. (1986). "The Useful Plants Of India". Publications and Information Directorate, Council of Scientific and Industrial Reaserch, New Delhi, 105-106 pp.

Ayub Shah, M.A., Garg, S.K. and Garg, K.M. (1997). Sub-acute toxicity studies on pendimethalin in rats. *Indian Journal of Pharmacology,* **29** : 322-324.

Bhattacharjee, S.K. (1998). "Hand Book of Medicinal plants", Pointer Publishers Jaipur, 71-72 pp.

Goyal, B.S., Garg, S.K. and Garg, B.D. (1986). Malathion induced hyperadrenal activity in WLH chicks. *Curr. Sci.,* **55** : 526-528 pp.

Hoffman, B.B. and Lef Kowitz, R.J. (1996). Catecholamines Sympathomimetics drugs and Adrenergic receptor Antagonists. In : Hardman J.G., Limbird L.E., Molinoff P.B., Ruddon R.W., Gillman A.G. (eds) Goodman and Gilmans. The Pharmacological Basis Of Thereputics. 9th ed. New York McGraw Hill, 199 pp.

Jayshekar, P., Mohanam, P.V. and Ratinam, K. (1997). Hepatoprotective activity of ethyl acetate extract of acacia catechu, *Indian Journal of Pharmacology,* **29** : 426-428 pp.

Kurinan, J.C. (1995). "Plants that Heal", Oriental Watchman Publishing House, Pune, 92-93 pp.

Rouiller, C.H. (1964). "Experimental Toxic Injury of the Liver". In the Liver, **2** : 335-476 pp.

Turner, R.A. (1965). "Screening Methods in Pharmacology", London, Academic Press, 61 pp.

Recent Advances in Medicinal Plant Research : Vision 21st Century, 2005
Editors : S.K. Prabhuji, G.P. Rao and S.K. Patil
Satish Publication House, New Delhi

Chapter 28

HYPOGLYCAEMIC, HYPOLIPIDEMIC AND ANTIOXIDANT EFFECT OF *WITHANIA SOMNIFERA* (L.) DUN FRUITS ON ALLOXAN INDUCED DIABETES IN EXPERIMENTAL RATS

R. Udayakumar[1], A. Ganapathi[2], R. Saravanan[1], R. Gopal[1], M. Sohanabanu[1] and G. Thirupurasundari[3]
[1]Department of Biochemistry, J.J. College of Arts and Science, Pudukkottai - 622 404, Tamilnadu, India
[2]Department of Biotechnology, Bharathidasan University, Tiruchirappalli - 620 024, Tamilnadu, India
[3]Department of Botany, Government Arts College, Thanthonimalai, Karur - 639 005, Tamilnadu, India

Hyperlipidemia is an associated complication of diabetes mellitus. In Egyptian folk medicine, many plants are used in the treatment of diabetes mellitus. *Withania somnifera* is attributed with curative properties against a number of diseases including cancer, diabetes, arthritis and diuretic. There was no report for *W. somnifera* fruits on diabetes mellitus. The experimental rats were divided into 4 groups, such as group 1 - normal control, group 2 - diabetic control (150 mg of alloxan/kg body weight by i.p.), group 3 - diabetic + drug treated (1 g of crude powder extract/kg body weight by p.o.), group 4 - diabetic + glibenclamide (0.6 mg/kg body weight by p.o.) treated and each group consisting of twelve rats. The levels of blood glucose, urine sugar, glycosylated haemoglobin (HbA_1C), serum enzymes like glutamate [1]pyruvate transaminase (SGPT), glutamate oxaloacetate transaminase (SGOT), acid phosphatase (ACP), alkaline phosphatase (ALP), and glucose-6-phosphatase in liver were significantly raised, whereas haemoglobin (Hb), serum total protein, albumin, liver glycogen, sodium and potassium in serum significantly decreased in diabetic rats. The levels of cholesterol, triglycerides and phospholipids were estimated in plasma, liver, kidney and lungs of all groups. The lipid profile was significantly raised in diabetic rats, when compared to normal control rats. Administration of *W. somnifera* fruit extract decreases the diabetic complications by reversing this trend more significantly. This study also reveals the status of antioxidants like superoxide dismutase (SOD), glutathione peroxidase (GPx), glutathione - S-transferase (GST), reduced glutathione (GSH), catalase (CAT) in liver, vitamin C and E in plasma decreased and thiobarbituric acid reacting substances (TBARS) in liver increased in diabetic rats when compared to normal control rats. The *W. somnifera* fruit extract treatment reversed this trend and subsequently significant decrease in the hyperglycaemic and hyperlipidemic complications were noticed. It may be concluded that *W. somnifera* fruit extract possess hypoglycaemic, hypolipidemic and antioxidant properties.

INTRODUCTION

Diabetes mellitus may be defined as a heterogeneous metabolic disorder involving carbohydrate, protein and fat metabolism due to a relative or an absolute lack of insulin. Many herbal products have been described for the care of diabetes mellitus (Chattopadhyay, 1999). Herbal medicine is still the main stay of about 75-80% of the world population, mainly in the developing countries, for primary health care because of better cultural acceptability, better compatibility with the human body and lesser side effects. Several species of plants have been shown to have hypoglycaemic action in animals and humans (Brinker, 1998). More than 400 species have been reported to display hypoglycaemic effects, but a few of them have been investigated (Bailey and Day, 1989). *Withania somnifera* (L.) Dun (Winter cherry), also called 'ashwagandha', belongs to the solanaceae family. The plant is known for its varied therapeutic uses in Siddha, Ayurvedic and Unani practices in India (Nadkarni, 1976). It is used for a very long time for all age groups and in both sexes and even during pregnancy without any side effects (Sharma *et al.*, 1985). *Withania somnifera* is commonly known for its usefulness in the treatment of arthritis in combination with other drugs. It is also credited to be hypoglycaemic and diuretic (Bhattacharya *et al.*, 1997a). This plant is attributed with curative properties against a number of diseases including cancer (Asthana and Raina, 1989). Clinical investigations with the *W. somnifera* root extract indicate that it exerts significant antiageing effect in normal healthy subjects (Kuppurajan *et al.*, 1980). Previous reports demonstrate the growth inhibitory, antitumour and radiosensitizing activities of *W. somnifera* on transplantable tumour sarcoma – 180 in mice (Devi *et al.*, 1992). *Withania somnifera* is reputed to promote physical and mental health and can be classified in modern terminology as an adaptogen (Bhattacharya, 1993). The plant has been used medicinally in folk medicine as an aphrodisiac and geriatric tonic. It is reported to have several pharmacological effects including anthelmintic, narcotic, radiosensitizer, antistressor, adaptogenic and cardio protective effects (Singh *et al.*, 1982; Devi, 1996; Dhuley, 1998; 2000; Archana and Namasivayam, 1999).

The pharmacological effect of the roots of *W. somnifera* is attributed to withanolides, a group of steroidal lactones (Budhiraja and Sudhir, 1987). Several withanolides and alkaloids have been extracted from the plant (Rizki, 1986). The aerial part of *W. somnifera* yielded 5-dehydroxy withanolide R and withasomniferin-A (Atta-ur-Rahman *et al.*, 1991). Withaferin A and 3-¼-hydroxy-2,3-dihydrowithanolide F shows antibacterial, antitumor, immuno modulating and anti-inflammatory properties (Budhiraja and Sudhir, 1987). Glycowithanolides withaferin-A and Sitoindosides VII- X isolated from the roots of *W. somnifera* significantly reversed ibotenic acid induced congenitive defecits in Alzheimer's disease model (Bhattacharya *et al.*, 1995). Increased generation of oxidative free radicals or impaired antioxidant defence mechanisms, have been implicated in the ageing process, neuro degenerative conditions including parkinsonism and Alzheimer's disease in chronic stress induced perturbed homeostasis, immuno depression, inflammation, diabetes mellitus, peptic ulcer and other diseased conditions (Maxwell, 1995). Major oxidative free radical scavenging enzymes and nonenzymes are superoxide dismutase (SOD), catalase (CAT), glutathione peroxidase (GPx), reduced glutathione (GSH), glutathione-S-transferase (GST), lipid peroxidase (LPO), vitamin C and vitamin E. Deficient functioning of these antioxidants leads to accumulation of toxic oxidative free radicals and consequent degenerative changes (Maxwell, 1995). *Withania somnifera* containing ayurvedic herbal formulation was found to attenuate streptozotocin-induced hyperglycaemia in rats, concomitant with augmented oxidative free radical scavenging activity in the pancreas (Bhattacharya *et al.*, 1997b). Earlier studies have reported the absence of any side effects of *W. somnifera* in animals (Aphale *et al.*, 1998). There was no report for *W. somnifera* fruits on diabetes mellitus. So, this study was conducted to assess the hypoglycaemic, hypolipidemic and antioxidant effects of *W. somnifera* fruit extract on alloxan induced diabetes in experimental rats.

METHODOLOGY

Animals

Male albino wistar - strain rats, body weight between 150 – 200 grams, were selected and housed in polypropylene cages. Rats were housed in a temperature - controlled room (25 ± 1 °C) and kept on a 12:12 light dark cycle. Food (Sri Saidurga feeds and foods, Bangalore) and water were available *ad libitum* throughout the experiment.

Chemicals

Alloxan was purchased from Sigma-Aldrich, St. Louis, Mo, USA. Cholesterol was purchased from Himedia, Mumbai, India and thiobarbituric acid (TBA), pyrogallol, hydrogen peroxide, 1-chloro-2,4-dinitrobenzene (CDNB), dithiodinitro benzoic acid (DTNB) and glutathione were purchased from E. merck, Mumbai, India. All other chemicals and reagents, used in the study, were of analytical reagent grade with high purity and were purchased from Loba chemie (P) Ltd., Mumbai, India, Nice chemicals, Kerala, India, Fischer chemicals, Chennai, India and Sd. fine chemicals (P) Ltd., Mumbai, India.

Plant material

Fruits of *W. somnifera* were collected from Nagappattinam District, Tamilnadu, India. The fruits were dried under shade and ground well to fine powder, the powder was sieved through a fine cloth and it was used as drug in the crude form along with distilled water.

Experimental design and Treatment

The rats were divided into four groups of twelve rats each. Diabetes was induced in rats two weeks before starting the treatment.

Group – 1 : Normal control

Group – 2 : Diabetic control

Freshly prepared alloxan in normal saline was administered in a single dose of 150 mg/kg body weight through intraperitoneally to overnight fasted rats. Animals were allowed to develop diabetes for 2 weeks.

Group 3 : Diabetic + drug treated rats

After two weeks of alloxan treatment the rats were treated with *W. somnifera* fruits powder (1g/kg body weight) dissolved in distilled water daily using an intragastric tube for 4 weeks.

Group 4 : Diabetic + glibenclamide treated rats

After two weeks of alloxan treatment the rats were treated with glibenclamide (0.6 mg/kg body weight) dissolved in distilled water daily orally for 4 weeks.

The rats were carefully monitored every day and weighed every week. The sugar levels of urine and blood of all rats were determined. After 4 weeks of treatment the rats were sacrificed by cervical dislocation. Blood was collected and processed for the estimation of glucose, haemoglobin (Hb), glycosylated haemoglobin (HbA$_1$C), total protein, albumin, serum glutamate pyruvate transaminase (SGPT), serum glutamate oxaloacetate transaminase (SGOT), acid phosphatase (ACP), alkaline phosphatase (ALP), sodium, potassium, vitamin C & E and plasma lipids. Tissues such as liver, kidney and lungs were collected in ice-cold containers, weighed accurately and used for lipid extraction, by the method of Folch *et al.*, (1957). The liver was homogenized in 10% using physiological saline. The homogenate was centrifuged at 15,000 X g for 30 minutes at 4^0 C and the supernatant was used for subsequent analysis.

Biochemical estimation

Blood glucose was estimated by Ortho Toluidine method (Sasaki *et al.*, 1972). Haemoglobin was estimated by Drabkin and Austin's method (1932), glycosylated haemoglobin (HbA$_1$C) by Sudhakar and Pattabiraman's method (1981), liver glycogen by the method of Morales *et al.* (1973), serum protein by the method of Lowry *et al.* (1951), albumin by the method of Doumas *et al.* (1977), glucose-6-phosphatase activity by the method of Koide and Oda (1959), acid phosphatase (ACP) and alkaline phosphatase (ALP) activities in serum were estimated by the method of King (1959), serum glutamate pyruvate transaminase (SGPT) and serum glutamate oxaloacetate transaminase (SGOT) activities were estimated by the method of Reitman and Frankel (1957) and sodium and potassium concentrations in serum were estimated by flame photometry. Vitamin C (Roe and Kuether, 1943; Roe, 1961) and vitamin E (Bakes and Frank, 1980) were estimated in plasma. The supernatent was used for the determination of superoxide dismutase (SOD) activity by the method of Marklund and Marklund (1974), catalase (CAT) activity by the method of Sinha (1972), thiobarbituric acid treating substances (TBARS) by the method of Nichans and Samuelson (1968), glutathione peroxidase (GPx) activity by the method of Rotruck *et al.* (1973), glutathionine-S-transferse (GST) activity by the method of Habig *et al.* (1974) and reduced glutathione (GSH) activity by the method of Ellman (1959). Total cholesterol and phospholipids in liver, kidney, lungs and plasma were estimated by the method of Zlatkis *et al.* (1953) and Zilversmit and Davies (1950) respectively. Triglycerides were estimated by the method of Foster and Dunn (1973).

Statistical analysis

Results were analysed using ANOVA, and a scheffe post hoc test was used to compare individual means. Significance was accepted at 5% level ($P<0.05$). All results are expressed as mean + SD of six samples from each group.

RESULTS AND DISCUSSION

Diabetes mellitus is associated with accumulation of biochemical, histological and functional changes and progressive increase in the change of diseases and death of an organism. Alloxan, a beta cytotoxin, includes 'Chemical diabetes' in a wide variety of animal species by damaging the insulin secreting cells of the pancreas (Bopanna *et al.,* 1997). This experimental animal model of diabetes mellitus is similar to type 1 diabetes, in which insulin secretion is defective. Literature sources indicate that alloxan diabetic rats are hyperglycaemia (Stanley *et al.,* 2000). Damage due to hyperglycaemia and lipid peroxidation appears to be an important determinative mechanism of diabetic complications.

Changes in initial and final body weight in control and experimental groups are shown in Table-1. Significant weight loss was observed in diabetic rats more than on normal control rats. Treatment of *W. somnifera* fruit extract improved the weight gain when compared to diabetic control rats. Moreover weight gain was less in the diabetic rats when compared to the normal control rats. Similar results were observed by Al-Shamaony *et al.* (1994).

Alteration in the blood glucose and urine sugar on treatment of diabetic rats with *W. somnifera* fruit extract and glibenclamide individually are given in Table-2. The blood glucose (261.49 mg/dl) and urine sugar (+++) were significantly increased in alloxan diabetic rats when compared to the normal

Table - 1 : Effect of *Withania somnifera* fruit extract on body weight

Groups	Initial weight (g)	Final weight (g)	Change in body weight (g)	Change in body weight (%)
Group 1 : Normal control	162 ± 3.90	210 ± 7.07	+48 ± 7.82	29.6
Group 2 : Diabetic control	165 ± 8.94	146 ± 8.98	-19 ± 5.51	11.5
Group 3 : Diabetic + *Withania somnifera* fruit extract treated	162 ± 7.16	184 ± 8.64	+22 ± 2.40	13.98
Group 4 : Diabetic + glibenclamide treated	170 ± 9.24	200 ±13.93	+30 ± 5.85	17.65

Values are expressed as mean ± SD of twelve rats from each group.

Table - 2 : Effect of *Withania somnifera* fruit extract on urine sugar, blood glucose, haemoglobin, glycosylated haemoglobin and liver glycogen

Groups	Urine sugar	Blood glucose (mg/dl)	Haemoglobin (g/dl)	Glycosylated Haemoglobin (mg/g Hb)	Liver glycogen (mg/g tissue)
Group 1 : Normal control	Nil	86.68 ± 1.1128^{a}	13.68 ± 0.3923^{ab}	0.92 ± 0.08877^{a}	3.64 ± 0.05797^{c}
Group 2 : Diabetic control	+++	261.49 ± 1.6454^{d}	12.32 ± 0.4036^{a}	2.06 ± 0.1493^{c}	2.06 ± 0.0662^{a}
Group 3 : Diabetic + *Withania somnifera* fruit extract treated	+	112.46 ± 4.3349^{b}	13.98 ± 1.1163^{b}	1.32 ± 0.09879^{b}	3.48 ± 0.06812^{b}
Group 4 : Diabetic + glibenclamide treated	+	120.38 ± 5.5567^{c}	13.80 ± 0.6986^{ab}	1.40 ± 0.0681^{b}	3.50 ± 0.0894^{b}

Values are expressed as mean ± SD of six samples from each group
Values not having common letters are significant at 5 % level ($P< 0.05$).

control rats (86.68 mg/dl). The observation of this study correlates with the previous research findings, that the blood glucose and urine sugar levels were elevated significantly in alloxan diabetic rats. Moreover, administering *W. somnifera* fruit extract to diabetic rats significantly reduced the blood glucose (112.46 mg/dl) and urine sugar levels. *Withania somnifera* fruit extract, possibly, brings about its hypoglycaemic action by potentiating the insulin effect, either by increasing the pancreatic secretion of insulin from the cells of islets of langerhan's or its release from bound insulin. *Withania somnifera* fruit extract exhibited reduction in blood sugar as observed in normal control indicating that *W. somnifera* fruit extract contains an effective regulator of blood sugar and thus useful in diabetes (Andallu *et al.*, 2000). In this context a number of other plants and plant products have also been observed to have hypoglycaemic effects (Prasanna, 2000). Haemoglobin content was significantly decreased (12.32 g/dl) in diabetic rats when compared to normal control rats (13.68 g/dl). Glycosylated haemoglobin (HbA_1C) was increased (2.06 mg/g Hb) in diabetic rats when compared to normal control rats (0.92 mg/g Hb). Liver glycogen was significantly decreased (2.81 mg/g) in alloxan induced diabetic rats when compared to normal control rats (3.64 mg/g). Administering of *W. somnifera* fruit extract tends to bring the values of haemoglobin (13.98 g/dl), glycosylated haemoglobin (1.32 mg/g Hb) and glycogen (3.48 mg/g tissue) to near those of the normal control rats (Table-2). A significant reduction in the level of haemoglobin was observed in diabetic rats when compared to normal control rats. But the glycosylated haemoglobin (HbA_1C) which is part of the total haemoglobin was significantly higher in diabetic rats when compared to the normal rats due to diabetes, the excess glucose present in circulation react with Hb to form HbA_1C. So, the decrease in total Hb during diabetes may be due to the formation of HbA_1C. Glycogen synthesis in the rat liver and skeletal muscles was impaired during diabetes (Huang *et al.*, 2000). Similar type of results were observed in this study. The level of Hb, glycosylated Hb and liver glycogen were reverted to near normal in the treatment of *W. somnifera* fruit extract when compared to untreated diabetic rats.

The levels of serum protein (7.20 g/dl), albumin (4.0 g/dl), sodium (126.0 meq/l) and potassium (3.81 meq/l) were observed in normal control rats. In the alloxan induced diabetic rats the levels of protein (6.40 g/dl), albumin (3.4 g/dl), sodium (86.0 meq/l) and potassium (2.12 meq/l) were decreased when compared to normal control rats. Treatment with *W. somnifera* fruit extract tends to bring the values of protein (7.0 g/dl), albumin (3.8 g/dl), sodium (108.0 meq/l) and potassium (3.67 meq/l) to near those of the normal control rats (Table-3). Renal disease is one of the most common and severe complications of diabetes. Distinct metabolic renal alterations are demonstrable in experimental diabetes leading to a negative nitrogen balance, enhanced proteolysis and lowered protein synthesis (Pathak and Dhawan, 1998). The total protein and albumin were found to be reverted

Table - 3 : Effect of *Withania somnifera* fruit extract on serum total protein, albumin, sodium and potassium.

Groups	Serum total protein (g/dl)	Albumin (g/dl)	Sodium (meq/l)	Potassium (meq/l)
Group 1 : Normal control	7.20 ± 0.4733[a]	4.0 ± 0.2828[a]	126.0 ± 4.2426[c]	3.81 ± 0.08832[d]
Group 2 : Diabetic control	6.40 ± 0.4733[a]	3.4 ± 0.3178[a]	86.0 ± 3.7947[a]	2.12 ± 0.02366[a]
Group 3 : Diabetic + *Withania somnifera* fruit extract treated	7.00 ± 0.2828[a]	3.8 ± 0.2733[a]	108.0 ± 6.8118[b]	3.67 ± 0.02828[c]
Group 4 : Diabetic + glibenclamide Treated	7.10 ± 0.5762[a]	4.0 ± 0.3797[a]	105.0 ± 3.1623[b]	3.54 ± 0.07797[b]

Values are expressed as mean ± SD of six samples from each group
Values not having common letters are significant at 5 % level ($P < 0.05$).

when compared to normal control rats (11.47 mg/dl and 10.54 mg/dl). Administration of *W. somnifera* fruit extract enhances the level of vitamin C and E of plasma (9.26 mg/dl) (10.73 mg/dl), respectively.

The level of liver lipid peroxidation in the *W. somnifera* fruit extract treated diabetic rats indicates that fruit extract can reduce lipid peroxidation significantly. The lipid peroxidation is one of the characteristic features of chronic diabetes (Maxwell *et al.*, 1997). The increased level of thiobarbituric acid reactive substances (TBARS) indicates the lipid peroxidation. The level of TBARS decreased significantly in the liver tissue of rats given *W. somnifera* fruit extract and glibenclamide individually when compared to the diabetic control rats where the levels were quite high. This indicates that *W. somnifera* fruit extract may inhibit oxidative damage of the liver tissue. The hypoglycaemic effect of *W. somnifera* fruit extract could be due to its property to increase the level of serum insulin and or the activities of catalase, superoxide dismutase and glutathione peroxidase, indicative of its antioxidant property (Bhattacharya *et al.*, 1997b). Pancreatic islet cells are having very low levels of free radicals scavenging enzymes, including SOD, and are therefore vulnerable to free radical toxicity (Mason and Chignell, 1982). Free radical oxidative stress has been implicated in the pathogenesis of a variety of diseases, resulting usually from defective natural antioxidant defences. Potential antioxidant therapy should, therefore include either natural antioxidant enzymes or agents which are capable of augmenting the function of these oxidative free radical scavenging enzymes (Bast *et al.*, 1991). Under *in vivo* condition, glutathione (GSH) acts as an antioxidant and its decreased is reported in diabetes mellitus (Illing *et al.*, 1951). The increased GSH content in the liver of the rats treated with *W. somnifera* fruit extract may be one of the factors responsible for the inhibition of lipid peroxidation. Superoxide dismutase and catalase are the two major scavenging enzymes that remove the toxic free radicals *in vivo.* Vucic *et al.* (1997) reported that the activity of SOD is low in diabetes mellitus. The *W. somnifera* fruit extract treated rats showed decreased lipid peroxidation associated with increased activity of SOD and catalase. The antioxidants like glutathione peroxidase (GPx) and glutathione-S-transferase (GST) in liver and vitamin C and E in plasma were enhanced in the *W. somnifera* fruit extract treated rats.

The results show the positive anti-peroxidative activity of *W. somnifera* fruit. *Withania somnifera* and their glycowithanolides have been reported to increase SOD, CAT and GPx activities in the rat brain (Bhattacharya *et al.*, 1995; Bhattacharya *et al.*, 1997b). In conclusion, our results have shown that crude powder extract of *W. somnifera* fruits given orally possess hypoglycaemic, hypolipidemic and antioxidant effects in alloxan induced diabetic rats.

REFERENCES

Al-Shamaony, L., Al-Khaznaji, S.M. and Twaij, I.A.A. (1994). Hypoglycaemic effect of Artemisia herba alba. 11. Effect of valuable extract on some blood parameters in diabetic animals. *J. Ethnopharmacol.*, **8** : 167-171.

Alta-ur-Rahman, Samina-Abbas, Dur-e-shahwar, Jamal, S.A., Choudhary, M I. and Abbas, S. (1991). New Withanolides from *Withania spp. J. Nat. Prod.*, **56** : 1000-1006.

Andallu, B., Radhika, B. and Dawar, R. (2000). Hypoglycaemic, diuretic and hypocholesterolemic effects of winter cherry (*Withania somnifera* L. Dun) root. *Ind. J. Exp. Biol.*, **6** : 607-609.

Aphale, A.A., Chhibba, A.D., Kumbhakarna, N., Mateenuddin, M. and Dahat, S.H. (1998). Subacute toxicity study of the combination of ginseng and ashwagandha in rats: A safety assessment. *Ind. J. Physiol. Pharmacol.*, **42** : 299-302.

Archana, R., Namasivayan, A. (1999). Antistressor effect of *Withania somnifera*. *J. Ethnopharmacol.*, **64** : 91-93.

Asthana, R. and Raina, M.K. (1989). Pharmacology of *Withania somnifera* (Linn) Dunal: a review. *Indian Drugs,* **26(5)** : 199-204.

Bailey, C.J. and Day, C. (1989). Traditional treatments for diabetes. *Diabetes care,* **12** : 553-564.

Bakes, H., Frank, O., De Angeles, B. and Rein Gold, S. (1980). Plasma tocopherol in man at various times after ingesting free or acetylated tocopherol. *Nutr. Rep. Int.*, **20** : 531.

Bast, A., Haenen, G.R. and Doelman, G.J. (1991). *Am J Med.*, **91** : 25.

Bhattacharya, S.K. (1993). In: *Traditional medicine*, (Ed by Mukherjee B), IBH publishing Co., New Delhi, 320.

Bhattacharya, S.K., Kumar, A. and Ghosal, S. (1995). Effects of Glycowithanolides from *Withania somnifera* on animal model of Alzheimer's disease and perturbed central cholinergic markers of cognition in rats. *Phytother Res.*, **9** : 110-113.

Bhattacharya, S.K., Satyam, S.K. and Chakrabarti, A. (1997a). Effect of Trasina an Ayurvedic herbal formulation, on pancreatic islet superoxide dismutase activity in hyperglycaemic rats. *Ind. J. Exp. Biol.*, **35** : 297–299.

Bhattacharya, S.K., Satyan, K.S and Ghosal, S. (1997b). Antioxidant activity of glycowithanolides from *Withania somnifera. Ind. J. Exp. Biol.*, **35** : 236-239.

Bopanna, K.N., Kannan, J., Gadgil, S., Balaraman, R. and Rathod, S.P. (1997). Antidiabetic and antihyperlipemic effects of neem seed kernel powder on alloxan-diabetic rabbits. *Ind. J. Pharmacol.*, **29** : 162-167.

Brinker, F. (1998). Herb contraindications and drug interactions. 2nd edn. Sandy. OR: Eclectic Medical Publications. 36-82.

Budhiraja, R.D. and Sudhir, S. (1987). Review of biological activity of Withanolides. *J. Sci. Ind. Res.*, **46** : 488-491.

Chatterjea, M.N. and Shinde, R. (1994). Metabolism of Carbohydrates. Part II. Text book of Medical biochemistry. Ist edn. Jay pee Brothers, Medical Publishers Pvt. Ltd., 421.

Chattopadhyay, R.R. (1999). A Comparative evaluation of some blood sugar lowering agent of plant origin. *J. Ethnopharmacol.*, **67** : 367-372.

Chinsky, M.S. and Shessy, H. (1957). *Am. J. Med. Soc.*, **223** : 400.

Devi, P.U., Sharda, A.C., Solomon, F.E. and Kamath, M.S. (1992). In vivo growth inhibitory effect of *Withania somnifera* (Ashwagandha) on a transplantable mouse tumor sarcoma –180. *Indian J. Exp. Biol.*, **30** : 169-172.

Devi, P.U. (1996). *Withania somnifera* dunal (Ashwagandha): potential plant source of a promising drug for cancer chemotherapy and radiosensitization. *Ind. J. Exp. Biol.*, **34** : 927-932.

Dhuley, J.N. (1998). Therapeutic efficacy of ashwagandha against experimental asparigillosis. *Immmunopharmacol. immunotoxicol.*, **20** : 191-198.

Dhuley, J.N. (2000). Adaptogenic and cardioprotective action of ashwagandha in rats and frogs. *J. Ethnopharmacol.*, **70** : 57-63.

Doumas, B.T., Watson, W.A. and Biggs, H.G. (1977). Albumin standards and the measurement of serum albumin with bromocresol green. *Clin. Chem. Acta.*, **31** : 87-96.

Drabkin, D.L. and Austin, J.M. (1932). Spectrophotometric constraints for common haemoglobin derivatives in human, dog and rabbit blood. *J. Biol. Chem.*, **98** : 719-733.

Ellman, G.L. (1959). Tissue sulphydryl groups. *Arch Biochem Biophy.*, **82** : 70-77.

Folch, J., Lees, M. and Stanely, G.H.S. (1957). A simple method for the isolation and purification of total lipids from animal tissues. *J. Biol. Chem.*, **256** : 497-509.

Foster, C.S. and Dunn, O. (1973). Stable reagents for determination of serum triglycerides by a colorimetric Hantzsch condensation method. *Clin. Chem.*, **19** : 338-340.

Habig, W.H., Pabst, M.J. and Jakoby, W.R. (1974). Glutathione S – transferase: the first enzymatic step in mercapturic acid formation. *J. Biol. Chem.*, **249** : 7130-7139.

Huang, X., Vagg, A., Hanson, M., Weng, J., Laurila, E. and Group, L. (2000). Impaired insulin stimulated expression of the glycogen synthase gene in skeletal muscle of type 2 diabetic patiens is acquired rather than inherited. *J. Clin. Endocrinol. Metab.*, **85** : 1584.

Kannel, W.B. and Mc Gee, D.L. (1979). Diabetes and cardiovascular risk factors, the Framingham study. *Circulation,* **59** : 8-13.

Khosla, P. (1995). Effect of *Trigonella joenum graecum* on serum lipids in normal and diabetic rats. *Ind. J. Pharmacol.*, **27** : 89-93.

King, J. (1959). Determination of serum alkaline and acid phosphatase. In: Practical clinical Enzymology. Dvan Nostrand, London.

Koide, H. and Oda, T. (1959). Pathological occurrence of glucose-6-phosphatase in serum in liver diseases. *Clin. Chem. Acta.* **4** : 554-561.

Kuppurajan, K., Rajagopalan, V., Janaki, K., Revathi, R. and Venkataraghavan, S. (1980). Effects of Ashwagandha on the process of ageing in human volunteers. *J Res. Ayurveda. and Siddha,* **1(1)** : 247-258.

Illing, E.K.S., Gray, C.H. and Lawrence, R.D. (1951). Blood glutathione and non-gulcose reducing substances in diabetes. *Biochem. J.,* **48** : 637.

Lowry, O.H, Roesborough, M.J., Farr, A.L. and Randall, R.J. (1951). Protein measurement with Folin-phenol reagent. *J. Biol. Chem.,* **193** : 265-275.

Marklund, S. and Marklund, G. (1974). Involvement of superoxide dismutase anion radical in the antioxidant of pyrogallol and a convenient assay for SOD. *Eur. J. Biol. Chem.,* **47** : 469-474.

Mason, R.P. and Chignell, C.F. (1982). *Pharmacol Rev.,* **33** : 189.

Maxwell, S.R.J., Thomason, H., Sandler, D., Le Guen, C., Baxtex, A.M., Thorpe, G.H.G., Jones, A.F. and Barnett A.H. (1997). Poor glycaemic control is associated with reduced serum free radical scavenging (antioxidant) activity in non-insulin dependent diabetes mellitus. *Annal. clin. Biochem.,* **34** : 638-644.

Maxwell, S.R.J. (1995). *Drugs,* **4** : 345.

Morales, M.A., Jabbagy, A.J. and Terenizi, H.R. (1973). Mutations affecting accumulation of glycogen. *Neurospora News,* **20** : 24.

Murray, R.K., Granner, D.K., Mayes, P.A. and Rodwell, V.W. eds. (2000). *Harper's biochemistry*-25th edn. Stanford CT: Appleton and Lange, 610-617.

Nadkarni, K.M. (1976). *Withania somnifera* (L.) Dunal. The Indian Meteria Medica. Vol (Popular Prakashan Pvt., Ltd) 1292-1294.

Navarro, C.M., Montilla, P.M., Martin, A., Jimenez, J. and Utrilla, P.M. (1993). Free radicals scavenger and antihepatotoxic activity of Rosmarinus. *Plant Med.,* **59** : 312-314.

Nichans, W.C, Jr. and Samuelson, B. (1968). Formation of Malondialdehyde from phospholipid arachidonate during microsomal lipid peroxidation. *Eur. J. Biochem.,* **6** : 126.

Pathak, A. and Dhawan, D. (1998). Effect of lithium on the levels of blood urea and creatinine in diabetic rats. *Med. Sci. Res.,* **26** : 855.

Prasanna, M. (2000). Hypolipidemic effect of fenugreek: a clinical study. *Ind. J. pharmacol.,* **32** : 34-36.

Reitman, S. and Frankel, S.A. (1957). Colorimetric method for the determination of serum glutamate oxaloacetic and glutamate pyruvic transaminases. *Am. J. Clin. Pathol.,* **28** : 56-63.

Rizk, A.M. (1986). The phytochemistry of the flora of Qatar. King Print, Richmond

Roe, J.H. and Kuether, C.A. (1943). The determination of ascorbic acid in whole blood and urine through the 2,4 dinitrophenyl hydrazine derivatives of dehydroascorbic acid. *J. Biol. chem.*, **147** : 2399.

Roe, J.H. (1961). *Standard methods in clinical chemistry* (Ed by David Seligson), Academic press, New York. **8** : 35.

Rotruck, C.T., Pope, A.L., Ganther, H.E., Swanson, A.B., Hafeman, D.G. and Hoekstara, W.G. (1973). Selenium : Biochemical role as a component of glutathione peroxidase. *Science,* **179** : 588-590.

Sasaki, T., Matsy, S. and Sonae, A. (1972). Effect of acetic acid concentration on the colour reaction in the O-toluidine boric acid method for blood glucose estimation. *Rinshokagaku,* **1** : 346-353.

Scott, M. Grundy. (1999). Diabetes and cardiovascular disease. *Circulation,* **100** : 1134-1146.

Sharma, S., Dahanukar, S. and Karandikar, S.M. (1985). Effect of long term administration of the roots of ashwagandha and Shatavari in rats. *Indian Drugs,* **29** : 133 -139.

Singh, N., Nath, R., Lata. A., Singh, S.P., Kohli, R.P. (1982). *Withania somnifera,* a rejuvenating herbal drug which enhances survival during stress (adaptogen). *Int. J. Crude. Drug. Res.,* **20** : 29.

Sinha, A.K. (1972). Calorimetric assay of catalase. *Annal. Biochem.,* **47** : 389.

Stanley Mainzen Prince, P. and Menon, V.P. (2000). Hypoglycaemic and other related actions of *Tinospora cardifolia* roots in alloxan – induced diabetic rats. *J. Ethnopharmacol.,* **70** : 9-11.

Sudhakar, N.S. and Pattabiraman, T.N. (1981). A new colorimetric method for the estimation of glycosylated haemoglobin. *Clin. Chem. Acta.,* **109** : 267-274.

Vucic, M., Gavell, M., Bozikov, V., Asheroft, J.H. and Rocic, B. (1997). Superoxide dismutase activity in lymphocytes and polymorphonuclear cells of diabetic patients. *Eur. J. Clin. Chem. Clin. Biochem.,* **35** : 517-521.

Zilversmit, D.B. and Davies, A.K. (1950). Micro determination of plasma phospholipid by TCA precipitation method. *J. Lab. Clin. Med.,* **35** : 155-160.

Zlatkis, A., Zak, B. and Boyle, A.J. (1953). A method for the direct determination of serum cholesterol. *J. Lab. Clin. Med.,* **45** : 486-492.

Recent Advances in Medicinal Plant Research : Vision 21st Century, 2005
Editors : S.K. Prabhuji, G.P. Rao and S.K. Patil
Satish Publication House, New Delhi

Chapter 29

INFLUENCE OF *ARTEMISIA PALLENS* WALL PLANT ON EXPERIMENTAL WOUNDS IN ALBINO RATS

V.P. Rasal, Ajay Kshirsagar★, Raj Bagali and Santosh K. Rai
Department of Pharmacology, K.L.E.S's College of Pharmacy, J.N.M.C Campus, Nehru Nagar, Belgaum - 590 010, Karnataka, India
★e-mail : ajay_ksagar@yahoo.com

Studies have been undertaken for the excision, resutured incision and dead space wound healing property of various extracts of *Artemisia pallens* Wall leaves.

Alcoholic and aqueous extracts of *Artemisia pallens wall* leaves were prepared. Wistar rats of either sex were tested for excision, dead space, resutured incision wounds taking aseptic precaution. Control animal received normal saline and other group received various extracts of *Artemisia pallens* plant. The various parameters monitored include wound contraction, time for complete epithelization and scar area and shape on complete epithelization for excision wounds, tensile strength of resutured incised wounds, breaking strength, quantification of granulation tissue and their histological studies for dead space wounds in control and extracts treated groups. The present study demonstrates

the wound healing potential of Artemisia *pallens* plant.
The percentage reduction in wound size, collagen content and thickness of the epithelium in alcoholic and aqueous extracts treated group was found to be significant than the control and other extracts.
The alcoholic and aqueous extracts of *Artemisia pallens* plant significantly favoured wound healing.

INTRODUCTION

Artemisia pallens Wall, commonly known as Davana, is reputed for its medicinal use as antihyperglycemic (Subramaniam *et al.*, 1996), anthelmentic, tonic and antipyretic. Davana improves the hormonal balance (Drurey and Wallonglon, 1980), relieves, excess gases and painful spasm (The Wealth of India, 1985). The present study describes the influence of ethanolic and aqueous extract of *Artemisia pallens* plant on open wound healing, resutured incision and dead space wound healing. Wound contraction and histopathological studies have also been carried out.

METHODOLOGY

Plant Material and Extraction

The plant *Artemisia pallens* was collected from Kolhapur region in the month of November-December. The plant was dried and cut into small pieces and crushed mechanically to obtain a coarse powder.

The powdered plant was extracted with ethanol using a Soxhlet's extraction apparatus and concentrated and dried under reduced pressure. The aqueous extract was prepared with 50 g of powdered herb soaked in 250 ml distilled water for one week at 60-70 °C. The extract was filtered followed by evaporation under reduced pressure.

Animals

Albino rats of wistar strain of either sex weighing 150-200 g were used. All the animals were kept under identical housing conditions of temperature (23 ± 1 °C), 12 h light/dark cycle and fed with standard pellet diet with water *ad libitum*.

Acute toxicity studies

Acute toxicity studies were carried out on the extracts as per OECD revised draft guidelines 423, received from CPCSEA, Govt. of India. The marked toxicity at the dose of 2000 mg/kg has been observed, hence $1/10^{th}$ of this lethal dose was taken as LD_{50} cutoff value of extract and effective dose (therapeutic dose) for subsequent wound healing activity was taken (200 mg/kg) for the treatment of groups, respectively.

Treatment

Animals were maintained under laboratory conditions before the experiment was carried out. Wounds were inflicted under light ether anesthesia. Animals were, then, housed individually in clean poly-propylene cages. The experimental protocols were approved by the Institute Animal Ethics Committee. The animals were randomly divided into three different groups of six animals each. A group of six rats were taken as control. The other groups (2 and 3) received treatment of ethanolic and aqueous extracts of *Artemisia pallens* plant respectively.

Wound Contraction Studies

Animals were anesthetized with ether and the back of the animals were shaved and washed with spirit. An impression was made on the dorsal thoracic central region, 5 mm away from the ears by using a round seal of 2.5 cm diameter as described by Mortan and Malone (1972).

The wound contractions were measured as a percentage reduction in the wound area every 4 days. The progressive decrease in the wound area was monitored periodically by tracing the wound margin on a tracing paper and the area assessed using a graph paper. The results were expressed as percentage reduction in wound size.

Resutured Incision wound

Under light ether anaesthesia, the animals were secured to an operation table in its natural position. Two paravertebral straight incisions of 6 cm were made through the entire thickness of the skin on either side of the vertebral column with the help of a sharp blade. Care was taken to see that the incisions were atleast 1 cm lateral to the vertebral column. After complete haemostasis, the wounds were closed by means of interrupted sutures placed at equidistant points about 1 cm apart using 4 – zero silk thread and round needle. Wounds were then mopped with cotton swabs soaked in 70 % alcohol and were caged individually. Removal of sutures was done on 8th post wounding day and breaking strengths of the wounds were determined on 10th post wounding day by continuous, constant water flow technique by using Ehrlich and Hunt (1969) method.

Granuloma studies (Dead Space Wounds)

In this model, the physical, mechanical, histological, biochemical changes in the granuloma tissue were studied. Under light ether anesthesia, the hairs in the axilla and groin were clipped out and a subcutaneous dead space wounds were inflicted in the same region, by making a pouch through a small nick in the skin.

Granuloma formation was induced by implanting either sterile cotton pellets or grass piths.

a) Two sterile cotton pellets weighing 10 mg (sterilized by autoclaving) were used to grow granuloma by the technique of D' Arcy *et al.* as described by Turner (1965).

b) As described above two cylindrical grass piths measuring 25 × 3 mm were also introduced in the subcutaneous pouch in each animal.

The sutured wounds were mopped with an alcoholic swab and animals were placed into their individual cages.

The granulomas were excised from the surrounding tissue on 10th post – wounding day under light ether anesthesia. Cotton pellet granuloma excised from dead space wounds were dried overnight at 60° C so as to obtain constant dry weight. Their weights were expressed as mg /100 g body weight.

Granuloma surrounding grass piths were excised and slit open by a longitudinal incision in one plane so as to obtain rectangular strips. The breaking strength of a strip of granuloma measuring about 15 mm in length and 8 mm in width (obtained by trimming the rectangular strip of granuloma tissue) was measured.

Histochemical Studies

A group of 30 wistar albino rats of either sex weighing 150 – 180g were selected for the study. Excision wound and treatment to all the animals were made in the same manner as described above. The regenerated tissues extracted from the open wounds were collected from each 6 animals group sacrificed at every 4th, 8th, 12th, 16th and 20th days and were divided into two equal parts. One part was placed in 10 % buffered formalin for histopathological studies and the other portion was used for histochemical studies for the estimation of collagen as per the method of Udupa (1970).

Histopathological studies

Sections from regenerated tissues were observed under light microscope for epithelization, fibroblast, collagen, cell infiltration (inflammation) and neovascularization. The results were visually quantified by numbering the results from 1 to 5 with 5 standing for maximum similarity and 1 standing for least similarity from the normal tissues around the wound area in both test and control wounds (Drurey and Wallonglon, 1980).

Statistical Analysis

The significance of the results were statistically evaluated using student's't' –test and ANOVA followed by Dunnett's 't' test.

RESULTS AND DISCUSSION

Wound Contraction Studies

Ethanolic and aqueous extracts of *Artemisia pallens* plant showed significant wound healing activity as compared to the control and other extracts. The rate of wound contraction was maximum on 8th day in the test as compared to the control group (Table 1).

Table - 1 : Mean raw wound area (mm^2) of original excision wound at different time intervals in control and treated groups

Day	Control (Normal Saline)	Ethanolic extract (200mg/kg)	Aqueous extract (200mg/kg)
0th day	490	497	503
4th day	394	370	385
8th day	175.5	151.5	165.5
12th day	101.5	75.3	91.8
16th day	57.7	40.4	48.2
20th day	26.3	4.6	17.5

Resutured Incision wounds

The tensile strength of ethanolic extract of *Artemisia pallens* treated group was significant. It was 419.6 ± 22.43 for ethanolic extract, 326.81 ± 17.39 for aqueous extract as compared to control - 263.35 ± 9.105 (Table-2).

Granuloma studies

a) Mean dry weight of 10 day old cotton pellet granuloma dry weight in control group (normal saline) was 35.5 ± 1.263 mg, whereas the ethanolic extract treated group was 47.16 ± 2.686 mg. It showed a significant increase in ($p<0.0014$)granuloma weight in ethanolic extract treated group as compared to control group. In case of aqueous extract (200 mg/kg) treated group it was 94.75 ± 4.266 mg, indicating that there was a significant increase ($p<0.0317$) of granuloma formation as compared to control group.

b) The mean breaking strength of granuloma pith in control group (normal saline) was 263.35 ± 9.105 gm, whereas the ethanolic extract treated group was 419.6 ± 22.43 gm, indicating that there was significant ($P<0.001$) increase in breaking strength of granuloma pith as compared to control. Similarly, the aqueous extract treated

Table - 2 : Tensile strength after 10th post wounding day (gram) on resutured incision wound in different treated groups.

Animal No.	Control (Normal Saline)	Ethanolic Extract (200mg/kg)	Aqueous extract (200mg/kg)
1.	237.4	329.4	352.4
2.	260.4	460.2	331.2
3.	271.3	470.3	323.4
4.	238.5	460.7	270.4
5.	292.4	389.5	390.5
6.	280.1	407.5	293.0
Mean ± SE	263.35 ± 9.105	419.6 ± 22.43	326.81 ± 17.39
t-value	—	6.454	3.234
p-value	—	0.001	0.0045

F-value à 20.85, $p<0.1904$
Dunnet's 't' test à $p<0.01$ for ethanolic extract, $p<0.05$ for aqueous extract.
Note : Significantly increased breaking strength in both extract treated groups.

group was 326.81 ± 17.39gm, indicating significant increase in tensile strength of incision wound ($p<0.004$), as compared to control.

Collagen Content Studies

The collagen content was estimated from regenerated tissues, in control as well as various treated groups. There was significant increase in collagen content on 4th, 8th, 12th, 16th and 20th days in ethanolic (200 mg/kg) and aqueous (200 mg/kg) extract treated groups as compared to the control group and both the extract of *Artemisia pallens* plant showed a statistical significance of collagen content ($p<0.02$). (Table-3).

Histopathological studies

a) **Epithelization :** From the result, it was observed that there was no much difference in epithelization on the 4th and 8th day between the test and Control groups. On 12th, 16th, 20th day the ethanolic and aqueous extract (200 mg/kg) treated groups showed significant increase in epithelization as compared to control wounds.

b) **Inflammation :** From the results obtained there was no considerable difference in the inflammatory response between the test and control wounds up to the 8th day. It showed a significant decrease on 12th day onwards in ethanolic and aqueous extract treated groups.

Table - 3 : Collagen content from regenerated tissues of excision wound (mg/gm)

Day	Control (Normal Saline)	Ethanolic Extract (200mg/kg)	Aqueous extract (200mg/kg)
4	8.59 ± 1.72	18.57 ± 3.4★	8.22 ± 0.700
8	13.61 ± 1.91	25.52 ± 3.86★	16.72 ± 0.769★
12	17.23 ± 2.37	32.2 ± 3.18★★	18.26 ± 0.568★
16	32.03 ± 0.83	39.81 ± 2.32★	35.65 ± 0.858★
20	38.27 ± 2.23	41.7 ± 0.68★	39.02 ± 1.042

Values are Mean ± SEM (4) readings.
p-values – p<0.05★, p<0.02★★ Vs. Control.

c) **Collagen Content :** Collagenization was significant in ethanolic and aqueous extract treated groups (200 mg/kg).

d) **Neovascularisation :** Histopathological studies showed significant (p<0.05) neovascularisation on the fourth day in the test group wounds.

e) **Cellularity (Fibroblast) :** There was no quantitative difference in fibroblast levels of test groups on any of the days tested (Table-4).

The process of wound healing occurs in four phases – 1) Coagulation, which prevents blood loss, 2) Inflammation of wound, 3) Repair, including cellular proliferation and 4) Tissue remodeling and collagen deposition (Smith, 1985). Any agent which accelerates the above process is a promoter of wound healing. Various reports are available on the plants promoting wound healing and good therapeutic potential as anti-inflammatory agents, due to the presence of active terpenes, alkaloids and flavanoids (Husain, 1990; The Wealth of India, 1985).

Alcoholic extract of *Artemisia pallens* plant prompted wound healing on oral administration in rats, are evidenced by wound contraction, resutured incision wounds, histochemical and histopathological studies. The increased tensile strength in the test group indicates wound healing activity of the ethanolic extract compared to the control group. This is further supported by significant increase in the mean breaking strength of granuloma pith.

The plant *Artemisia pallens* Wall has been reported to contain sequiterpene lactones, known as davanone and trace compounds (mainly monoterpenes) (The Useful Plants of India, 1994; Akhila and Tewari, 1986). The presence of terpenes and sesquiterpenes might be responsible for its wound healing activity (Evans, 1980; Sharma *et al.*, 1990). It has been

Table - 4 : Results of histopathological studies

Drug	Parameters	Day-4		Day-8		Day-12		Day-16		Day-20	
		C	T	C	T	C	T	C	T	C	T
Ethanolic Extract (200mg/kg)	Epithelization	0.25 ± 0.25	0.75 ± 0.25	0.75 ± 0.27	1.25 ± 0.25	1.25 ± 0.25	2.25 ± 0.25★	2.25 ± 0.25	4.25 ± 0.75★★	2.75 ± 0.75	4.5 ± 0.25★★★
	Inflammation	4.25 ± 0.25	4.75 ± 0.25	3.75 ± 0.25	3.25 ± 0.25★★	2.5 ± 0.5	1.5 ± 0.5★★	2.25 ± 0.25	2.25 ± 0.25	1.25 ± 0.25	1.25 ± 0.75
	Collagen	1.00 ± 0.00	2.25 ± 0.25	1.25 ± 0.25	2.5 ± 1.5★	1.75 ± 0.25	3.5 ± 0.5★★★	2.25 ± 0.25	3.75 ± 0.25★★★	2.5 ± 0.5	4.0 ± 0.00★★★
	Neovascularity	3.25 ± 0.25	4.00 ± 0.00★★	3.50 ± 1.5	3.25 ± 0.25	2.75 ± 0.25	3.0 ± 1.00	2.25 ± 0.75	2.5 ± 0.5	1.75 ± 0.25	2.25 ± 0.25
	Cellularity	3.25 ± 0.25	4.00 ± 0.00	3.00 ± 0.00	3.75 ± 0.25	2.25 ± 0.25	2.75 ± 0.25	1.75 ± 0.25	1.25 ± 0.75	1.5 ± 0.5	1.0 ± 0.00
Aqueous Extract (200mg/kg)	Epithelization	0.25 ± 0.25	0.75 ± 0.25	0.75 ± 0.75	1.0 ± 0.00	1.25 ± 0.25	2.0 ± 0.00	2.25 ± 0.25	4.0 ± 0.00★★★	2.75 ± 0.075	4.25 ± 0.00★★★
	Inflammation	4.25 ± 0.25	4.50 ± 0.00	3.75 ± 0.25	3.5 ± 0.5★★	2.5 ± 0.5	1.75 ± 1.25★★	2.25 ± 0.25	2.0 ± 0.5★	1.25 ± 0.25	1.25 ± 0.5
	Collagen	1.00 ± 0.00	2.0 ± 1.0	1.25 ± 0.25	2.25 ± 0.25★	1.75 ± 0.25	3.25 ± 0.25★★★	2.25 ± 0.25	3.5 ± 0.5★★★	2.5 ± 0.5	3.75 ± 0.25★★★
	Neovascularity	3.25 ± 0.25	3.75 ± 0.25★★	3.5 ± 1.5	3.5 ± 0.5	2.75 ± 0.25	3.25 ± 0.25	2.25 ± 0.75	2.75 ± 0.25	1.75 ± 0.25	2.5 ± 0.5
	Cellularity	3.25 ± 0.25	4.00 ± 0.00	3.0 ± 0.00	3.75 ± 0.25	2.25 ± 0.25	3.0 ± 1.00	1.75 ± 0.25	1.75 ± 0.25	1.5 ± 0.5	1.25 ± 0.25

Values 5 refers for maximum similarity and 1 refers for least similarity of wound from the normal tissue.
Values are mean ± SEM readings (4); p-values are $p<0.05$★, $p<0.02$★★, $p<0.01$★★★ V/s. control. [C = Control, T = Test group]

demonstrated that the aqueous and ethanolic extracts (oral and topical) of *Artemisia pallens* improved the rate of wound healing in rats as evidenced by wound contraction and histopathalogical studies. Aqueous and ethanolic extracts increased the collagen content and thickness of epithelium also. The treated wound epithelialised faster and the rate of wound contraction was higher as compared to the control.

The wound healing activity of aqueous and ethanolic extracts of *Artemisia pallens* has also been supported by histochemical studies. The wound healing property appears to be due to the presence of its active principle sequiterpene which accelerates the healing process and confers breaking strength to the healed wounds.

ACKNOWLEDGEMENT

We are thankful to Dr. F.V. Manvi, Principal, K.L.E.S's College of Pharmacy, Belgaum for providing the necessary facilities.

REFERENCES

Akhila, A. and Tewari, R. (1986). "Chemistry of davana oil : A review". *Current Research Medicinal Aromatic Plant,* **8** : 128-138.

Drurey, R.A.B. and Wallonglon, E.A. (1980). Carleton's histological techniques 5th edition. 57-67, 77-78.

Ehrlich, H.P. and Hunt, T.H. (1969). Effects of cortisone and anabolic steroids on the tensile strength of healing wounds. *Ibid,* **170** : 103-206.

Evans, P. (1980). *Physiotherapy,* **20** : 256.

Husain, A. (1990). "Essential oil, plants and their cultivation-Davana" Farm Bulletin ND-12, (1990), Central Institute of Medicinal and Aromatic Plants, Lucknow, India, 17-19.

Mortan, J.J.P. and Malone, M.H. (1972). Evaluation of vulnerary activity by an open wound procedure in rats. *Arch. Int. Pharmacodyn,* **196** : 117-126.

Sharma, S.P., Aithal, K.S., Srinivasan, K.K. *et al.* (1990). *Fitoterapia,* **61** : 263.

Smith, R. (1985). "Recovery and Tissue Repair". British Medical Bulletin, **41(3)** : 295-301.

Subramoniam, A., Pushpangada, P., Rajasekharan, S., Evans, D.A. and Valsaraj, R. (1996). "Effects of Artemisia pallens Wall on blood glucose levels in normal and alloxan-induced diabetic rats". *Journal of Ethnopharmacology,* **50(1)** : 13-17.

Turner, R.A. (1965). Screening methods in Pharmacology. New York Academic Press, 61-153.

The Useful Plants of India (1994). Publications and Information Directorate, New Delhi, 55.

The Wealth of India (1985). Raw Materials, Vol. I (A), Revised edition, New Delhi, 439-441.

Udupa, K.N. (1970). Advances in research in Indian Medicine. Banaras Hindu University, Varanasi, 270-303.

www.irisherbal.com/ritual.html

Recent Advances in Medicinal Plant Research : Vision 21st Century, 2005
Editors : S.K. Prabhuji, G.P. Rao and S.K. Patil
Satish Publication House, New Delhi

Chapter 30

ANTIFEEDANT ACTIVITY OF *CHROMOLAENA ODORATA* ON LARVAE OF TEAK DEFOLIATOR, *HYBLAEA PUERA*

P. Ramana[1★], S.K Patil[1] and K.S. Rao[2]
[1]*College of Forestry, Sirsi - 581 401, Karnataka, India*
[2]*Director, Institute of Wood Science and Technology, Bangalore, India*
★e-mail : ramtarunsirsi@yahoo.com

Chromolaena odorata has strong antifeedant properties against larvae of *Hyblaea puera*. This natural pesticide of plant origin can be extensively used in pest control measures of other pests after careful screening and identification of active ingredients of this bioresource.

INTRODUCTION

The teak (*Tectona grandis* Linn. f.) is a most valuable timber tree in India. It is being attacked by two serious lepidopteron pests namely the defoliator, *Hyblaea puera* (Cramer) (Hyblaeidae) and skeletoniser, *Eutectona machaeralis* Walker (Pyraustidae) which cause epidemic defoliation in teak. In Kerala, *Hyblaea puera* infestation is a routine annual phenomenon and its defoliation reported to cause 44 per cent loss in the volume increment of teak (Nair *et al.*, 1985).

Insect antifeedants have been the subject of considerable interest. In recent years, they have been given special importance because of their toxicity to *Lepidopteron* pests (Sahayaraj, 1998). Many insect pests have been increasingly demonstrating ability to become resistant to conventional insecticides. Interest has therefore grown in the development of alternative strategies, including the use of antifeedants (Meshram *et al.*, 1994).

HISTORICAL REVIEW

Chromolaena odorata L. *is* a commonly occurring perennial obnoxious weed in the Western Ghats of India. It is a colonizer in the moist deciduous, semi-evergreen and evergreen forests of Western Ghats. The plant is easily recognized by its appearance and its characteristic odour. Dense thickets of the weed are a common sight and sometimes it scrambles small trees in moist areas of Western Ghats. The weed is capable of growing throughout the year (Anon, 1952). Some species are reported to contain active phytochemicals (Baruah and Pathak, 1993; Hai Mohemmad *et al.*, 1995; Ramana *et al.,* 1998; Ramana *et al.*, 2003).

Many crude plant extracts containing secondary compounds, some with antifeedant principles, have been tested against insect species (Jacobson *et al.*, 1978). There are many odoriferous herbs, which contain a variety of secondary compounds that possess antifeedant properties (Berenbaum, 1985). A number of plant species have been evaluated for antifeedant and other related properties with varied degree of success against many insect species of agriculture and forestry importance (Ahmed, 1996).

The present study was undertaken to evaluate the antifeedant efficacy of the various extracts of *Chromolaena odorata* on larvae of teak defoliator *Hyblaea puera.*

METHODOLOGY

Preparation of plant extracts

Whole plants of *Chromolaena odorata* were collected from the fields in and around College of Forestry, Sirsi during September-October. The plant material was shade dried and powdered. The powdered plant material (110 gm) was subjected to Soxhlet extraction separately and successively with petroleum ether (60-80 °C), butanol, methanol and distilled water. The crude plant extracts were concentrated to dryness in a flash-evaporator under reduced pressure and controlled temperature. All the extracts were stored in a refrigerator.

Antifeedant Activity

The antifeedant activity of various crude extracts was assessed by exposing fresh teak leaves treated with extractive solution to the third instar larva of *Hyblea puera* in no choice test (Ahmed *et al*, 1996). Since the third instar larvae of *Hyblea puera* are voracious feeders, this is the ideal stage for antifeedant assay. The test solutions of each extract was prepared by dissolving suitable quantity of crude plant extract to get a 4.0 and 2.0 per cent concentrated solutions, respectively. The fresh leaf discs of teak were cut with the help of a round cutter and were measured with help of digital leaf area meter (4.91 cm^2 area) and dipped in test solutions of the extracts for four seconds. The treated leaf got dried at room temperature and was placed on a moistened filter paper in the petridishes (15 cm dia.). One, third instar larva, pre-starved for two hours, was introduced at the center of the leaf. Control was run in the separate petridishes by placing measured, respective solvent viz. petroleum ether, butanol, methanol and distilled water, treated teak leaf disc along with a larva. After 24 hours, exposed leaves were again measured with a leaf area meter to find out the consumption of the leaf by the larva. The experiment was replicated five times. The data was statistically analysed and mean percent damage and protection in treatments over control were worked out by taking damage in control as standard.

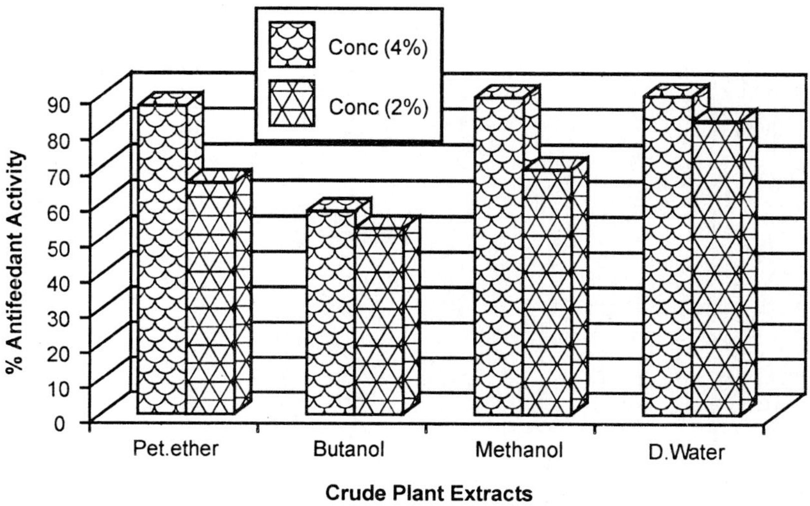

Fig. 1 : Antifeedant activity of *Chromolaena odorata* extracts.

Table - 1 : Antifeedant activity of extracts of *Chromolaena odorata* on third instar larvae of *Hyblaea puera*

Crude Plant Extracts	Control solvent/ Concentration of Plant Extracts	Mean leaf area consumed after 24 hrs* in cm^2	Mean leaf area consumed (%) after 24 hrs*	Mean reduction in feeding (%) after 24 hrs*
Petroleum Ether Extract	Petroleum ether	4.91	100	-
	4%	0.63	12.97 (21.13)	87.03 (68.87)
	2%	1.69	34.43 (35.91)	65.57 (54.09)
Butanol Extract	Butanol	4.45	100	-
	4%	1.88	42.31 (40.57)	57.69 (49.43)
	2%	2.10	47.16 (43.34)	52.84 (46.61)
Methanol Extract	Methanol	4.91	100	-
	4%	0.52	10.61 (19.00)	89.39 (71.00)
	2%	1.51	30.72 (33.65)	69.28 (56.35)
Distilled Water Extract	Distilled Water	4.91	100	-
	4%	0.50	10.14 (18.53)	89.86 (71.37)
	2%	0.84	17.09 (24.43)	82.91 (65.57)
	C.D. at 5%	1.24	2.56	

*Each value is the mean of five replications and values in parenthesis are the angular transformed values

RESULTS AND DISCUSSION

The results of antifeedant activity of the extracts of *C. odorata* are summarized in Table-1 and Fig. 1. All the four extracts have exhibited antifeedant activity. The petroleum ether, methanol and distilled water extracts have exhibited high antifeedant activity, while butanol extract has shown moderate antifeedant activity. Highest antifeedant activity was exhibited by both the distilled water extract and methanol extracts with distilled water extract exhibiting highest antifeedant activity at both high and low concentrations. The petroleum ether extract also showed high antifeedant activity against the pest. The butanol extract has exhibited moderate antifeedant activity at both the concentrations.

The present investigation strongly suggests that *C. odorata* extracts have high antifeedant activity. The results are in agreement with the earlier reports on biological activities of *C. odorata* (Ramana *et al.,* 1998, 2003). The natural pesticides of plant origin have least toxicity to animal and plant life, being biodegradable and do not leave toxic residues hence plant based pesticides can be extensively used in pest control measures. Solvents also play an important role in extracting active ingredients of specific nature. Further purification needs to be done before any such conclusions are drawn. Work is under way to identify compounds responsible for this bioefficacy.

REFERENCES

Ahmed, M., Dayal, R. and Borai, P. (1996). Evaluation of insecticidal properties of *Dalbergia stipulacea* against poplar defoliator, *Clostera cupreata* (Lepidoptera: Notodontidae). *Ann. For.*, **4(1)** : 47-51

Anonymous (1952). Wealth of India: Raw materials (CSIR, New Delhi), Vol. III: pp 223.

Baruah, R.N. and Pathak, M.G. (1993). Fatty acid composition of *Chromolaena odorata* flower fat. *Indian J. Nat. Prod.,* **9(1)** : 17-18.

Berenbaum, M. (1985). Recent Advances in Phytochemistry. GA Cooper-Driver; T. Swain and E.Com. (Eds.), Plenum, New York 19 : 139-169.

Hai Mohammad, A., Saha Koush, K. and Ahmad Mesbah, U. (1995). Chemical constituents of *Eupatorium odoratum* Linn (Compositae). *J. Bangladesh Chem Soc.*, **8(2)** : 139-142.

Jacobson, M., Reed, D.K., Crystal, M.M., Moreno, D.S. and Soderstrom, E.L. (1978). Chemistry and biological activity of insect feeding deterrents from certain weed and crop plants. *Entomol. Exp Appl.*, **24** : 448-457.

Meshram, P.B., Kulkarni, N. and Joshi, K.C. (1994). Antifeedant activity of certain plant products against teak skeletanizer *Eutectong mochaerali* Walk (Lepidoptera: Pyralidae). *Ann. Ento.,* **12** : 53-56.

Mukhtar, Ahmed, Rameshwar, Dayal, Mishra, R.K. and Dobhal, P.C. (1997). Antifeedant potency of some plants produce against defoliating pest of poplar, *Clostera cupreata. Indian Forester.*, **123(9)** : 821-826.

Nair, K.S.S., Sudheendra Kumar, V.V., Varma, R.V. and Chacko, K.C. (1985). Studies on the seasonal incidence of defoliators and the effect of defoliation on volume increment of teak. KFRI Research Report, 30. Kerala Forest Research Institute, Peechi, Kerala. pp 78.

Ramana, P., Neelannavar, T.N., Patil, S.K. and Rao, K.S. (2003). Phytochemical and antifungal activity of *Eupatorium odoratum* Linn. *Ind. J For.*, **26(1)** : 19-22.

Ramana, P., Patil, S.K. and Shivanna, H. (1998). Influence of *Chromolaena odorata* extracts on biocontrol agents. *My Forest.*, **34(4)** : 927-931.

Sahayaraj, K. (1998). Antifeedant effect of some plant extracts on the Asian armyworm *S. liturea* (Fabricius). *Curr. Sci.*, pp 523-525.

Recent Advances in Medicinal Plant Research : Vision 21st Century, 2005
Editors : S.K. Prabhuji, G.P. Rao and S.K. Patil
Satish Publication House, New Delhi

Chapter 31

BIOPROSPECTING OF INDIGENOUS FLORA OF MEDICINAL PLANTS — A GLIMPSE FROM BAY ISLANDS

T.E. Sheeja, D. Chattopadhyay*, Asit B. Mandal and R.B. Rai
Central Agricultural Research Institute, Port Blair - 744 101
**ICMR Virus Unit, Kolkata - 700 010*

Andaman and Nicobar islands situated between 10° 30′ and 13° 42′ latitudes and 92° 14′ and 94° 16′ E longitudes represent a vast archipelago covering 8249 sq km area amidst Bay of Bengal. The islands harbour a number of endemic and extra Indian species of medicinal plants used extensively by the tribals to cure diverse ailments. Two extra Indian plant species of tribal importance viz. *Alstonia macrophylla* and *Mallotus peltatus* not found in mainland India were analysed for their bioactivity involving leaf and stem bark. The active principles were extracted, purified, fractionated and tested involving microbial systems as well as animal models. Methanolic extract of *Alstonia macrophylla* leaves showed *in vitro* antibacterial activity against 147 strains of bacteria and one dermatophytic fungus. It also displayed strong antileishmanial activity. *Mallotus peltatus* inhibited three fungal strains. Significant antipyretic activity was observed in case of yeast induced pyrexia in both the cases. The crude methanol extracts from both the species also displayed psychopharmacological and analgesic activities comparable to that of standard drugs Paracetamol and Aspirin. Phytochemical analyses showed the presence of tannins, reducing sugars, flavonoids, saponins, steroids and ursolic acid, β-sitosterol and its glucoside in both the species. The folk lore in respect of ethnomedicinal values was scientifically validated and used for the development of four herbal leads viz. (1) Herbal antimicrobial, anti-inflammatory and antihistaminic formulation (2) Mouth wash (3) Vaginal contraceptive and (4) a formulation to treat oral submucus fibrosis (OSF) involving different fractions of bioactive compounds which are under process of patenting. These wonder drugs may be developed from the forest floor of Bay Islands.

INTRODUCTION

In the recent past, bioprospecting has got a new dimension in therapeutics. Especially the economically poor but biologically rich developing countries have started revisiting their indigenous bioresources for green medicines. Isolation and characterization of bioactive principles having single activity or "magic bullets" against deadly diseases like AIDS, cancer etc. from traditional medicinal plants would be of immense importance. About 119 pure chemicals extracted from higher plants are being used in health care system world wide. Extensive research on traditional medicines could identify several drugs that led to substantial foreign exchange earnings in countries like China, Japan, Madagascar and Thailand.

Andaman and Nicobar Islands harbour a number of invaluable medicinal plants, which remain mostly unexplored. Those may be exploited judiciously for economic prospects of the islands. A large number of them are extensively used by the tribals in curing various ailments. Those plants are found to be time tested in the indigenous system of health care, however, lacks scientific validation. A systematic study was thus planned to assess the bioactive characteristics of two important medicinal plant species used extensively by Onge and Nicobarese tribals of this Union Territory. The plants viz. *Alstonia macrophylla* locally known as Chattiyun and *Mallotus peltatus* locally known as Kamela are used for treating stomache (Dagar and Dagar, 1991), skin diseases and urinary infections (Bhargava, 1983; Chakraborty and Vasudeva Rao, 1988). Studies were conducted to evaluate the antimicrobial, antileishmanial, antipyretic, anti-inflammatory and psychopharmacological activities of both stem bark and leaf samples from these two plant species.

METHODOLOGY

The leaves and stem bark of the plants were collected from the tropical rain forests of middle and south Andamans during 1999-2000. The voucher specimens were identified with the help of Botanical Survey of India, Port Blair.

Extraction

The samples were cleaned, dried and powdered. They were screened through a 40 mesh sieve and extracted in cold with 95% methanol for 72 h at room temperature. The whole extract was filtered and solvent were evaporated to dryness under reduced pressure in a rotary vaccum evaporator (Eyela make) at 40-45°C. The concentrated leaf extract were partitioned between n-butanol and water in 3:1 ratio, while aqueous portion was lyophilized to dryness and the solvent of n-butanol part was removed under reduced pressure. Preliminary phytochemical test of leaf extract was done following Pollock and Stevens (1965). The extract was stored in desiccators, dissolved in propylene glycol for further study. Phytochemical tests were done following the method outlined by Pollock and Stevens (1965) after solvent extraction. The major and minor compounds were detected by TLC, Co-TLC, superimposable IR, mass spectroscopy and NMR. The isolated compounds were characterized by 2D NMR and mass spectroscopy.

Animal model used

Swiss albino mice and Wistar albino rats of either sex weighing 180-200 g were used. The animals were housed in standard metal cages at ambient temperature and provided with standard diet and water *ab libitum*.

Microorganism used

Consisted of 317 strains of bacteria including 314 human isolates, five yeast and five dermatophytic fungi collected from Calcutta School of Tropical Medicine, Department of Pathology, Peerless Hospital and B.C. Roy Research Centre, Kolkata and the reference strains were a gift from London Hospital Medical College, London, UK. The bacterial strains were grown in blood agar or Mac Conkey agar plates at 37°C and maintained in Sabourand glucose agar slants.

Antimicrobial assay

Sensitivity tests were performed by disk diffusion methods (NCCLS, 1993). Extract impregnated discs (concentrations of 0-2000 μg/ml for bacteria and 0-128 μg/ml for yeasts and fungi) were placed aseptically on sensitivity plates with appropriate controls. The plates were incubated at 37°C overnight for bacteria and 30°C for fungi. The clear zone was measured to assess sensitivity. MIC was determined by agar and broth dilution methods (Chattopadhyay *et al.*, 1998).

The effect of crude extracts on bacteria and fungi was determined by incubating the culture and adding the extract involving mid logarithmic phase culture and drawing at fixed intervals to determine OD_{540} and colony count (Chattopadhyay, 1998). The growth of dermatophytic fungi was determined by weight basis, after drying the sample at 60°C for 20 h for 3 days (Chattopadhyay *et al.*, 2001). The results were analysed statistically by Students t-test (Woodson, 1987).

Antiprotozoal activity

This was studied involving *Leishmania donovani* AG 83 by using tritiated thymidine uptake assay.

Psychopharmacological activity

General behaviour profile was assessed following Dixit and Varma (1976). Exploratory behaviour was analysed by Y-maze test (Rushton *et al.*, 1961; Mandal *et al.*, 2001) and head dip test (Dorr *et al.*, 1971). Muscle relaxant activity was analysed by the traction test, rotarod test and 30°C inclined screen test (Rudzik *et al.*, 1973; Dunham and Miyan, 1957; Radal, 1960). Phenobarbitone sodium induced sleeping time (Fastier *et al.*, 1957) were analysed. The statistical analysis was done by Students t-test following Woodson (1987). Analgesic activity was analysed following Koster *et al.* (1959) for acetic acid induced writhing test, tail immersion test (Ghosh, 1984) and tail flick test (Ghosh, 1984). Antipyretic activity was studied by monitoring normal body temperature and induction of yeast induced pyrexia as described by Smith and Hambourger (1935) and modified by Murugesan *et al.* (2000).

Anti-inflammatory activity

Carageenan induced rat paw oedema (Winter *et al.*, 1962) and dextran induced rat paw oedema (Parmar *et al.*, 1978 and Kavimani, 1996) and cotton pellet induced granuloma (D'Arcy *et al.*, 1960 and Winter and Porter, 1957) were studied.

RESULTS

Phytochemistry

Phytochemical analysis of the crude methanol extract showed presence of tannins, reducing sugars, flavonoids, saponins, steroids in leaf and bark of all three species. Ursolic acid and β-sitosterol glucosides were abundant. The compounds were analysed by NMR and IR spectroscopy. The n-butanol part of *A. macrophylla* and *M. peltatus* were further fractionated and the study revealed that a synergestic effect of these fractions is responsible for attributing sensitivity towards microbial strains.

Antibacterial activity

The methanol extract of *A. macrophylla* displayed appreciable *in vitro* inhibitory activity against 95 of 314 strains of bacteria. The n-butanol part of the extract had more activity in nine strains of *Proteus mirabilis*. Seven *Pseudomonas vulgaris* and five of *E. coli* strain had a MIC of 64 mg/ml (inhibition zone dia : >12 mm), while seven of *Streptococcus faecalis* and 27 of *Staphylococcus aureus* had MIC of 1000 μg/ml (inhibition zone dia : >10 mm) against the crude extract. Overall 17.73% of gm -ve and 47.42% gm +ve strains were found to be sensitive to the extract. When susceptible species were analysed, growth of the organisms was found to decrease with more concentration of the extract and completely inhibited at their MIC values (Table-1).

The methanol extract of *M. peltatus* leaf showed a moderate *in vitro* antibacterial activity against four bacteria and one fungal isolate. The n-butanol fraction of the methanol extract was found to be more active. 147 isolates were found to be inhibited at a concentration of 128-1000 μg/ml. The MIC revealed that 63 gm +ve isolates had MIC of 128-256 μg/ml (inhibition zone dia : 11-14 mm), while 47 including 22 *E. coli* and seven Proteus species had MIC of 512-1000 μg/ml (inhibition zone dia : 10-12 mm). The extract of *M. peltatus* on two susceptible bacteria at different concentration showed that growth of these organisms decreased with increasing concentrations of the extract and was found to be completely inhibited at their MIC values (Table-2).

Antifungal activity

Dermatophytic fungi that parasitize keratinized tissues were used. The MIC values of *A. macrophylla* leaf extract showed that three fungal strains (viz. *Trichophyton rubrum, T. mentagrophytes* var. *metagrophytes* and *M. gypseum*) were found to be inhibited had MIC of 128 μg/ml and were more sensitive to the extract with a zone of inhibition of 10-12 mm. The extracts showed fungistatic activity at lower concentrations, but lethal effect at higher concentrations.

Antiprotozoal activity

A. macrophylla leaf extract showed appreciable activity against *Leishmania donovani* AG 83 at a concentration of 15 μg/ml as evident from ^{3}H thymidine uptake study (Fig. 3). This hold enormous promise in drug development for Kala-azar, a dreadly disease prevalent in Asian countries.

Psychopharmacological activity

The methanolic extract of MPL influences general behavioural profiles as evident from the spontaneous activity, touch, sound and pain responses (Table-3). The extract significantly potentiated the phenobarbitone sodium

Table - 1 : Antibacterial activity of *Alstonia macrophylla* leaf extracts

Bacterial species	No. of Strains	MIC of crude leaf extracts (μg/ml)							MIC of Amoxycillin (μg/ml)						
		64	128	256	512	1000	2000	>2000	0.25	0.50	8	64	128	256	>1000
Escherichia coli	75	05	01	02	02	06	08	51	-	-8	02	02	05	15	43
Klebsiella pneumoniae	25	-	-	-	-	01	02	22	-	-	-	-	11	04	10
Salmonella typhi VI	30	-	-	-	-	-	02	28	-	-	-	-	01	05	24
Shigella dysenteriae	15	-	-	-	-	-	03	12	-	-	-	01	02	02	10
Vibrio cholerae	15	-	-	-	-	-	01	14	-	-	-	-	-	12	13
Citrobacter sp.	15	-	-	-	-	-	09	06	-	-	01	10	02	02	-
Proteus mirabilis	15	09	01	-	-	01	03	01	-	-	-	-	-	01	14
Proteus vulgaris	15	07	-	01	02	01	01	03	-	-	-	-	01	01	13
Pseudomonas aeruginosa	15	-	-	-	-	-	-	15	-	-	-	-	-	-	15
Staphylococcus aureus	72	-	-	-	02	27	31	i2	-	24	10	12	14	08	04
S. saprophyticus	10	-	-	-	-	05	02	03	02	01	-	07	-	-	-

Table - 1 : Continued ......

Bacterial species	No. of Strains	MIC of crude leaf extracts (µg/ml)							MIC of Amoxycillin (µg/ml)						
		64	128	256	512	1000	2000	>2000	0.25	0.50	8	64	128	256	>1000
Streptococcus faecalis	10	-	-	-	01	06	03	-	01	01	02	02	01	03	-
Bacillus subtilis	05	-	02	01	01	01	-	-	-	03	-	01	01	-	-
Staphylococcus aureus NCTC 65761					+					+					
Escherichia coli NCTC 10418				+							+				
Pseudomonas aeruginosa 10662					+										
Total	317	21	04	04	08	48	65	167	0.03	37	15	35	38	43	146

+ = inhibitory concentration; - = no inhibition; Inoculum size used : 10^4 cfu/spot for all the organisms except *Staphyloccus aureus*, where the inoculum size per spot was 10^6 cfu (NCCLS, 1993). The results represent average values of triplicate tests (repeated three times) after 24 h to 72 h of incubation at 37°C.

Table - 2 : Antibacterial activity of *Mallotus peltatus* leaf and stem bark extracts

Bacterial species	No. of Strains	MIC of leaf extracts (μg ml^{-1})					MIC of stem bark extracts (μg ml^{-1})			MIC of Amoxycillin (μg/ml)						
		128	256	512	1000	>2000	1000	2000	>2000	0.25	0.50	8	64	128	256	>1000
Escherichia coli	75	-	-	12	07	56	-	17	58	-	8	0.2	02	05	15	43
Klebsiella pneumoniae	25	-	-	-	04	21	-	05	20	-	-	-	-	11	04	10
Salmonella typhi VI	30	-	-	-	02	28	-	04	26	-	-	-	-	01	05	24
Shigella dysenteriae	15	-	-	01	04	10	-	04	11	-	-	-	01	02	02	10
Vibrio cholerae	15	-	-	-	02	13	-	02	13	-	-	-	-	-	02	13
Citrobacter sp.	15	-	-	-	01	14	-	-	15	-	-	0.1	10	02	02	-
Proteus mirabilis	15	-	-	05	04	06	-	03	12	-	-	-	-	-	01	14
Proteus vulgaris	15	-	-	02	03	10	01	02	12	-	-	-	-	01	01	13
Pseudomonas aeruginosa	15	-	-	-	-	15	-	-	15	-	-	-	-	-	-	15
Staphylococcus aureus	72	21	25	22	04	-	62	10	-	-	24	10	12	14	08	04
S. saprophyticus	10	0.4	02	02	02	-	06	04	-	02	01	-	07	-	-	-

Table - 2 : Continued .....

Bacterial species	No. of Strains	MIC of leaf extracts ($\mu g\ ml^{-1}$)					MIC of stem bark extracts ($\mu g\ ml^{-1}$)			MIC of Amoxycillin ($\mu g/ml$)						
		128	256	512	1000	>2000	1000	2000	>2000	0.25	0.50	8	64	128	256	>1000
Streptococcus faecalis	10	05	03	-	02	-	05	05	-	01	01	02	02	01	03	-
Bacillus subtilis	05	01	02	02	-	-	03	02	-	-	03	-	01	01	-	-
Staphylococcus aureus NCTC 65761 NCTC 65711	-	-	-	-	+	-	+	-	-	+	-	-	-	-	-	-
Escherichia coli NCTC 10418	-	-	+	-	-	-	-	+	-	-	-	+	-	-	-	-
Pseudomonas aeruginosa 10662	-	-	-	-	-	+	-	-	+	-	-	-	-	-	+	-
Total	317	31	32	46	35	173	77	58	182	0.3	37	15	35	38	43	146

+ = inhibitory concentration; Inoculum size used : 10^4 cfu/spot for all the organisms except *Staphyloccus aureus*, where the inoculum size per spot was 10^6 cfu (NCCLS, 1993). The results represent average value of triplicate tests (repeated three times) after 24 h to 72 h of incubation at 37°C. Diameter of inhibition zones varies from 10 to 14 mm ± S.D. at MIC concentrations.

The results are the mean values of triplicate tests repeated three times after 96 h of incubation at 30°C under shaking at 200 rpm (Ref. : Antifungal Drugs, St. Georgiev V. ed., *Ann N.Y. Acad Sci.* 1988, 544 : 1-590). The growth of the dermatophytes was determined on dry weight basis, after drying the sample at 60°C for 20 h.

Table - 3 : Effect of the methanolic extract of A. macrophylla leaf on general behaviour in mice and rats

Behaviour	Methanol extract (mg kg^{-1})				Chlorpromazine (5 mg kg^{-1})	Propylene glycol (5 ml kg^{-1})
	100	200	300	Fr A + B 50		
Spontaneous activity	+	++	++++	++++	++++	-
Alertness	+	+++	+++	++	++	-
Awareness	+	++	+++	++++	++++	-
Sound response	+	+++	+++	+++	++++	-
Touch response	++	+++	++++	++++	++++	-
Pain response	+	+++	++++	++++	++++	-

- = no effect; + = slight depression; ++ = moderate depression; +++ = strong depression; ++++ = very strong depression; n = 10; Fr. A = fraction A (β-sitosterol); Fr. B = fraction B (ursolic acid).

Table - 4 : Effect of the methanolic extract of *Mallotus peltatus* leaf on general behaviour profiles in mice and rats

Behaviour	Methanol extract (mg kg^{-1})					Chlorpromazine (5 mg kg^{-1})	Propylene glycol (5 ml kg^{-1})
	100	200	300	Fr A 50	Fr B 50		
Spontaneous activity	+	++	+++	++++	++	++++	-
Alertness	+	++	+++	+++	+++	+++	-
Awareness	+	++	++	+++	+++	+++	-
Sound response	+	++	++++	+++	++	++++	-
Touch response	++	+++	+++	+++	+++	++++	-
Pain response	+	+++	++++	+++	+++	++++	-

- = no effect; + = slight depression; ++ = moderate depression; +++ = strong depression; ++++ = very strong depression; n = 10; Fr. A = fraction A (ursolic acid).

Table - 5 : Analgesic effect of the methanol extract of *Alstonia macrophylla* leaf (MEAML) and *Mallotus peltatus* leaf (MEMPL) and their fractions by physical and chemical methods in mice.

Treatment	Dose (mg kg^{-1})	Tail flick test (Reaction time in second)	Tail immersion test (Reaction time in second)	Number of writhing (% inhibition)
Propylene glycol	5 ml kg-1	2.25 ± 0.14	2.32 ± 0.16	41.50 ± 0.50
Aspirin	100	-	-	10.10 ± 0.43 (75.66)
Pethidine	5	4.20 ± 0.18	4.48 ± 0.12	-
MEAML	100	2.75 ± 0.04	2.64 ± 0.03	31.50 ± 0.34 (24.09)
	200	3.21 ± 0.08	3.18 ± 0.02	20.90 ± 0.34 (49.63)
	300	3.98 ± 0.12	3.72 ± 0.18	10.50 ± 0.34 (74.69)
Fraction A	50	3.78 ± 0.14	3.54 ± 0.14	12.50 ± 0.04 (69.87)
Fraction B	50	3.43 ± 0.02	3.81 ± 0.04	10.20 ± 0.03 (75.42)
Fraction C	50	3.41 ± 0.04	3.61 ± 0.06	13.20 ± 0.22 (68.19)
MEMPL	100	2.50 ± 0.13	2.48 ± 0.02	32.70 ± 0.36 (21.20)
	200	3.02 ± 0.06	3.05 ± 0.02	20.70 ± 0.43 (51.12)
	300	3.73 ± 0.10	3.65 ± 0.16	10.80 ± 0.34 (73.97)
Fraction A	50	3.69 ± 0.21	3.65 ± 0.16	10.40 ± 0.02 (74.93)
Fraction B	50	3.01 ± 0.12	3.02 ± 0.12	12.20 ± 0.04 (70.60)

Values are mean ± S.E.; N = 10; $p<0.001$ vs Control; Student's t-test

Table - 6 : Effect of *Alstonia macrophylla* leaf extract on normal body temperature

Treatment (mg kg^{-1})		Rectal temperature (°C) before and after treatment					
		0 h	1 h	2 h	3 h	4 h	5 h
Control (5 ml kg^{-1})		36.5 ± 0.08	36.6 ± 0.07	36.4 ± 0.03	36.4 ± 0.08	36.5 ± 0.07	36.4 ± 0.05
A. macrophylla leaf	100	35.6 ± 0.08	35.6 ± 0.08★	35.3 ± 0.09★	35.0 ± 0.04★	35.4 ± 0.05★	35.6 ± 0.05★
	200	36.0 ± 0.04	35.9 ± 0.04★	35.4 ± 0.04★	34.9 ± 0.01★	35.3 ± 0.09★	35.5 ± 0.09★
	300	35.9 ± 0.08	34.9 ± 0.06★	34.2 ± 0.05★	34.2 ± 0.05★	35.3 ± 0.05★	35.3 ± 0.08★

Each value represents mean ± SE of six rats. Control = Propylene glycol ★P<0.001 as compared to the control values of corresponding hour.

Table - 7 : Effect of *Mallotus peltatus* leaf extract on normal body temperature

Treatment (mg kg^{-1})		Rectal temperature (°C) before and after treatment					
		0 h	1 h	2 h	3 h	4 h	5 h
Control (5 ml kg^{-1})		36.5 ± 0.08★	36.6 ± 0.07	36.4 ± 0.03	36.4 ± 0.08	36.5 ± 0.07	36.4 ± 0.05
M. peltatus leaf	100	36.9 ± 0.04	36.8 ± 0.05★★	36.3 ± 0.04★★	36.2 ± 0.08★★★	36.5 ± 0.08★★★	36.8 ± 0.07★★★
	200	36.0 ± 0.07	35.9 ± 0.06★	35.3 ± 0.07★	35.2 ± 0.08★	35.3 ± 0.07★	35.4 ± 0.07★
	300	36.2 ± 0.08	36.0 ± 0.08★	35.3 ± 0.09★	35.1 ± 0.06★	35.3 ± 0.06★	35.5 ± 0.06★

Each value represents mean ± SE of six rats. Control = Propylene glycol ★P<0.001 as compared to the control values of corresponding hour.

Table - 8 : Effect of *Alstonia macrophylla* leaf extract on yeast-induced pyrexia in rats.

Treatment (mg kg^{-1})	Rectal temperature (°C) before and after treatment					
	0 h	1 h	2 h	3 h	4 h	5 h
Propylene glycol (5 ml kg^{-1})	37.3 ± 0.03★	39.6 ± 0.03	39.6 ± 0.02	39.5 ± 0.01	39.5 ± 0.01	39.5 ± 0.02
Paracetamol 150	37.3 ± 0.01	39.6 ± 0.04★	38.3 ± 0.01★	38.3 ± 0.02★	37.7 ± 0.01★	37.6 ± 0.02★
AML 100	37.3 ± 0.06	39.6 ± 0.01★	38.9 ± 0.02★	38.6 ± 0.03★	38.2 ± 0.03★	37.9 ± 0.01★
AML 200	37.3 ± 0.01	39.6 ± 0.01★	38.9 ± 0.01★	38.3 ± 0.01★	38.0 ± 0.02★	37.8 ± 0.01★
AML 300	37.3 ± 0.01	39.6 ± 0.02★	38.7 ± 0.01★	38.4 ± 0.01★	38.0 ± 0.04★	37.7 ± 0.05★

★Each value represents mean ± SE of six rats. Control = Propylene glycol ★$P<0.001$ as compared to the control values of corresponding hour.

Table - 9 : Effect of *Mallotus peltatus* leaf extract on yeast-induced pyrexia in rats.

Treatment ($mg\ kg^{-1}$)	Rectal temperature (°C) before and after treatment					
	0 h	19 h	20 h	21 h	22 h	23 h
Propylene glycol ($5\ ml\ kg^{-1}$)	37.3 ± 0.03★	39.6 ± 0.03	39.6 ± 0.02	39.5 ± 0.01	39.5 ± 0.01	39.5 ± 0.02
Paracetamol 150	37.3 ± 0.01	39.6 ± 0.04★	38.3 ± 0.01★	38.3 ± 0.02★	37.7 ± 0.01★	37.6 ± 0.02★
MPL 100	37.3 ± 0.02	39.6 ± 0.01★	38.7 ± 0.07★	38.3 ± 0.02★	38.2 ± 0.03★	37.9 ± 0.01★
MPL 200	37.3 ± 0.02	39.6 ± 0.03★	38.8 ± 0.04★	38.3 ± 0.06★	38.2 ± 0.04★	37.4 ± 0.09★
MPL 300	37.3 ± 0.04	39.6 ± 0.01★	38.6 ± 0.04★	38.2 ± 0.03★	38.0 ± 0.01★	37.2 ± 0.05★

★Each value represents mean ± SE of six rats. Control = Propylene glycol ★$P<0.001$ as compared to the control values of corresponding hour. MPL = *M. peltatus* leaf extract.

induced sleeping time, possibly through a CNS depressant action (Fastier *et al.*, 1957) or through tranquilizing action (Mandal *et al.*, 2001). Further investigation on psychopharmacological functions revealed a reduction in exploratory behaviour and lack of motor co-ordination and muscle relaxant activity. This is probably attributed to the action of either β-sitosterol or triterpenes alone or their combination. Both AML and MPL showed moderate to strong antipsychotic activity owing to the action of ursolic acid or due to its combined action with β-sitosterol. The presence of alkaloids viz. picrinine (CNS depressant), picralsonin (beriberi) and 0-benzoyl vincamajine (Asolkar *et al.*, 1992) in *A. macrophylla* may be the reasons for this activity.

Significant decrease in exploratory behaviour was also observed with AML and MPL in doses of 100 mg/kg (Table-4). In the head dip test also a decrease in number of head dips was observed with AML and MPL at doses of 100 mg/kg and above. Significantly potentiated phenobarbitone sodium induced sleeping time was observed at 100 mg/kg of the combination of fractions in case of both *A. macrophylla* and *M. peltatus*.

The muscle relaxant activity analysis showed that AML at 100 mg/kg and above significantly reduced motor co-ordination of the tested animals. In 30° inclined screen test, the extract showed a significant loss of co-ordination and muscle tone of all tested animals. The results were found to be similar in case of both extracts.

Analgesic activity

A. macrophylla at 300 mg/kg showed an inhibition comparable to that of standard drug Aspirin (75.66%), while for MPL, it was 73.97% at 300 mg/kg dose. Both the plant extracts significantly increased the induction time required to produce writhing movements compared to that of Aspirin. Methanolic extract of *A. macrophylla* at 300 mg/kg significantly increased the reaction time compared to Pethidine (4.48), while for *M. peltatus* it was 3.65 second. The respective fractions also showed a reaction time of 3.65 and 3.02 seconds (Table-5).

Antipyretic activity

A. macrophylla leaf extract at 100 mg/kg significantly reduced the body temperature upto 3h (Table-6). Similarly reduced yeast induced pyrexia was also observed. Methanol extract of *M. peltatus* also displayed significant antipyretic effect in yeast provoked elevation of body temperature in rats and the effect were comparable to that of Paracetamol. MPL also reduced the normal body temperature significantly (Table-7).

Both AML and MPL extract at 100 mg/kg caused significant lowering of body temperature upto 3h following extract administration (Table-8, 9). The effect was maximum at 200 and 300 mg/kg. In case of yeast induced pyrexia subcutaneous injection of 10 ml/kg of 15% w/v yeast suspension could elevate the body temperature to 39.6°C and MPL at 100, 200 and 300 mg/kg decreased this temperature. The earlier studies have revealed presence of β-sitosterol as a major compound (Chattopadhyay *et al.*, 2002), which is reported to have good antipyretic effects (Gupta *et al.*, 1996).

Based on different properties involving diverse systems from the above species four herbal formulations *viz.* an antimicrobial, anti-inflammatory and antihistaminic formulation, vaginal contraceptive, mouth wash and a preparation for treating oral submucal fibrosis (OSF) were developed and are being processed for patent grant.

REFERENCES

Bhargava, N. (1983). Ethnobotanical studies of the tribes of Andaman and Nicobar Islands, India. 1. *Onge. Economic Bot.*, **37** : 110-119.

Chakraborty, T. and Vasudeva Rao, M.K. (1988). Ethnobotanical studies of the Shompen of Great Nicobar Islands. *J. Eco. Taxon. Bot.*, **12(1)** : 43-54.

Chattopadhyay, D., Mukherjee, T., Pal, P., Saha, B. and Bhadra, R. (1998). Altered membrane permeability as the basis of bactericidal action of methdilazine. *J. Antimicrob. Chemother,* **42** : 83-86.

Chattopadhyay, D., Arunachalam, G., Mandal, A.B., Sur, T.K., Sur Bhadra, R., Mandal Subhash, C. and Bhattacharya, S.K. (2002). Antimicrobial and anti-inflammatory activity of *Mallotus peltatus* leaf extract. *Ethnopharmacol.*, **82(2-3)** : 229-237.

Chattopadhyay, D., Maiti, K., Kundu, A.P., Chakrabarty, M.S., Bhadra, R., Mandal, S.C. and Mandal, A.B. (2001). Antimicrobial activity of *Alstonia macrophylla* : A folklore of Bay Islands. *J. Ethnopharmacol.*, **77** : 49-55.

D'Arcy, P.D., Howard, E.M., Muggleton, P.W. and Townsend, S.B. (1960). The anti-inflammatory action of Griseofulvin in experimental animals. *Pharmacy and Pharmacology,* **12** : 659.

Dagar, H.S. and Dagar, J.C. (1991). Plant folk medicines among the Nicobarese of Katchal Island, India. *Economic Bot.*, **45(1)** : 114-119.

Dixit, V.K. and Varma, K.C. (1976). Effect of essential oil of leaves of *Blumea lacera* DC on central nervous system. *Indian Pharmacol.*, **18** : 7-11.

Dorr, M., Stienberg, H., Tomkiewiez, M., Joyee, D., Porosolt, R.D. and Summerfield, A. (1971). Persistence of dose related behaviour in mice. *Nature,* **231** : 121-123.

Dunham, N.W. and Miya, T.S. (1957). A note on simple apparatus for detecting neurological deficit in rats and mice. *American Pharmacology,* **46** : 208-209.

Fastier, F.N., Spenden, R.N. and Hendrieka, W. (1957). Prolongation of chloral hydrate sleeping time by 5-HT and by certain other drugs. *British Pharmacology,* **12** : 251-259.

Ghosh, M.N. (1984). Fundamental of Experimental Pharmacology, 2nd Edn. Scientific Book Agency, Kolkata, India. p. 153.

Gupta, M.B., Nath, R., Srivastava, N., Shanker, K., Kishor, K. and Bhargava, K.P. (1996). Anti-inflammatory and antipyretic activities of β-sitosterol. *International J. Immunopharmacol.*, **18(12)** : 693-700.

Kavimani, S., Vetrichelvan, T., Ilango, R. and Jaykar, B. (1996). Anti-inflmmatory activity of the volatile oil of *Toddalia asiatica*. *Indian J. Pharmaceutical Sciences,* **58** : 67-70.

Koster, R. and Anderson, N. (1959). Acetic acid for analgesic screening. *Fed. Proc.*, **18** : 412.

Mandal, S.C., Dhara, A.K. and Maiti, B.C. (2001). Studies of psychopharmacological activity of *Andrographis paniculata* extract. *Phytotherapy Research,* **15** : 253-256.

Murugesan, T., Mandal, S.C., Bhakta, T., Das, J., Pal, M. and Saha, B.P. (2000). Evaluation of anti-pyretic potential of *Jussiaea suffrutucosa* Linn. extract in rats. *Phytomedicine*, **7** : 231-234.

National Committee for Clinical Laboratory Standards, 3rd Ed., Approved Standard M7-A3, NCCLS, Villanova, PA, 1993.

Parmar, N.S. and Ghosh, M.N. (1978). Anti-inflammatory activity of gossypin bioflavonoid isolated from *Hibiscus viofolius* Linn. *Indian Pharmacol.,* **10** : 277-293.

Pollock, J.R.A. and Stevens, R. (1965). In : Dictionary of organic compounds, Vol. 5, 4th edition, Eyre and Spottiswoode Publishers Ltd., London.

Randall, P. (1960). Studies of the neurological effects using simple experimental model. *J. Pharmacol. Exp. Theor.*, **135** : 543-545.

Rudzik, A.D., Hester, J.B., Tang, A.H., Staw, R.N. and Friis, W. (1973). The Benzodiazepines, Raven Press, New York, 285-297.

Rushton, R., Steinberg, H. and Tinson, C. (1961). Modification of the effects of an amphetamine barbiturate mixture by the past experience of rats (Y-shaped runway). *Nature*, **192** : 533-535.

Smith, P.K. and Hambourger, W.E. (1935). *J. Pharmacol.* Articles from *Exp. Theor.*, **54** : 346.

Winter, C.A. and Porter, C.C. (1957). Effect of alteration in side chain upon anti-inflammatory and liver glycogen activities in hydrocortisone esters. *J. American Pharmacol. Soc.*, **46** : 515.

Winter, C.A., Risley, E.A. and Nuss, G.W. (1962). Carrageenin induced oedema in hind paw of rat as assay for anti-inflamatory drugs. *Experimental Biology and Medicine,* **111** : 544-547.

Woodson, R.F. (1987). Statistical methods for the analysis of bio-medical data, probability and mathematical statistics. Wiley, Chichester, pp. 313-316.

Recent Advances in Medicinal Plant Research : Vision 21st Century, 2005
Editors : S.K. Prabhuji, G.P. Rao and S.K. Patil
Satish Publication House, New Delhi

Chapter 32

EXTRACTION AND ISOLATION OF FURANOCOUMARINS FROM THE SEEDS OF 'BABCHI' (*CULLEN CORYLIFOLIUM* L.)

C.S. Raghav★ and Poonam Suneja
National Bureau of Plant Genetic Resources, Pusa Campus, New Delhi - 110 012
★e-mail : raghav63@yahoo.com

Various tools and techniques of isolation of furanocoumarins from the seeds of *Psoralea corylifolia* L. have been developed. Results on relative efficiency of individual technique along with its merits and demerits have been discussed in detail. The isolation of furanocoumarins has been done in four major steps i.e., seed processing, seed soaking, extraction and isolation. Each step has been further sub-divided into different parameters. The best parameter out of each step has been selected for easy, quick and high yield of furanocoumarins. All the techniques are gravimetric, simple and quick.

FURANOCOUMARINS

Furanocoumarins are a group of compounds, which occur in various plants species. They are toxic compounds. The chemical structure of furanocoumarins derives from the condensation of a furanic ring with the coumarin nucleus and there could be numerous isomers (Fig.). However, the parent compound of the most important group of furanocoumarin derivatives is psoralen. Among the other isomers, only isopsoralen and allopsoralen have importance. Isopsoralen, also known as angelicin, is a naturally occurring substance while allopsoralen is not found in nature but has been synthesized. They are used to treat vitiligo or leucoderma. Various furanocoumarins based allopathic drugs are available in the market, which emphasize its commercial importance. Furocoumarins are primary photodynamic agents, which have property to fluoresce when exposed to visible light. These sensitizers, when present in the skin and get exposed to visible light, absorb a quantum of energy and become activated. This energy may further pass on to other molecules such as amino acids, histidine, tryptophane, tyrosine, which in turn become activated and subsequently involve in further chemical reactions. As a result of these chemical reactions inflammatory changes are noticed in the skin. Furanocoumarins on exposure to sunlight or UV light produce erythema of varying degree, followed by pigmentation that appears after a latent period. Two forms of furanocoumarin condensation occur in nature. The first forms linear molecules by linking the 3'2' furan bond to the 6,7 coumarin bond. The 3'2' bond may also link to the 7,8 coumarin bond to produce angular furanocoumarins. In general, the linear furocoumarins (psoralens) are more phototoxic than the angular furocoumarins (angelicins). The only exception to this is the angular furocoumarin known as pimpinellin found in members of the genus *Heracleum* (Umbelliferae). The most severe reactions occur with 5-methoxypsoralen (5-MOP; bergapten) isolated from *Citrus bergamia* (Rutaceae), the bergamot orange, and 8-MOP (xanthotoxin) isolated from *Fagara xanthoxyloides* (Rutaceae). Psoralen is more phototoxic than 8-MOP and 5-MOP. The most commonly detected linear furocoumarins are 8-methoxypsoralen (8-MOP, xanthotoxin), 5 methoxypsoralen (5-MOP, bergapten) and psoralen itself.

BABCHI (*CULLEN CORYLIFOLIUM* L.)

Babchi is an annual herb belonging to family Papalinaceae. The seeds of *Cullen corylifolium* L. are well known for their therapeutic uses in various kinds of ailments (Anonymous, 1969). Recently, the plant has been recognised in agriculture sector for its uses as pulse protectant (Chander and Ahmad, 1982) against insect infestation, as antifeedent for the larvae of pest of castor (*Ricinus communis*) (Mohanty *et al.*, 1988) and as growth promoter in silk worms (*Bombyx mori*) (Murugan *et al.*, 1998). Research findings indicate that the plant extract could be a cheap and abundant source of insect hormones and can be used to increase silk yield in commercial silk worm rearing (Rajashekhargouda *et al.*, 1997). The major groups of compounds identified in the seeds and other parts of the plant include furanocoumarins, coumestans, flavones/chalcones, terpenes, saponins/ sterols, hydrocarbons and fats (Siddiqui and Ansari, 1995). Out of these, the furanocoumarins including psoralen and its angular isomer, isopsoralen, psoraliden and bakuchicin are of immense importance in the medical science as well as in agriculture.

NATURAL SOURCES

Furanocoumarins are occurred in various plants of 4 families such as Rutaceae, Papilionaceae, Moraceae and Apiaceae. Furanocoumarins are found in various parts of plants such as leaves, seeds, fruits, roots and rhizomes. In Rutaceae family four main linear furanocoumarins are found in Ruta genus. They are psoralen, 5-methoxy-psoralen (5-MOP, also known as bergapten), 8-methoxy-psoralen (8-MOP, xanthotoxin), and 5-8-methoxy-psoralen (5-8-MOP, isopimpinellin). The plants of Moraceae family contain bergapten (5-MOP) and psoralen. Among the large Papilionaceae family, only 2 plant genere are known to contain furanocoumarins. These genere are *Psoralea* and *Coronilla*. These plants contain the most simple linear and angular furanocoumarins that are repectively psoralen and angelicin. Furanocoumarins are the most widespread in the Apiaceae family. So far these molecules have been identified in more than 35 species. Most of these plants contain 4 linear furanocoumarins that are 5-MOP, 8-MOP, 5-8-MOP, and imperatorin.

HISTORY OF USES

Plants have been known to cause hyperpigmentation since antiquity. As early as 2000 B.C. in Egypt, the juice of *Ammi majus* (Apiaceae), false Bishop's Weed, was rubbed on patches of vitiligo after which patients were encouraged to lie in the sun. Even today, Egyptian herbalists sell Aatrillal, a yellowish-brown powder made from Ammi majus seeds for the treatment of leukoderma. Indian medicine used boiled extracts of leaves, seeds, and roots of *Cullen corylifolium* (Papilionaceae) a South African native known as the bavachee. Topical or systemic figs followed by sun exposure was also used to treat 'white leprosy', as vitiligo is still called in India. Chinese herbal medicine even today recommends that leaves of Angelica archangelicum (Apiaceae) be used. Our name for the chemical family of phototoxic ingredients, psoralens, was derived from Cullen corylifolium. In 1897, reports of dermatitis following contact with parsnips, *Pastinaca sativa*, (Apiaceae) and/or *angelica, Angelica archangelicum*, came from both the United States and England. Unfortunately, neither author recognized the necessity of UVR for the reaction. In 1916, Freund observed characteristic hyperpigmented lesions resembling pendants that he attributed to eau de cologne containing bergamot oil. He, too, did not recognize the necessity of UVR for the cutaneous changes. In 1932, Oppenheim reported that the combination of meadow grass exposure and UVR could result in a striate arrangement of itching blisters 24-48 hours later. In 1938, Kuske showed that plant furanocoumarins caused photosensitization. In 1939, Jensen and Hansen reported that UVR wavelengths between 320 and 380 nm (UVA) caused the maximal reactions.44 Klaber introduced the term 'phytophotodermatitis' in 1942 to emphasize the necessity of plants and light to cause the reaction. It was an Egyptian, Professor Abdel Monem El Mofty, Department of Dermatology, Cairo University Medical School, who observed plants used in Egyptian folk medicine and began the development of modern photochemotherapy (PUVA) for vitiligo and psoriasis. In the 1940s, he used crystalline methoxsalen (8-MOP, xanthotoxin) followed by sunlight exposure to treat vitiligo. The Department of Dermatology at Harvard Medical School first successfully treated psoriasis with PUVA in 1974. Today, PUVA is used in the therapy of many other diseases including palmoplantar pustulosis, mycosis fungoides, atopic dermatitis, and generalized lichen planus.

STRUCTURE AND ACTIVITY

Based on the condensation of a furanic ring with the coumarin nucleus, furanocoumarins may be divided in two groups, linear furanocoumarins and angular furanocoumarins (fig.). The linear furanocoumarins are naturally occurring aromatic compounds. They are benzo-2-pyrone compounds with a furan ring fused at the 6,7 position. There are two types of benzopyrones, coumarins (benzo-alpha- pyrones) and chromones (benzo-gamma-pyrones). They differ only in the position of the carboxyl group (Fitzpatrick *et al.*, 1966). A furan ring can be condensed with a coumarin molecule in twelve different ways and each of the resulting compounds can become the parent of a family of psoralen-like derivatives. Those with linear tricyclic structure are active and with angular structure are least active or not active.

The phototoxic effect of furocoumarins relies on their ability to absorb photons. It has been documented that active furanocoumarins show a maximum of excitation in the region 340-380 mμ and a maximum of fluorescence in the region 420-460 mμ. The molecular mechanism of photosensitization by the active furanocoumarins consists of three steps:

- molecular complexing
- photobinding of furanocoumarins to DNA and bases
- cross-linking between strands of DNA.

When furanocoumarins are added to an aqueous solution of nucleic acids a molecular complex is formed, which is observed by changes in the viscosity of DNA solutions, increased solubility of furanocoumarins in aqueous DNA solutions, and shifts of the absorption maxima (9-12 nm) of furanocoumarins to longer wavelengths. From the above observations it is inferred that the furanocoumarins intercalate between two pyrimidine bases of DNA. However, since both biologically active and inactive furanocoumarins form complexes with DNA, the biological activity cannot be explained by complexing alone.

Photobinding of furanocoumarins to DNA and bases occur following absorption of a photon. In the classic mechanism, the molecule forms via ground state complexes that enables them to intercalate with pyrimidine bases, cytosine, uracil, and thymine thus forming covalent monoadducts, or by triplet excited states of furanocoumarins mediated by singlet oxygen resulting in superoxide and hydoxyl radical generation, or by formation of sensitized free radicals. However, more recently photochemical events have been known to occur with amino acids and proteins, and photoreactions with membrane constituents. The reactive sites occur, at least partly, from the C4-photocycladdition. The reactive sites for linear furanocoumarin are the 4', 5'-double bond on the furan ring and the 3,4-double bond on the pyrone ring by initial involvement of the 5,6-double bond of thymine. Thus, two monoadducts can be formed. The 3,4-monoadduct however does not absorb further UVA. In contrast, the 4'5'-monadduct can be further stimulated by UVA.

Cross-linking between opposite strands of DNA occurs with absorption of additional quanta of radiant energy at 365 nm. Thus, further irradiation stimulates the 4'5'-monadduct to form a cross-link between pyrimidine bases in the opposite strand of DNA. The efficiency of forming cross-links varies from furanocoumarin to furanocoumarin, intensity of the light source, the route of administration, and the species tested. Only a fraction of 8-MOP molecules are intercalated with DNA under normal physiological conditions. Double-stranded crossslinks in molecules other than DNA has been acknowledged in RNA and of RNA and DNA duplexes. At the cellular level with the different types of photochemical reactions involving psoralens and their ability to interact with RNA, DNA, proteins, and lipids, it is not surprising that psoralens induce point/gene mutations and chromosomal aberrations in a variety of organisms. These mutations are probably more responsive to crosslinking than monadducts since the former is found to be more toxic.

EXTRACTION AND ISOLATION

Various modern instrumental techniques of estimation and isolation of these compounds have been evolved in the recent past (Lu *et al.*, 1999; Spencer *et al.*, 1987), but very high isolation cost associated with these techniques has always been a limiting factor in the use of these techniques by the

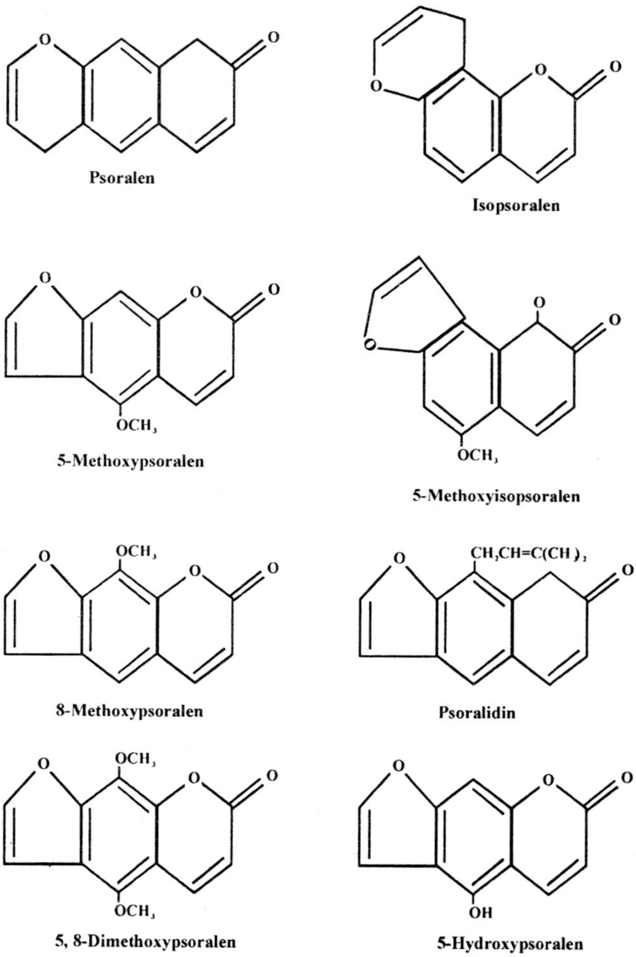

Fig : Structure of important furanocoumarins

common users. They need a handy, easy and cheap solution to their problems. They also don't need these compounds in absolutely pure state. The gravimetric method of isolation of these compounds is very important not only to these users but also to the small-scale entrepreneurs. The isolation of furanocoumarins is based on the retention or elution of the desired compound from a medium (silica gel, aluminium oxide etc.) against a polar or non-polar solvent. This process helps to enrich the desired compound by discarding other unwanted impurities. There is an array of solvents, which when used in altered sequence or in the form of mixture change their polarity and thereby their solubilizing efficiency towards a compound. How quickly and effectively a desired compound is isolated depends upon selection of various parameters and factors like solvent, mixtures of various solvents in varying proportions, heating, non heating or cooling, sequence of extraction or elution with solvent or solvent mixture etc.

Authors have developed the techniques of extraction and isolation of furanocoumarins, wherein a comparative study in terms of time, yield and simplicity of these sets of methods and materials has also been conducted to find out the most suitable method for isolation of furanocoumarins.Accordingly, all the experimental techniques of isolation of furanocoumarins were divided into four major steps, which are described below.

1. **Sample processing :** The seeds (100g. to 1 kg.) were cleaned, dried and ground. The observations were recorded on moisture losses at different intervals in two seasons i.e., summer and rainy, in order to standardize optimum drying time in a particular season.

2. **Seed soaking :** Dried ground seeds were soaked in five different solvents (water, petroleum ether, benzene, chloroform and methanol) each for 6, 12, 24, 48 and 72 hr. in order to find out the best combination of solvent and soaking time. Simple and acidic hydrolysis of the seed was also done. After a fixed time of soaking the seed extract of each sample was collected, the seed was repeatedly washed with the same solvent and the collected extract was concentrated and the furanocoumarins were isolated from each sample separately using the same methodology for all.

3. **Extraction :** The acidic methanol soluble part of the soaked seed, which showed maximum furanocoumarins, was concentrated and extracted with hot and cold petroleum ether and its mixtures (in various proportion) with chloroform, pure chloroform, diethyl ether and benzene, separately to find out the suitable solvent/mixture extracting maximum furanocoumarins from the methanol soluble part in a given quantity of solvent and time. The furanocoumarins were isolated using the same methodology for all the samples.

4. **Isolation :** Two methods of isolation of furanocoumarins from the pure chloroform (which was found best extractant) extract were tried.

1. Repeated extraction with chloroform and petroleum ether mixture (in various proportions) followed by pure hot petroleum ether.
2. Column chromatography, using various solvents and their mixtures as mobile phase and silica gel as stationary phase.The furanocoumarins yield from each altered step/solvent was recorded.

5. **Identification :** The identification of the isolated furanocoumarins was done using Thin Layer Chromatography (TLC) technique against standard furanocoumarin compounds.

RESULTS AND DISCUSSION

The recorded data and visual observations on each step are presented as follows.

1. Seed Processing : The air-dried seed was oven dried at 60° for 7 hr. a day in the month of August till the weight loss was constant. At 4th day the weight loss was maximum, which was constant at 5th day of drying. However, when dried continuously for 24 hr. the weight loss was maximum and no further loss was recorded after 24 hr. (Table-1). In the month of May drying of 2 days @ 7 hr./day gave the highest weight loss, which was less than the period continuous drying of 24 hr. in the month of August. It indicates that in the month of August the moisture content of the seed was more due to which it took more drying time. While in the month of May the seed moisture was already less, therefore, the drying time also reduced even below 24 hr. over required drying period in the month of August. The drying loss was 6.6% higher in the rainy season as compared to summer season. This may also be due to presence of high moisture content in the seed in rainy season.

The oven dried seed was ground in an electric grinder through a mesh (size 64 holes/inch^2). This could facilitate easy and quick penetration of solvent in the seed to solubilize and release the furanocoumarins from the seed tissues. While grinding, it was observed that the just dried seed was difficult to grind as it turned in to a sticky lump in the grinding machine. This was probably on account of dilution and release of oil present in the pericarp of seed due to heating effect. Therefore, the grinding of hot seeds,

Table - 1 : Drying loss in seeds of *Psoralea corylifolia* in two different seasons.

Sample size 1 kg. seed

Drying period	**Weight loss (%) in two seasons**	
	Summer (May)	**Rainy (August)**
1st day (9.30 a.m. - 4.30 p.m.	2.3	7.8
2nd day (9.30 a.m. - 4.30 p.m	3.0	8.7
3rd day (9.30 a.m. - 4.30 p.m.	3.0	9.2
4th day (9.30 a.m. - 4.30 p.m.	-	9.6
5th day (9.30 a.m. - 4.30 p.m.	-	9.6
Continuously for 24 hr. (9.30 a.m. - 9.30 a.m.)	3.0	9.6

just taken out of the oven, should be avoided. It should be cooled down to the room temperature before grinding.

2. Seed soaking : The soaking of seed is a very important part of isolation as it loosens the seed tissues and thus, makes the release of compounds easy. There may be many factors, which can influence the loosening and release process. Some of them are, selection of solvent or solvent mixture, soaking time, stirring or non stirring, pH change, heating, non heating or cooling or any combination of any of these factors. The important factors relevant to the present study were applied at various stages and their respective efficiency and effects were tested. The results are discussed in the following text.

Soaking solvent and time : The seeds were initially soaked in five different solvents (water, petroleum ether, benzene, chloroform, methanol) in order to test the relative efficiency of various solvents in relation to time of soaking (Table-2). It was found that acidic methanol released the maximum furanocoumarins from the seeds at all the time intervals as compared to any other solvent. It also yielded maximum quantity of oleo-resin. The highest yield of furanocoumarins was recorded at 72 hr. with all the solvents. However, the rate of recovery varied from solvent to solvent at different time intervals.

The data given in table-2 indicate that recovery of furanocoumarins was 0%, 1.20%, 3.61%, 60.24% with water, 11.54%, 53.85%, 78.85%, 92.30% with petroleum ether, 55.51%, 63.78%, 75.20%, 83.86%, with benzene, 62.16%, 74.22%, 80.62%, 92.50%, with chloroform, 75.24%, 95.36%, 98.33%, 99.76% with non-acidic methanol, and 76.86%, 95.70%, 99.43%, 99.90% with acidic methanol at 6, 12, 24, and 48 hr., respectively as compared to their respective yield at 72hr., which was recorded highest in all the cases. It is clear from the data that in case of methanol over 95% recovery was recorded only at 12 hr. soaking, while in case of all other solvents this percentage of recovery could be possible only between 48 and 72 hr. of soaking. The increased furanocoumarins yield with acidic methanol over non-acidic methanol may be due to acid hydrolysis and subsequent release of glycosidic (bound) psoralen and isopsoralen from the seed tissues, which is otherwise not freed with any of the solvents. The presence of glycosidic psoralen in the seeds of *Cullen corylifolium* has been reported by Krishnamurthi (1969).

The colour, quality and yield of extracts obtained with various solvents were also varied with solvent type. The yield of acidic methanol extract was 88%, 48%, 36%, 28% and 20% higher than the yield obtained with water, petroleum ether, benzene, chloroform and non-acidic methanol extract, respectively. The colour and quality of various solvent extracts were also different in some or the other way. The extract of water was very light in

Table - 2 : Effect of soaking solvents and time of soaking on extract and furanocoumarins yield (%)

Soaking solvent	Extract yield (%)	Extract colour	Furanocoumarin yield (%) at various soaking times (hr.)				
			6	12	24	48	72
Water	3	Straw colour pigment	nil	0.001	0.003	0.050	0.083
Petroleum ether	13	Brown viscous oil	0.006	0.028	0.041	0.048	0.052
Benzene	16	Brown oleoresin	0.282	0.324	0.382	0.426	0.508
Chloroform	18	Brown oleoresin	0.340	0.406	0.441	0.506	0.547
Methanol	20	Dark brown oleoresin	0.632	0.801	0.826	0.838	0.840
Acidic methanol	25	Black coloured oleoresin	0.804	1.001	1.040	1.045	1.046

colour, which quickly separated on the silica gel coloum. The extract of petroleum ether was absolutely oily in nature and brown in colour. The extracts of benzene and chloroform were brown oleo-resinous. Methanolic extract was dark brown in colour, while the acidic methanol extract was black and sticky oleo-resin, which delayed the separation of furanocoumarins through column chromatography as compared to any other solvent extract.

3. Extraction : Once the acidic methanol extract of soaked seed, which contained 99% furanocoumarins after 24 hr. soaking, was obtained and concentrated, it was further extracted with various solvents separately in order to isolate the furanocoumarins.

(i) **Extraction with petroleum ether :** Concentrated black coloured viscous acidic methanol extract of soaked seeds when further extracted with petroleum ether it could not give any furanocoumarins. However, when extracted with hot petroleum ether it gave 20% furanocoumarins in the first fraction of 300 ml, but simultaneously the residue turned so sticky that further release of furanocoumarins became difficult and lingered on till lots of hot petroleum ether used in extraction. Even after that the major quantity of furanocoumarins could not be recovered from the parent extract. This chelating effect may be due to formation of a sticky layer over the parent extract as a result of solubility of fixed oil and partial solubility of other oleo-resinous substances in the petroleum ether at the time of extraction.

(ii) **Extraction with chloroform :** When the concentrated acidic methanol extract was further extracted with chloroform it extracted 100% furanocoumarins from the parent extract leaving behind 80% insoluble residue, which contained no furanocoumarins as tested on TLC. It also consumed lowest solvent and time as compared to any other solvent. For example, from a 50 gram sample lot a total of 12.5 gram acidic methanol extract was obtained, which required only 350 ml chloroform in 7 equal lots to extract the total furanocoumarins from this extract. On contrary, this quantity was more than 2 litres in case of hot petroleum ether.

(iii) **Extraction with mixture :** Extraction of concentrated acidic methanol extract was tried with mixture of chloroform and petroleum ether in varying proportions. Out of these, extraction with 60:40% (v/v) mixture (60% chloroform + 40% petroleum ether) followed by hot extraction with the same mixture was found suitable for more than 70% recovery of furanocoumarins with 500 ml of mixture for a 50 gram sample lot. The recovery of remaining less than 30% furanocoumarins from the parent extract was possible only with repeated hot extraction with another litre of mixture.

Though it extracted furanocoumarins with least unwanted pigments, the low recovery from it could not make the mixture comparable with pure chloroform extraction.

4. Isolation and purification : The concentrated chloroform extract was weighed as 18-20% of the total seed weight. It contained mainly three components viz., the 100% furanocoumarins of the seeds, 12-14% fixed oil and the pigment. All further steps in this process were taken to separate furanocoumarins from the oil and coloured pigment. Two approaches were adopted for this.

(1) further extraction of chloroform extract with mixtures of chloroform and petroleum ether in varying proportions ranging from 20:80 to 80:20 (v/v)was tried. Out of these mixtures, the extraction with mixture of 25:75 (v/v) (chloroform : petroleum ether) ratio followed by hot extraction with the same mixture was found the most suitable as it discarded maximum coloured pigment extracting 70-95% furanocoumarins depending upon the sample size. It was observed that the bigger (½ -1 kg.) the sample size the lower the recovery. The reason for low recovery in bigger lots was perhaps the thick layer of parent extract in the flask, which could not be penetrated completely in an insufficient extracting solvent. This was supported by the fact that when the size of flask was increased, the recovery also increased with the same amount of solvent. For further purification of crystals this concentrated mixture extract was further extracted with pure hot petroleum ether. This extracted almost pigment free crystals alongwith oil. The oil associated with the furanocoumarin crystals was separated by chilling it in refrigerator over night. The chilled oil was decanted through a cotton plug and the crystals deposited in the bottom of flask were collected separately. Thin Layer Chromatography (TLC) of the over night chilled decanted petroleum ether of the flask showed absence of furanocoumarins in the separated oil.

(2) direct column chromatography with

- (i) petroleum ether followed by chloroform + petroleum ether (25:75) (v/v) mixture
- (ii) petroleum ether followed by pure chloroform
- (iii) directly with chloroform
- (iv) benzene

In case of (i) initial oily fraction which contained no furanocoumarins was discarded. Subsequent fractions rich in furanocoumarins were combined and further purified. Use of mixture as a subsequent elutor facilitated more colour retention on the column giving more pure crystals, but it took longer time to elute the furanocoumarins from the column.

In case of (ii) similar to (i) initial furanocoumarins free oily fraction was collected separately. The subsequent chloroform soluble fractions rich in furanocoumarins were collected and purified. This quickly removed the furanocoumarins from the column, but with more pigment along with crystals, which needed repeated subsequent chromatography for purification.

In case of (iii) oil plus furanocoumarins fraction came simultaneously and the coloured pigment retained on the column. In this case also retention of coloured pigment on the column was very less however, the recovery of furanocoumarins was very fast as compared to all other eluents. The quantities used to remove the furanocoumarins from the column was also very less as compared to any other eluent. To remove the oil from the mixture it was concentrated and the chloroform was removed from it completely. Then a little petroleum ether was added to it and chilled in refrigerator for over night. This separated the oil from furanocoumarin crystals settled in the bottom of flask. The coloured crystals were further purified on column to obtain pure off white crystals. In this case more than 95% furanocoumarins were recovered very quickly and easily. In a silica column (40 g. silica gel, sample size 50 g. seed) 1st and 2nd fraction of 150 ml chloroform each eluted 70% and 30% furanocoumarins, respectively from the column.

In case of (iv) the retention of colour was very good but the recovery of furanocoumarins was very low and long lasting. lots of solvent was required for complete recovery of furanocoumarins from the column.

5. Identification : The furanocoumarins isolated through various methods and solvents were identified through Thin Layer Chromatography (TLC) technique against standard compounds of this group.

REFERENCES

Anonymous (1969). Wealth of India, Vol. VIII, pp 295-298.

Bensasson, R.V., Chalvet, E.J. Land and Ronfard-Haret, J.C. (1984). Triplet radical anion and radical cation spectra of furanoocoumarin. *Photochem. Photobiol.*, **29** : 287-291.

Bordin, F., Carlassare, F., Busulini, L. and Baccichetti, F. (1992). Furanocoumarin sensitization induces DNA-protein cross-links. *Photochem. and Photobiol.*, **58(1)** : 133-136.

Brown, S.A. (1978). Biochemistry of the coumarins. In Recent Advances in Photochemistry, Swain, T., Harborne, J.B. and Sumere, C.H. Eds. Plenum Press, NY **12** : 249-284.

Cadet, J., Vigny, P. and Midden, W.R. (1990). Photoreaction of furanocoumarins with biomolecules. *J. Ohotochem. and Photobiol. B: Biol.*, **6** : 197-206.

Chander, H. and Ahmed, S.M. (1982). Extraction of medicinal plants as pulse protectants against *Callosobruchus chinensis* l. infestation. *Journal of Food Science and Tech.*, **19(2)** : 50-52.

Dall'Acqua, F. and Martelli, P. (1990). Photosensitizing action of furanocoumarins on membrane components and consequent intracellular events. *J. Photochem. Photobiol. B: Biol.*, **8** : 235-25.

Dall'Acqua, F., Marciani, S. and Rodighiero, G. (1969). The action spectrum xanthotoxin and bergapten for the photoreaction with native DNA. *Z. Naaturforsch,* **24(b)** : 667-688.

Fitzpatrick, T.N., Arndit, K.A., El Mofty, A.M. and Pathak, M.A. (1966). Hydroquinone and psoralens in the therapy of hypermelanosis and vitiligo. *Archs Derm.*, **93** : 589.

Lu, Feng, Liu Li, Li Ling, Wu, Yutian, Lu F, Liu LL, Li L and Wu YT (1999). Determination of psoralen and isopsoralen in *Psoralea corylifolia* L. by supercritical fluid chromatography. *Acta Pharmaceutica Sinica.*, **34(4)** : 301-303.

Mohanty, K.K., Chakraborty, D.P. and Roy, S. (1988). Antifeedant activity of oil fraction of seed of some leguminous plants against *Diacrisia obliqua. Indian J. Agric. Sci.*, **58(7)** : 579-580.

Murugan, K., Jeyabalan, D., Kumar, N.S., Nathan, S.S., Shivaprakasam, N. and Pandey, A. (Eds.) Soccol, C.R. and Josh, V.K. (1998). Growth promoting effects of plant products on silk worm- a biotechnological aproach. *J. Sci. & Indus. Res.*, **57** : 10-11.

Parrish, J.A., Fitzpatrick, T.B., Tanenbaum, L. and Pathak, M.A. (1974). Photochemotherapy of psoriasis with oral methoxsalen and longwave ultraviolet light. *New Engl. J. Med.*, **291** : 1207-1214.

Rajashekhargouda, R., Gopalan, M., Jayaraj, S. and Natarajan, N. (1997). Field performance of plant extracts on mulberry silk worm (Bombyx mori L.) *Entomon.*, **22(3-4)** : 235-238.

Schmitt, I.M., Chimenti, S. and Gasparro, F.P. (1995). Psoralen-protein photochemistry - a forgotten field. *J. Photochem. Photobiol. B: Biol.*, **27** : 101-107.

Scott, B.R., Pathak, M.A. and Mohn, G.R. (1976). Molecular and genetic basis of furanocoumarin reactions. *Mutat, Res.*, **39** : 29-74.

Siddiqui, A.A. and Ansari, S.H. (1995). Phyto-chemical and pharmacological investigation on *Psoralea corylifolia. Hamdard Medicus.*, **38(3)** : 109-115.

Spencer, G.F., Tjarks, L.W. and Powell, R.G. (1987). Analysis of linear and angular furanocoumarins by dual column High Performance Liquid Chromatography. *J. Agric. Food. Chem.*, **35(5)** : 803-805.

Recent Advances in Medicinal Plant Research : Vision 21st Century, 2005
Editors : S.K. Prabhuji, G.P. Rao and S.K. Patil
Satish Publication House, New Delhi

Chapter 33

DEMAND POTENTIALS OF MEDICINAL AND AROMATIC CROPS-CHANNELISING MARKETING SYSTEM : AN OPTION IN NATIONAL ECONOMY

V. Suryanarayana[1], G.O. Manjunath[2] and S.K. Patil[2]
[1]*Department of Forest Protection, College of Forestry, Sirsi - 581 401, Karnataka*
[2]*Department of Forest utilization, College of Forestry, Sirsi - 581 401, Karnataka*

Globally, there is an increasing interest of herbal usages in the lively hood set-ups. The demand for MAC's in India to meet the domestic and export market is increasing steeply. A mismatch in demand-supply levels amounting to more than 40,000 tonnes due to high dependency on forest resources is established. Cultivation of MAC's has high potentials over agricultural crops with regard to net profits. It has a wide scope with a special concern of promoting social harmony balance, employment opportunities, self-employment motives and up-surgence in national economy. Currently, from the point of fast depleting forest resources of MAP's, quality concern and national economic drive "Commercial cultivation of MAC's" is a need of the day. Government assisted and promoted national network framed by a combined abilities of planners, economists, policy makers, scientists, industrialists and progressive farming communities can be the unique solution in commercializing MAC's. Mass drive commercialization can be achieved by introducing them in different cropping systems and in afforestation programme. A network of organizing and mobilizing a common platform of growers, NGO/GO/self help groups, local traders, regional traders, processors and stake holders under the umbrella of equitable bio-partnership is the fore most need in channelising market issues in India. A ready-reckoner of resource data base or a web site discharging the updated production and marketing issues on MAC's can be a potential weapon in creating mass motivation among rural communities by building up their confidence levels. Further, multiple linked chain of community based organization and MAC's based companies in a value based chain or production to consumption and marketing continuum (PC& M) is to be formulated in production and supply of quality material to industry and to assure benefits to growers. In nutshell, MAC's can be an integral part of natural resource management contributing for economic growth; environmental protection and trade provided marketing channel system is perfectly stream lined.

INTRODUCTION

Globally, there is an increasing interest of herbal usages in the lively hood set-ups. The toxicities associated with synthetic medicines, cosmetics and perfumeries have been well realized in recent years. Over 60 per cent of all the pharmaceuticals are plant based. The world market for plant derived chemicals, pharmaceutics, nutraceutics, fragrances and colour ingredients alone exceeds several billions/year. In South Asian countries, medicinal and aromatic crops (MAC's) play a significant role in subsistence of rural economy through phyto-pharmaceutics and cosmetic industries. India in specific has better prospects, being one of the mega-biodiversity centers of the World with 80 per cent of its population as rural based, who have rich potentials of indigenous skills. WHO recommends the inclusion of herbs in natural health care by looking into expenditure of 40-50 percent of total health budget on drugs of westernized pharmaceutical base. With the rising populations, reduced economic levels and rural based poor population, our country finds inability to afford the high cost of synthetic medicines. The situation warrants us to adopt herbal-based lively hood, which is cheap, safe and this can be made available in plenty through efforts of adopting skills of cultivation, collection, storage and processing. It forms a broad base in poverty alleviation drive by uplifting the prospects of our MAC's status at International forum.

HISTORICAL REVIEW

Plants have been used by healers and health rejuvenators since time immemorial. Now, herbal based medication is widely practiced, inspite of so much advancement in the field of allopathic medicine. As per the survey conducted by FRLHT (a Bangalore based NGO), about 880 botanicals are traded in India with an estimation of 600,000 registered practitioners and crores of local healers who are dependent on medicinal plants grown in wild. It has over 6000 manufacturing units with a turn over of Rs. 4,200 crores. Eighty percent of world's population is using herbal medicines in their health care systems. During 1990's the International trade of natural product medicines was US $ 50 billion and annual increase rate was 30 percent (Sarin, 2003). Our country has a lion share of 75 to 80 per cent of herbal raw materials with respect to export potentials (Rawat and Uniyal, 2003).

An annual domestic trade of herbal raw drugs, in India, is accounted to 1,98,000 tonnes with an annual trade value of 4000 to 4200 crores (Handa, 1996). The demands for MAC's in India to meet the domestic and export market is increasing steeply and will be more than 15 to 16 percent by 2005 (CRPA, 2001). The world trade figures suggest that, India is next to China by exporting 32,600 tonnes of medicinal raw materials worth US$ 46 m annually (Lange, 1997; Dhar *et al.*, 2000; Manjikhola and Dhar, 2002).

The mismatch in demand-supply potentials has been attributed to dependence on forest resources for MAP's up to an extent of 90 percent (FAO data base, 2003). In meeting the growing demands of herbal products at national and global levels, natural forest solely cannot fulfill the needs unless it is supplemented by some other means like commercial cultivation coupled with value addition of the produce (Khanuja *et al.*, 2004). Especially it pushes an economic growth in resource-constrained areas suffering from limited education opportunities, lack of money generating jobs and lack of infrastructure (Karki, 2002). Baby P. Skaria *et al.* (2003) have described that MAP's trade in our country is highly un-systematic and not a quality oriented system.

Marketing network has been studied and a flow chart of information was suggested which include a flow of produce from the collector to consumers through a series of middlemen (Maikhuri *et al.*, 2003).

METHODOLOGY

The present study has been conducted with the aim of collecting information related to gaps in supply – demand levels of medicinal and aromatic crops, to arrive, whether MAC's sector is remunerative or not. This was under taken in proposing an ideal model of marketing network system. For fulfilling these ideologies, a team of scientists took up surveys and met the cultivators of MAC's, the local traders, industrialists and local health practitioners, besides, visiting related Government and Non-government organizations. A series of discussions were made with them in respect to the problems and benefits in raising MAC nurseries, sale of seedlings, cultivation, post-harvest processing, transportation and marketing of their produce. An ideology has been flown in crisping the data under different headings to highlight the aspects and prospects. A serious effort was made in drafting the commonly prevailing marketing network in MAC's sector. A special focus was given in gathering information on constraints in MAC's commercialization as India is possessing 25 percent of the global floral diversity coupled with very rich Ayurvedic cultural heritage. A thorough study was also undertaken to prepare flow chart of marketing network from producer to consumer, which is currently operating.

With this background it was worked out in proposing a modified marketing network so as to reduce the number of changing hands (middlemen) or indirect links from producer to consumer. Care was also taken to incorporate the other relevant information's such as platform of bio-partnership to promote the transparency which can break down the main marketing constraint of hidden marketing agenda. In addition, a focus was also given to link-up multidisciplinary conceptuals in the proposed model rather than concentrating only on vicious circle of mediators.

RESULTS AND DISCUSSION

As the medicinal plants in wild has now entered the stage of extensive cultivation, we feel, it is the right time to replace the name MAP's with MAC's with reference to MP's, which are already under cultivation. The study was conducted with an idea of generating information on global market demands of MAC's and supply potentials of our country, to answer whether commercialization of MAC's sector is worth and if so how to go about, a vision to lacunae in the present marketing network and a means of correcting it, was also generated. The gathered information is, then, arranged under different headings as explained below :

(1) Growing demand rates and gaps in demand-supply potential

All India Co-ordinated Project on Ethno-biology has suggested that over 4000 ethnic communities are using around 8000 plant species in our country. On the economic context, out of 8000 species, more than 1000 species are believed to be in the streamline of commercial trade as raw drugs. Estimated annual domestic trade in herbal raw drugs is about 1,98,000 tonnes. Furthermore, over 6000 manufacturing units are presently functioning in this country with a turn over of Rs. 4,200 crores per year. The estimated International trading of herbal products is US $ 62 billion per year and is growing at the rate of 7 percent. Even the worlds essential oil production at raw material level is estimated over US$ 90 crores annually. The projected demand of for MP's alone is US$ 5 trillion by 2050. India has a supply potential of 32,000 tonnes per year valuing US$ 46M. The trading in India is far below the level of export level of China in monetary as well as quality aspects. So, the share of India in the global export market of MP's trade falls below 0.5 percent. The demand for MAC's in India to meet domestic and export market is increasing steeply and will be more than 15-16 percent by 2006. Inspite of possessing rich herital knowledge on herbs and natural resources, besides man power, currently the gap between demand and supply is more than 40,000 tonnes and expected to rise by 1.52 to 4 lakh tonnes valuing US$ 54M by 2006. The mismatch in demand-supply potentials may be attributed to several factors, viz., high dependence on forest resources (to an extent of 90 per cent and 95% of which is by premature harvesting), low investment on research and development, inadequate techniques with respect to production and processing, lack of concern in exporting MAC's to International quality standards, high chances of adulteration and bad packing, low sale price and opaqueness in marketing system.

(2) Can MAC's be a remunerative sector?

Usage of plant based medicines in meeting 90 and 40 percent of rural and urban medicare system respectively is a standing example for projecting MAC's as a remunerative sector. The cultivators of MAC's and important agricultural crops were met as indicated under methodology and the information regarding cost of cultivation and the income generated per acre basis was collected. From the generated information the net profit was estimated and presented in Table-1.

The results support in proving cultivation of MAC's to be highly beneficial than agriculture crop production in enhancing rural economy. *Withania somanifera* a dry land medicinal crop supercedes *Sorghum* by generating a three-fold increase in the net profit estimates. Similarly, the medicinal crop of, *Pogostemon patchouli,* growing under irrigated conditions, can generate a net profit of over two times more than that of paddy or potato which are again the irrigation dependent agricultural crops and are the staple food of greater demand. Besides economic benefits, the other advantages which make them highly remunerative are hardiness, non-browzability, wide adaptability and pilfer proof. From this, it can well be justified that MAC's are the potential option before us in blooming national economy.

(3) Can MAC's be a successful venture in Indian farming system?

MAC's have multifaceted spectral impacts in glorifying socio-cultural-economical perspectives. Our country do not need its introduction as a new venture, as the use of MAP's in health care and dietary systems is already

Table - 1 : Net profit estimates of some important medicinal and aromatic crops (MAC's) over agricultural crops.

Name	Duration (Months)	Economic part	Expend. (Rs/Ac.)	Gr. income (Rs/Ac./harvest)	Net Profit (Rs/Ac./harvest)
Medicinal/aromatic crop					
Raulwolfia serpentina (Sarpagandha)	20	Roots	20,000	1,06,250	86,250
Pogostemon patchouli (Patchouli)	5	Leaf	15,000	40,000	25,000
Withania somnifera (Aswagandha)	5-6	Roots	3,000	10,000	7,000
Agricultural crop					
Oryza sativa (Paddy)	4-5	Seeds	6,000	18,000	12,000
Solanum tuberosum (Potato)	3-4	Tuber	9,000	20,000	11,000
Sorghum bicolor (Jowar)	4	Seeds	2,000	4,500	2,500

embedded in Indian blood as a socially accepted traditional culture. In the present scene of ethically un-acceptable allopathic chemical productions, MAC's sector can cater the needs of domestic and plant based industries with environment friendly concept. Further, the vast range of agro climatic conditions ranging from alpine/mild temperate to tropical regions with abundant rains and sunshine makes it an ideal place for commercialization of MAC's. India being strategically well placed geographically with an easy access to many countries with well-connected conveyance systems, it has very good prospects in the emerging International markets by serving as a potential supplier of MAC's produce. It does not specify gender specificity as it encourages women entrepreneurship with a main focus on social harmony, balance and equity. Such developed binding can further strengthen self-help groups for promoting self-employment as a mass motive in rural sector. Since all MAC's require special keenness from harvest to post harvest processing, employing the local women is the best option in taking up the activities like cleaning, grading, drying, packing *etc.,*

In a base study conducted at Bhatkal of Uttara Kannada district of Karnataka, Vetivar based cottage industry is working since a decade with low working capital through indigenous technology. It is concentrating on Vetivar oil production from the roots of Vetivar grass and in making chappals, hand purse, vanity bags, hand fans, hats and other handicrafts. Recently preparation of hanging baskets has been introduced which is fetching high demand, as they are being useful in growing orchids and other ornamental plants. In the recent past, the industry has gained more popularity in this region and is shining due to the dedications of local women groups who have been specially employed for the task.

From the above information MAC's can be taken granted as a viable preposition in the Indian farming system as a successful venture as it fully utilizes men cum women force and skilled cum unskilled force thereby increasing living standards.

(4) Is it really a poverty alleviation drive?

The concept is to be viewed from MAC's role in bridging the issues of sustainable economic development, affordable health care and conservation of vital bio-diversity. The conversion of socio-cultural traditions and indigenous knowledge into lively hood means and as economic opportunities with enterprise-based application can be a base for up-surging rural economy. In the present state of enormous demand for MAC's at global level, there is an ample scope for India, which is a rich repository of MAP's diversity. Right from cultivation-processing-extraction to product formulation and fractionalization to blending perfumes requires both skilled and unskilled manpower. This, in turn, has a great potential in opening the floodgates of employment opportunities to rural communities. In the base study conducted at Shiralakoppa and Sagar of Shimoga district of Karnataka, small-scale industries preparing pickles from the medicinal herbs and non-alcoholic beverages from fruits and herbs of medicinal values through the inherited indigenous knowledge, this attempt has generated much employment opportunities to many unemployed rural youth in industrial and marketing sector and helping in alleviating rural poverty, which was earlier rampant in some places. Moreover, such industries are getting more popularity year after year in many places of the state from the health point of view.

(5) Why the commercialization is need of the day?

MAP resources are fast depleting through unsustainable exploitation by the unchecked growing populations. Our nation is shrinking in the near past by failing to generate employment to millions of youths and it is dwindling with a lack of concern in the conservation of natural resources. From the above point of view, a base study was under taken on the, " Over exploitation of *Nothapodytes nimmoniana,* an endemic species of Western Ghats of Uttara Kannada district of Karnataka with a concern of conservation". The observations indicated that, it is available only in wild and the wood of which is known to possess active principles of anticancer property with a tremendous demand in the domestic as well as International market. This potential leads to its continuous exploitation through illegal felling and marketing by many unemployed youth of this region which has resulted the species to reach the endangered status. In saving the species from further exploitation, which may lead to its disappearance, commercial cultivation of this species should be geared up immediately for its conservation and in meeting global market demands.

The vast, secret and unregulated trade in MAP's mainly from wild which continues to grow in the absence of a serious policy, attention is required to the material where the origin of particular drug is assigned to more than one plant due to which adulteration and the substitutes affect the market both directly and indirectly. So, wild collections reduce the quality of herbal product and this, in turn, disfigures Indian status in International market forum. The growing national and global demands of herbal products are enormous and, natural forests solely cannot fulfill the needs unless it is supplemented by some other means. So, from the point of conservation, employment generation, quality concern and as a national economic drive "Commercial cultivation of MAC's" should be the need of present day.

(6) How to go about commercialization?

For any crop to be commercialized, quality planting material, genetic vigour, high yielding potentials, pest tolerance trait, ready reckoner of agro-techniques and suitable market support are essential. In the context of unscientific and ruthless exploitation of forest resources for MAP's, incentives for the development of MAP's gardens may be taken up as an initial step in large-scale promotion. To start with the expansion of area under MAC's, it should be introduced under agro or farm forestry combinations such as certified farming, mixed or inter or companion or multi cropping systems through organic farming set ups. Further, it should be promoted in marginal and wastelands under afforestation programmes. This needs to be triggered up with top most priority without leaving a moment for rethinking. Government assisted and promoted national network, framed by a combined efforts of planners, economists, policy makers, scientists, industrialists and progressive farming communities can be the unique solution in commercializing MAC's.

From the point of cultivation with conservation concern, a base study was conducted in the Western Ghats of Uttara Kannada of Karnataka by selecting important endemic species or endangered species like *Saraca asoka*, *Nothapodytes nimmoniana* and *Cosnium fenestratum*. For commercialization, standardization of propagation, agro-techniques, genetic improvement and channelizing the marketing system are required.

So, "Commercialization", is the only weapon, which can check the fast eroding MAP's-based traditional bank. Practicing cultivation on a business platform by a chain of small and micro entrepreneurs is the best proposition in its success.

(7) MAP's marketing situation

Mass drive commercialization of MAP's in farming systems is solely dependent on marketing. Marketing channel in MAP's was started with the government supported license holders/cooperatives who served as collectors of MAP's from forest reserves. Later the inability of license holders to cover wide areas for collections, they started collecting the material from the gatherers (Village people) by acting as a trader. As the MAP's products availability was very less, it made them inaccessible to have direct dealing with the processor. So, addition of middlemen to channel took place that was associated in collecting the MAP produce from the gatherers/growers and moving it to the other stage of marketing. The exploitation of gatherers by middlemen was so systematic with a 'debt binding' relationships. Such system has turned into a tradition and because of this, the same set up has been forced to adopt in the present day trade of commercially cultivated crops. So, the channel of present marketing system, in general, has been depicted as a flow chart in Fig. 1.

From the flow chart it can well be predicted that, the price indexing is being monopolylised by local traders in day-to-day market transactions with very weak dissipation of trade secrets. It implies that the concept of transparent marketing and buy-back issues are still in infancy, which needs to be rejuvenated. The actual revenue that is expected to reach the grower, instead of reaching him, the substantial part of it is leaked in continuum of

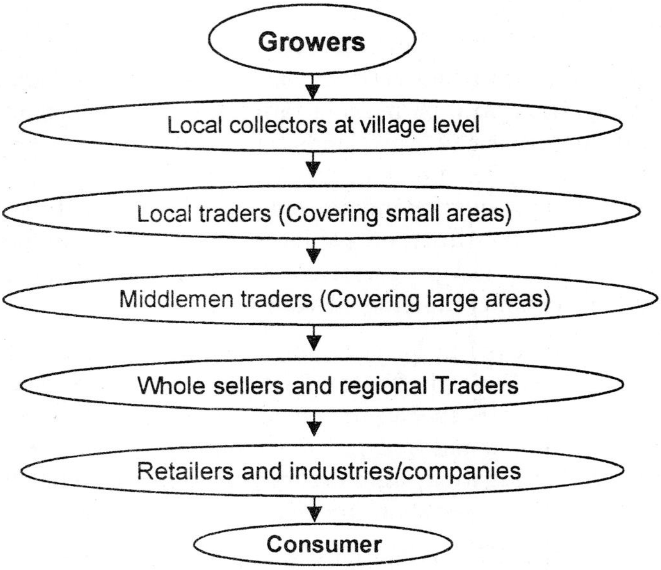

Fig. 1 : Flow chart showing present marketing situation of MACs

series of maximum number of changing hands from grower to the consumer with an ultimatum of victimizing the grower. So, non-existence of organized marketing infrastructure, un-assured pricing structures and lack of transparency in trade transactions, the farmers are reluctant in incorporating MAP's in their cropping system. These factors have been attributed for restricted expansion of area under MAC's, that too limited to few states. There is no systematic data on actual production, employment generation, technology development and its transfer, pricing and financial flows even for a single species of MAP's. This clearly indicates that a holistic approach is lacking in this sector.

(8) How to channalise marketing network?

Looking the barriers in the general marketing network of present days, the foremost requirement is to have restructuring. The factors like un-assured or guaranteed marketing, involvement of too many traders in the trade issues, lack of transparency in trade, lack of governments policy, lack of information bulletin availability on day to day domestic cum International trade and lack of plat form for interaction for streamlining the bio-partnership with a mediation of government organizations, have been considered in proposing a restructured marketing network system. With an ideology of breaking mediator's menace in the marketing system, three alternative marketing channels have been proposed by keeping all the pros and cons into consideration as depicted in Fig. 2.

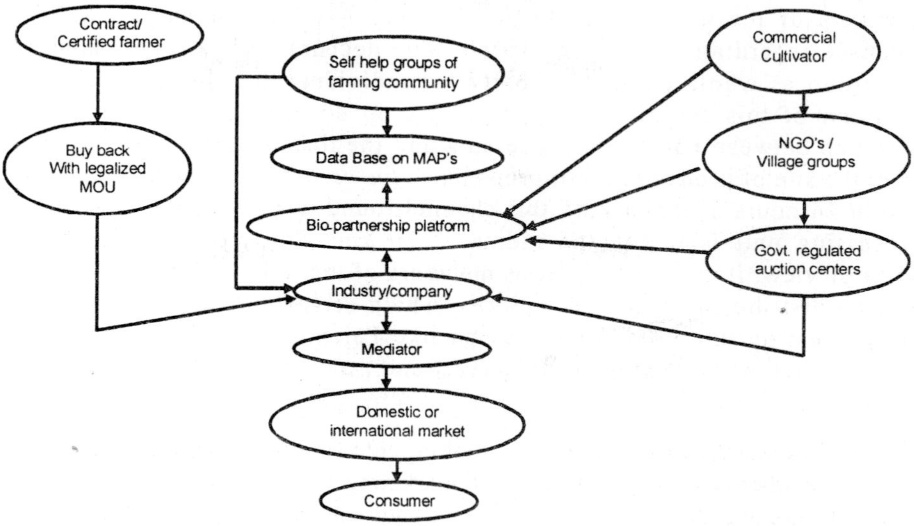

Fig. 2. Proposed flow chart marketing channels of MAC's

(a) Marketing channel I : In the traditional buy back system, always the understanding between the grower and industrialist with the conditions of providing the planting material on cost basis and purchasing the produce on oral terms without a legal binding. But, actually speaking in many cases like Aloe *vera*, *Safed musli* and Patchouli in this region the buyback binding was violated by the buyer, by restricting their association till the supply of planting material without a further turn up. In order to overcome this problem, the present proposal has been foot forth. As per this system, the farmer should be progressive with large land holdings having suitability to raise MAC as desired by the buyer. The buyer (Industry or Pharmaceutical company) should be a registered license holder who is authenticated in selecting the growers on his own choice with the responsibilities like, supply of good planting material, periodical supervision and guidance, arrangements in lifting of harvested produce, and giving scheduled payments. The main approach behind the system is legalizing the MOU between the grower and industry and thereby any violation in the dealings can be rooted out. As, this system enables the direct contact of farmer with the industry without a scope for any mediator, the farmer is highly benefited.

(b) Marketing channel II : This system is proposed with no loss or no profit basis by looking to the non-preferability of small landholders in the channel 1. In this system irrespective of land holdings, farmers come together and make a union with an ideology of growing suitable MAC's of greater demand uniformly in a scheduled period of time over large areas on community basis. Here, self help group unions at taluka level is encouraged and some volunteers among them will serve as active members. All the members of the unions will be provided with ready reckoner of MAC's database regarding cultivation aspects. The decisions are always taken by the group as whole without a liberty for individual decision making. The adoption of this system for a particular crop depends on the permissible time gap between harvest to processing and the distance of processing unit from the site of production. For crops like *Aloe vera*, which needs processing within 24 hours of its harvest, this channel found to be feasible, only if the processing unit is within the radius of 20 km. However, from practical point of view, it is not possible, as members of the union, growing the crop will be distributed in different places, which are at varied distances from the processing unit. But for the crops like Patchouli and *Rouwolfia*, it is highly suited as the gap between harvest and processing is slighly relaxed unlike *Aloe vera*.

(c) Marketing channel III : By looking into rate of increase in area under MAC's, another proposal of channelised marketing has been projected to operate in near future at individual commercial cultivators level. In this system proactive assistance of NGO's/Village groups by taking role as collectors on nominal fee structure is very important in the success of the system. So, collected produce gets channelised to auction centers, where it is being sold on auction basis under the supervision of government authority followed by its onward passage to Industry. In streamlining this channel, equitable bio-partnership (A network of organizing and mobilizing a common platform of growers, NGO/GO/self help groups, Officers of auction centers and industrialists) is the foremost need. So, a multiple linked chain of community based organization and MAC's-based companies in a value based chain or production to consumption and marketing continuum (PC&M) is to be formulated in production and supply of quality material to industry and to assure benefits to growers. To make the set up highly effective, all the associates of marketing system should be made access to a ready-reckoner of resource data base or a website carrying the information on MAC's, their uses, sources of quality planting material, region-wise buyers, day to day price indices, demand status at domestic and International markets, species-wise growing areas and their yield levels, simple agro-techniques, sustainable harvesting techniques, processing techniques and quality assessment standards should be made available from grower to the traders. In addition to this, establishment of grower co-operatives, market information and communication technology helps in information exchange and knowledge sharing. The adoption of this system depends on the permissible time gap between harvest to processing. This is suitable for the crops like Patchouli and *Rouwolfia*, which don't have a very short time frame for their processing after harvesting and shade drying. So, these act as a projected channel for years to come in near future.

To conclude, MAP's are the integral part of natural resource management, contributing for economic growth, environmental protection and trade provided marketing channel system is perfectly streamlined. Potential contribution of MAC's can be substantial and can be made highly remunerative in capital-poor but resource-rich India, if investment and efforts are substantially increased.

REFERENCES

Baby P. Skaria, Samuel, Mathew and Joy, P.P. (2003). Constraints and strategies for the development of medicinal and aromatic crops in India. *In : Proceedings of National Seminar on New perspectives in spices, medicinal and aromatic plants* held on 27-29, 2004 at Goa. pp 422-429.

CRPA (2001). Demand study for selected medicinal plants: Vol. I and II. Centre for Research, Planning and Action for Ministry of health and family welfare, GOI, Department of ISM & H & WHO, New Delhi.

Dhar, U., Rawal, R.S. and Upreti, J. (2000). Setting priorities for conservation of medicinal plants as a case study in the Indian Himalaya. *Bilogical Conservation,* **95** : 57-65.

FAO Database (2003). Website, www.fao.org.

Handa, S.S. (1996). Medicinal plants- Priorities in Indian medicinal studies and implications. *In: Cultivation and utilization of medicinal plants, Supplement (Handa and Kaul, eds),* RRL, Jammu, Niscom, New Delhi. pp 33-51.

Karki, M.B. (2000). Commercialization of natural resources for sustainable lively hood; the case of forest products: paper presented at the international conference on growth, poverty alleviation and sustainable management in the mountain areas of south Asia organized by German Foundation for International Development (DSE) and International Center for integrated mountain development (ICIMOD) held from Jan.31-Feb.4, 2000 at Katmandu, Nepal.

Khanuja, S.P.S., Alok Kalra, Singh, A.K. and Singh, J. (2004). Green economics. Survey of Indian Agriculture-2004. Published by Hindu daily newspaper, Mysore.

Lange, D. (1997). Trade figures for botanical drugs worldwide. *Medicinal Plant Conservation Newsletter,* **3** : 16-17.

Maikhuri, R.K., Rao, K.S., Kusum Chauhan, Kandari, L.S., Prasad, P. and Rajashekaran, C. (2003). Development of marketing of medicinal plants and other forest products-Can it be a pathway for effective management and conservation. *The Indian Forester,* **121(2)** : 169- 178

Manjikhola, S. and Dhar, U. (2002). Conservation and utilization of *Arenabia benthami* (Wall. ex G. Don) Johnston-A high value Himalayan medicinal plant. *Current Science,* **83** : 484-488.

Planning Commission (2001). Report of the task force on medicinal plants in India. Planning Commission, Government of India, Yojan Bhawan, New Delhi.

Rawat, R.B.S. and Uniyal, R.C. (2003). National Medicinal Plants Board Committed for over all development of the sector. *Agribios,* **1(8)** : 12-17.

Sarin, Y.K. (2003). Medicinal plant raw materials for Indian drug and pharmaceutical industry-An appraisal of resources. *Indian Forester,* **129(1)** : 3-24.

Recent Advances in Medicinal Plant Research : Vision 21st Century, 2005
Editors : S.K. Prabhuji, G.P. Rao and S.K. Patil
Satish Publication House, New Delhi

Chapter 34

UTILIZATION PATTERN, COMMERCIAL VALUE AND RED LIST CATEGORIZATION OF IMPORTANT INDIGENOUS TREE SPECIES OF MEDICINAL IMPORTANCE IN UTTARA KANNADA OF WESTERN GHATS (KARNATAKA)

G.O. Manjunath[1], V. Suryanarayana[2], S.K. Patil[1] and V.M. Dabgar[1]
[1]*Dept. of Forest Utilization, College of Forestry, UASD, Sirsi - 581 401, Uttara Kannada, Karnataka*
[2]*Dept. of Forest Protection, College of Forestry, UASD, Sirsi - 581 401, Uttara Kannada, Karnataka*

A study was done in Uttara Kannada part of Western Ghats of Karnataka on documentation of indigenous tree species of medicinal importance and local health traditional practitioners. Study revealed the existence of 28 tree species of indigenous nature falling under one of the Red listed categories. Among them *Dipterocarpus indicus*, *Dysoxylum malabaricum, Persia macrantha, Nothapodytes nimmoniana* and *Saraca asoka,* were recorded as endangered. Uttara Kannada has an heritage of possessing more than 200 local health practitioners, concentrated mainly in parts of Honnavar and Ankola followed by Sirsi, Kumta and Siddapur taluks. In this district, dependence on tree species is primarily for stem bark and seeds followed by wood and roots. Specifically, there is a growing pressure of commercial exploitation of *Dipterocarpus indicus, Dysoxylum malabaricum, Persia macrantha, Nothapodytes nimmoniana* and *Saraca asoka* for stem bark or wood as they have active principles for anti-cardiac, anti-cancer, anti-rheumatic and anti-menstrual disorders. Maximum populations of indigenous tree species were observed in moist deciduous to evergreen patches of Honnavar, Ankola, Siddapur, Joida and Sirsi taluks. The situation warrants to gear up conservation work through *in-situ* and *ex-situ* approaches.

INTRODUCTION

Uttara Kannada part of Karnataka is known for its richest treasure of medicinal plant biodiversity. Total geographical area of this region accounts to 10,291 Km^2 and of which 7,807 Km^2 is under forest cover. It stands first among 27 districts of Karnataka by covering nearly 21 per cent of the total state's forest cover (Anonymous, 2001). A great awareness on conservation issues of biodiversity can be well sensed here in spite of meagre exploitation levels. Enriched indigenous medicinal knowledge among local communities, rich cultural heritage of including medicinal plants in their dietary systems and concern of conserving them in their kitchen gardens have contributed for the existence of medicinal flora which has vanished in many parts of Karnataka. In addition, the district has high potential base of ayurved, which can be figured out by the presence of more than 200 local health practitioners. Their working system is purely a herital custom. They render services in the name of god with a main motto of well being of the local people. The concern of **"Living with greenery"** developed a non-inclination of Uttara Kannada people to allopathic aids.

With the passing years, the traditional knowledge is being slowly eroded which may reach a critical edge in near future. So far, the indigenous drugs have not been studied carefully and systematically. Till date, momentum of traditional knowledge is ill documented, without an essence of perseverance. Just the knowledge flow from one generation to another generation is partial and that too is restricted to verbal communications. In many cases, lapse in discharging actual information due to hidden secrecy among tribal based practitioners is prime struck in constant flow of complete knowledge. Even though efforts have been done in status documentation of herbs and shrubs in Uttara Kannada, the status documentation of trees of medicinal importance is in infancy.

In present days, due to a steep increase in human and animal populations, a high dependence pressure is being exerted upon forests. This is leading to gradual depletion of indigenous tree species of medicinal importance. Once they are destroyed, they cannot be recouped as we see in case of many of the herbs and shrubs. Increased trend of commercialization of medicinal plants, high rate of unemployment, limited income generating alternatives for lively hood for marginal and landless farmers and poverty have changed the life style in this region. In nutshell, over exploitation and unsustainable harvesting for livelyhood is limited to a small fraction of population. But, the maximum fraction of population is deliberately involved in earning more money without a pinch of concern on **"Sustainable medicinal plant (Tree) reserve"** through the most unethical motives. Hence the study was taken up on documentation of indigenous tree species of medicinal importance, their utilization patterns and Red listed categorizations.

HISTORICAL REVIEW

With a lot of work in respect to usage, commercial value and red list categorization of medicinal plants, ample of informations are now available. For many species to vanish, a gap in the flow of knowledge from one generation to another is the main cause as per the works conducted by Guptha and Guleria (1992). The possibility of above factors in shifts of socio-economic and cultural activities of rural communities have also been appraised by other workers in their case studies (Kala, 2003; Williams *et al.*, 2000). The usage patterns of medicinal plants have also been well documented from Uttaranchal, Arunachal Pradesh, Himalayas and north eastern parts by various workers (Sarin, 2003; Haridasan *et al.*, 2003; Manjikhola and Dhar, 2002)). In the process of selection, RET status of each of the species was taken as a criterion and it was gathered mainly through literature (Ravikumar and Ved, 2000). However the literature with regard to utilization pattern of perennial tree species is lacking.

METHODOLOGY

The indigenous tree species of medicinal importance, which are being subjected to commercial exploitation and falling in one or the other status of threatening, were selected. For the study, 28 tree species of maximum commercial value in Uttara Kannada have been taken into consideration. Later, for easy communication with the local people, the information on their common names was collected through literature, local people and local traders. With an intension of generating meaningful and unbiased data, the present study was carried out in different taluks of Uttara Kannada District of Karnataka. In each of the 10 taluks, five local health practioners were met and interactions were made regarding their herbal therapy. In addition, as per the preformatted proforma, details on parts of trees being used and their availability in the forest were also gathered. In order to assess and ascertain commercial exploitation, the details on commercial raw drugs and trade value of the selected species were collected by consulting local traders and local people.

With the prime and ultimate aim of working out suitable conservation strategy, a survey has been conducted to create an information bureau on tree species wise specific forest patches. This has been undertaken with a theme of protecting them to make it as a buffer zone for perfect achievement of conservation.

RESULTS AND DISCUSSION

The indigenous tree species of Red listed category was searched out in literature pertaining tc Uttara Kannada region and further it was confirmed with the local experts. From the present study, it is documented that 28 tree species, which are indigenous, found to fall in one or the other category of threatening. Overall, 5 species, viz., *Dipterocarpus indicus*, *Dysoxylum malabaricum, Persia macrantha, Nothapodytes nimmoniana* and *Saraca asoca,* fall under endangered category, 19 species fall under vulnerable category and remaining 4 fall under low risk category (Table-1). We are fortunate enough that no species has reached to critically endangered status in the indigenous tree species of commercial exploitation. The possible attributes for many species to fall into the endangered category may be attributed to the increasing global demand levels for roots and wood, biotic menaces, low regeneration capacity and low success in vegetative propagation. In the similar line of work related to enlistment of threatened medicinal plants in peninsular India, some of these reasons have been quoted (Balakrishne Gowda *et al.*, 1997). From the commercial point of view, interviews with local healers and local traders revealed that the plant parts like stem, roots, seeds and fruits possess medicinal values and are being extracted. A total of 55 local health practitioners, selected at random basis in different taluks of Uttara Kannada, have revealed that commonly *Aegle marmelos, Saraca asoca, Garcinia gummigutta, Garcinia indica, Dysoxylum malabaricum, Santalum album, Canophyllum inophillum, Mesua ferria* are being commonly used in their medicare systems. In the present study a notable observation was made that the number of local health healers was maximum in areas of lower Ghats comprising Honnavar, Ankola and Kumta taluks followed by their normal populations in up-Ghat areas covering Sirsi and Siddapur taluks. But towards Haliyal, Yellapur and Dandeli areas, local health practice is mostly restricted to tribal populations. Over all, the basic services offered by them were in the interest of local people on the custom of receiving a coconut in the name of god. The tradition of taking money for rendering services was very remote. In the similar line of works, the opinions drawn, fall in the line of our observations (Aronald and Perez, 2000). In this region, dependence on indigenous tree species is documented

Table - 1 : Continued .....

Sl. No.	Tree species	Local name	Parts used	Usage	Commercial value	Status of RET
19.	*Oroxylum indicum*	Alangi	Roots, leaves and fruits	**Roots :** One of the ingredients of Dasamoola, used as anti-inflammatory, dyspepsia, carminative, anti-periodic, asthma, bronchitis, dysentery. **Leaves :** To cure ulcers and headache. **Fruits :** To cure bronchitis, bleeding piles and to kill intestinal worms.	Local and regionally traded. Root bark is traded as Shynaaka mool chaal.	Vulnerable
20.	*Persia macrantha*	Kulur mavu	Stem bark	To treat asthma, constipation and rheumatism.	Regionally traded.	Endangered
21.	*Santalum album*	Srigandha	Heart wood	Treating burning sensation, skin diseases, leprosy, jaundice, bronchitis dysentery, gastric irritability, intermittent fever and general debility.	Local, regional, nation-wide and globally traded.	Vulnerable

Table - 1 : Continued .....

Sl. No.	Tree species	Local name	Parts used	Usage	Commercial value	Status of RET
22.	*Saraca asoca*	Ashoka mara	Stem bark, leaves, flowers	**Stem bark** - To treat indigestion, fever, burning sensation, ulcers, and menstrual disorders dysentery. **Leaves** - Used as blood purifier. **Flower** - Treating burning sensation and bleeding piles.	Locally, regionally and nationally traded	Endangered
23.	*Symplocos cochinchinensis*	Changa	Stem bark	**Stem bark** : To treat uterine complaints, vaginal diseases and menstrual disorders, eye diseases, diarrhoea and dysentery.	Locally and regionally traded	Low risk near threatened.
24.	*Teminalia arjuna*	Bilimatti	Stem bark, fruits	**Stem bark decoction** : Treating heart diseases, bone fractures, skin diseases, white discharge, giddiness, fever, leprosy. **Fruits:** To heal ulcers.	Stem bark is traded as Arjuna chall.	Low risk near threatened.
25.	*Vateria indica*	Bili dhoopa	Stem resin	To treat cough, asthma, leprosy, wounds, ulcers, gonorrhea, dysentery, anemia, chronic rheumatism.	Local and regionally traded as Dhupa	Vulnerable

Table - 1 : Continued .....

Sl. No.	Tree species	Local name	Parts used	Usage	Commercial value	Status of RET
26.	*Symplococcus racemosus*	Lodhrra	Stem bark	Used to treat haemorrhage, menstrual disorders, liver diseases, skin disorders.	Local, regional and nationally traded as Lodha.	Vulnerable
27.	*Garcinia morella*	Arasina gurgi	Fruit, stem, seeds	**Fruit juice** - curing rheumatism, constipation, dysentery and intestinal infections. **Seed oil** - curing ulcers and skin diseases. **Stem exudates** - curing leprosy.	Local and regionally traded.	Vulnerable
28.	*Mesua ferrea*	Naga Sampige	Flowers, seeds	**Dry flowers** - curing bleeding piles, dysentery. **Seed oil** - for rheumatism, skin diseases and ulcers.	Local and regionally traded.	Vulnerable

and it was for stem bark from 13 species followed fruits and seeds from 7-8 species and for roots from 5 species. Growing demand of herbal medicines and commercial exploitation are found to establish a perfect base, which in near future may result in extinction of many of the enlisted tree species. It is observed during the present study that there is a growing pressure on some of the species in these areas like *Dipterocarpus indicus*, *Dysoxylum malabaricum*, *Persia macrantha*, *Nothapodytes nimmoniana* and *Saraca asoca.* They are known to possess one or more active principles which may be anti-cardiac, anti-cancer, anti-rheumatic, anti-menstrual disorders and anti-bleeding piles as presented in Table-1. As bark and wood chips are the main parts of medicinal value in these species and that too they are performing well in treating the dreadly diseases, their population in the forests are fast depleting. If the same trend of unethical exploitation continues for money minting, it may pose a serious threat to their existence in near future. With a motto of identifying potential areas of their population, a general survey has also been undertaken. The study revealed that the maximum populations were pocketed in moist deciduous to evergreen patches of Honnavar, Ankola, Siddapur, Joida and Sirsi taluks. It is a clear indication that their populations are in specific areas of high bio-diversity prevalence. Under the prevailing circumstances of over harvesting or felling of trees for wood and roots, which are at the verge of entering into critically endangered status in coming years, the above study was highly warranted. From the identified areas there is a need to gear up sincere efforts of conservation, on war footing, through *ex-situ* and *in-situ* approaches.

REFERENCES

Anonymous (2001). State of forest Report. Forest survey of India, Dehradun. p. 130.

Aranald, J.E.M. and M.R. Perez (2001). Can non-timber forest products match forest conservation and development objectives?. *Ecological Economics*, **39** : 437-447.

Balakrishne Gowda, Nissar, M.M., Seetaram, Y.N. and Ramesh, S.R. (1997). The threatened plants of peninsular India. *My forest*, **33(1)** : 327-334.

Guptha, T.R. and Guleria, A. (1992). Non-wood forest products in India: Economic potentials. Oxford and IBH publishing company, New Delhi. P. 147.

Haridasan, K., Anupam Sharma, H., Bhuvan, L.R. and Bist, N.S. (2003). Medicinal plant sector in a Arunachal pradesh an over view. *Indian Forsetor*, **129(1)** : 37-47.

Kala, C.P. (2003). Commercial exploitation and conservation status of high value medicinal plants across the borderline of India and Nepal in Pithogarh. *Indian forester,* **129(1)** : 80-84.

Manjikhola, S. and Dhar, U. (2002). Conservation and utilization of *Arenabia benthami* (Wall. ex G. Don) Johnston-A high value Himalayan medicinal plant. *Current Science*, **83** : 484-488.

Ravikumar, K. and Ved, D.K. (2000). 100 Red-listed medicinal plants of conservation concern in southern India. FRLHT, Bangalore. P. 467.

Sarin, Y.S. (2003). Medicinal plant raw materials for indian drug and phramaceutical industry. An appraisal of resources. *Indian Forsetor*, **129(1)** : 3-24.

Williams, V.L., Blackwil, K. and Witkowski, E.T.F. (200). Unrevealing the commercial market for medicinal plants and plant parts on the wit water strand, South Africa. *Economic Botany*, **54** : 310-327.

Recent Advances in Medicinal Plant Research : Vision 21st Century, 2005
Editors : S.K. Prabhuji, G.P. Rao and S.K. Patil
Satish Publication House, New Delhi

Chapter 35

MARKETING OF SELECTED NTFP'S : A CASE STUDY OF SIRSI AND SIDDAPUR TALUKAS OF UTTARA KANNADA DISTRICT, KARNATAKA

A. Krishna and P. Ramana★
College of Forestry, Sirsi - 581 401, Karnataka, India
★E-mail : ramtarunsirsi@yahoo.com

An attempt has been made to study the market structure of important NTFP's in Sirsi and Siddapur talukas of Uttara Kannada district during 2002-2003. Forest dwellers largely depend on NTFP's for their survival and growth. NTFP's marketing is a process which starts with an NTFP harvest (procure) that can be sold in an organized/ unorganized market. It involves activities, comprising the process of collecting the products and services from producer to the end users. Simple value additions increase the selling price at primary, collector level and industrial level. Among the NTFP's studies, the *Myristica* seeds fetch lowest price, whereas arils of *Myristica* fetch highest price. There was greater reduction in price of Uppage during the period 1997-2003. In general, two marketing channels viz. substantial use and sale to private traders/middlemen, were identified in the study area. Lack of market and market information has been found to make the primary collectors to resort on thirsty middlemen to dispose the NTFP's at a throw away price. The processor, who is making substantial efforts to add value (industry), is getting a lion share of the end user price. There was declining trend in demand and prices of selected NTFP's during the year 2002-2003.

INTRODUCTION

Non Timber Forest Products (NTFP's) are the most sustainable source of revenue in lieu of the present scenario of restricted harvesting or no harvesting of trees, apart from Eco-tourism. Most important marketing functions, involved in the marketing of NTFP's, are gathering/assembling, cleaning, processing, storage, transportation and the distribution. The main functionaries involved in the marketing channel of NTFP's are primary collectors (tribal and rural people), the small agents/middle men, the forest contractors, the processing industries and the end users/consumers. The marketing channel of most of the NTFP's has not been properly defined which is another reason for market inefficiency. Market information is very specific to the locality, which cannot be generalized and especially so in trade of forest produces.

HISTORICAL REVIEWS

Over 50% of the forest revenue and 70% of income from exports come from NTFP's and they provide 50 % of income for 20-30% of the rural and tribal people in India (Theagaraja, 1994; Negi and Bhalla, 2002). In India, more than 30 million people are employed in NTFP sector (Masih *et al*, 2001). However, there is a lack of information on many aspects of NTFP's in general and about marketing aspect in particular. A large number of NTFP's are still handled by the forest contractors and only few NTFP's are nationalized. Despite acquiring a certain degree of marketing skills, the primary collectors have remained at the lowest rung of NTFP's marketing either working as labourer on daily wages or being underpaid collectors. The main reason being the market inefficiency created due to lack of market information. Since the NTFP's are often sold in weekly informal primary tribal markets, informations about prices, product flow, demand and supply pattern are less known (Masih *et al.*, 2001).

Hence, this study is an attempt to systematically document the market information/structure of few important NTFP's such as Katha, Kokam, Uppage, Nutmeg, Zumma, Litter and Soapnut in Sirsi and Siddapur markets during 2002-2003.

METHODOLOGY

1. Preliminary survey of market and literature was undertaken to identify important NTFPs in the study area.
2. The primary collectors were identified and enquired about the collection and marketing of NTFP's.
3. The market prices and sources of NTFP's were known from local shopkeepers.
4. Marketing information on NTFP's was collected both from forest contractors and other market functionaries.

RESULTS AND DISCUSSION

Uppage

The fruits of *Garcinia cambogia* (= *G. gummi-gutta*, Family : *Clusiaceae* = *Gutiferae*) are locally called as Uppage. Fruits are reddish-yellow in colour; it has thick rind with ridges. The fruit matures during June to August. Fruit contain nearly 30% of Hydroxy Citric Acid (HCA), which is used as a bioslimming agent. Traditionally, the fruit has been used as a substitute for tamarind in the study area and also in Kerala. It is also used in chemical industries and as medicine for rheumatism. Vinegar and black liquor is prepared from fruit and is used in fish food preparation.

The fruit is mainly supplied to Kerala, where there was a huge demand for it. The prices of fruit per kg were Rs. 90-100 in 1998-1999, Rs. 70-90 in 1999-2000; Rs. 30-35 in 2001-02 and Rs. 25-30 in 2002-03.The sudden drop in the market price of fruit in recent years has disturbed the local NTFP's market and the economic condition of the collectors.

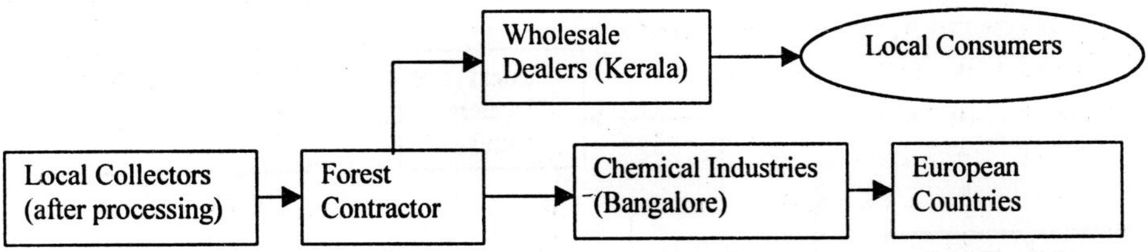

Marketing Channel for Uppage

Kokam

Kokam is the vernacular name for the soft drink prepared out of the rind of fruit of *Garcinia indica* (Family : *Clusiaceae*). Fruits mature during April-May, one kg of fruit yield about 300 g of rind, which in turn yield about 1 litre of Kokum juice. Kokam juice, in concentrated form, is marketed under different trade names in the local market. Often the local soft drink companies market it without trade name. Generally it is a mixture of kokam juice (40 %), sugar and other additives. Each bottle of 150 ml costs Rs. 3-4, the rind powder is also marketed in the brand name of 'Sapna'. Local market has a little contribution of Kokam butter, which is exported to Italy and other European countries.

The price of fresh fruit per kg was Rs. 8-10 during 1994-95, Rs. 30-40 during 2000-01 and Rs. 15-25 during 2002-03.

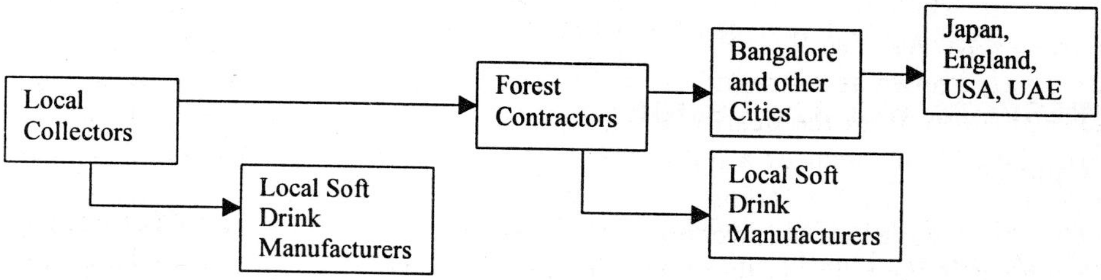

Marketing Channel for Kokam

Katha

Heartwood of *Acacia catechu* (Family : *Mimosaceae*) is the principal species used in manufacturing Katha. The chief constituent of the heartwood is catechin (Katha) and catechu tannic acid (cutch). *Acacia catechu* is commonly found in the tropical deciduous and thorn forests of northern India. In south India, it is replaced by *Acacia chundra*, which is mostly distributed in Tamil Nadu and Andhra Pradesh and in northern part of Karnataka. The yield of katha is very meagre i.e., 4.5-5 percent. This implies that to obtain 5 kg. of katha, nearly 100 kg. of heartwood is used.

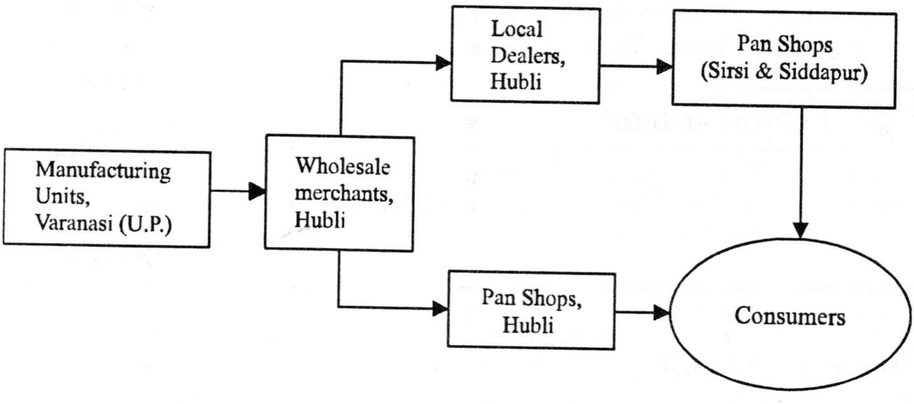

Marketing Channel for Katha

Katha is available in two forms in the market, viz., semisolid and in cube or biscuit form. The katha has several uses such as ingredient of chewing pan, as an astringent and as coolant for burn injuries.

Katha in large quantity is brought from manufacturing units at Varanasi (U.P. to Hubli (Karnataka). In Hubli market katha is available at Rs. 560-580 per kg in biscuit form and Rs. 115 per kg in cube form. In Sirsi market the same costs Rs. 600 per kg. and Rs. 120 per kg, respectively.

Dharaku

Tree leaf, whether green or dry (litter), is an important component of Areca nut (*Areca catechu*; Family : Palmaceae) farming in the Canara circle. Dry litter popularly known as "dharaku" in this region is generally collected in the months of April-May from the nearby forests and marketed to the areca nut planters at Rs. 200-225 per load of a bullock cart or Rs. 20-25 per head load. The price may generally be the fallout of the negotiation between buyer and the seller. Dharaku is collected by sweeping the forest floor. The dry litter is also spread in the cattle yard that is then used as manure. A certain proportion of litter is also collected from the Betta lands for domestic use.

Soapnut

The fruit of *Sapindus trifoliatus* and *S. laurifolia* (Family *Sapindaceae*) are popularly known as Soapnut. The pericarp of soapnut contains a chemical substance known as saponin, which is an essential ingredient of soaps, detergents and other industrial products. Fruits have been traditionally used as detergents for washing clothes, hair, jewels, tarnished ornaments and fabrics. Fruits have medicinal value and are used for curing asthma, diarrohea and tuberculosis. The fruit is marketed generally in three different grades. In fact, the collectors sell the un-graded raw seeds to contractors at Rs. 4-5 per kg. The contractor, after drying and cleaning, divides them into three grades. Few collectors grade the seeds and sell at higher prices (Table 1). First grade fetches the highest price. Primary collectors are more benefited by selling grade I fruits than by selling grade III fruits. The different grades are:

Table - 1 : Prices of different grades of Soapnut

Fruit Grades	Price received by Collectors (Rs.)	Selling price by contractors (Rs.)	Price Difference (Rs.)	Percent appreciation in value (Average)
I Grade	8.50-9.50	15.00-16.00	6.5	72.22
II Grade	7.50-8.50	14.00-15.00	6.5	81.25
III Grade	7.00-7.50	13.00-14.00	6.0-6.5	86.21

I Grade : Large size, brown colour and shining fruits.

II Grade : Medium size, brown colour fruit.

III Grade : Small size, blakish brown colour and without shine.

Nutmeg

Seeds and arils of *Myristica malabarica*, *M. dactyloides*, *M. fatua* and *M. fragrans* (Family : *Myristicaceae*) are having high commercial importance in the local market. The first two are indigenous and the last two are introduced. A very low quantity of aril is obtained from each fruit. The average nut to aril ratio after drying is only 0.16. Seeds are sold at lower prices but the arils fetch a very high price (Table-2). Hence, the tribal and local people collect almost all available seeds and aril and sell it in the weekly market or to businessmen or the local spice shop owners. The aril, then, is sold at 176-217 percent higher prices.

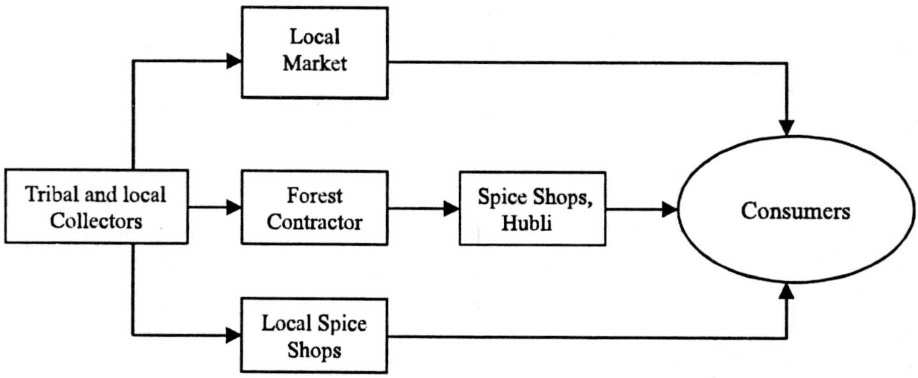

Marketing Channel for Nutmeg

Table - 2 : Market Prices of seeds and arils of different *Myristica* species

Species	Seeds (Rs. per seed)	Price received by Collectors (Rs. per kg)	Arils selling price by collectors (Rs. per kg)	Percent appreciation in value (Average)
M. malabarica	0.20-0.25	150.00-175.00	400.00-500.00	176.92
M. dactyloides	1.50-2.00	200.00-225.00	650.00-750.00	217.65
M. fatua	1.00-1.50	250.00-275.00	700.00-750.00	176.19
M. fragrans	1.50-2.00	250.00-300.00	750.00-950.00	209.09

Zumma

Zumma is the vernacular name for the seeds of *Zanthoxylum rhetsa* (=*Fagelia metsa*, Family : *Rutaceae*). The pericarp is rich in essential oil called Mullilam oil. It is used in the indigenous system of medicine for the treatment of cholera. Fruit has been widely used in preparing fish masala and fish curry. The fruits ripen in the months of September-November. The primary collectors harvest and sell fresh fruit in the local markets at Rs. 8-10 per kg. They sell them to the local shopkeepers and to the forest contractors. The forest contractors, after drying and cleaning, sell them at Rs. 25-30 per kg. to perfume extraction units located at Pune and Mumbai.

It has been observed during the study that in most cases NTFP's are sold by the primary collectors without any value addition that could have been otherwise elevated the earnings of the primary collectors. Simple value additions in the form of extraction, storage, de-seeding and drying of some NTFP's can be done very easily without having any sophisticated technology, thereby enhancing the selling price at primary collectors level. It has also been found that some prime collectors are industries, adding value to their collected NTFP's before they sell off. Lack of market and market informations force, the primary collectors to resort to thirsty middlemen to dispose the NTFP's at a throw away price.

The distribution network in the study area for some selected NTFP's by and large consist of primary collector, forest contractors/village shopkeepers, traders at taluka, district, state and national level distributors. Wholesale NTFP traders at leading market places including Sirsi, Hubli, Bangalore and Mumbai etc., manufacturing units located at different places, their agents play the important role and the consumers are the end users. Currently the middlemen have determined the prices of selected NTFP's in the study area. Poor primary collector almost has no say in price determination process (Ganapathy *et al.*, 2000; Surayya *et al.*, 2003). The prices of produce may depend upon the season, the prices of the NTFP, which may be cheaper during a particular season may fetch three to four times higher price in other seasons.

The demand for selected NTFP's has been declining for the past few years and there has been a drastic decline in the prices of selected NTFP's also in the local market.

ACKNOWLEDGEMENTS

The authors are thankful to local NTFP's collectors, Forest Contractors and retail shop owners for providing information.

REFERENCES

Ganapathy, M.S., Laxminarayana, M.T. and Surendra, S. (2000). Marketing of NTFP's in Kollegal Taluk, Chamarajanagar District, Karnataka. *My Forest*, **36(1)** : 41-46.

Masih, S.K., Sharma, C.B. and Sharma, M.C. (2001). NTFP's and their price trends in primary tribal markets. *Journal of Non-Timber Forest Produce*, **7(1/2)** : 1-8.

Negi, Y.S. and Bhalla, Pankaj (2002). Collection and marketing of important medicinal and aromatic plants in tribal areas of Himachal Pradesh, *Indian Forester*, **128(6)** : 641-649.

Surayya, T., Mishra, M. and Mishra, R.P. (2003). Marketing of selected NTFP's: A case study of Korapur, Malkangiri and Rayagada districts, Orissa state. *Journal of Non-Timber Forest Produce*, **10(3/4)** : 186-194.

Theagaraja, K.S. (1994). Non-Wood Forest products, problems and Prospects, *My Forest*, **30(1)** : 33-35.

Recent Advances in Medicinal Plant Research : Vision 21st Century, 2005
Editors : S.K. Prabhuji, G.P. Rao and S.K. Patil
Satish Publication House, New Delhi

Chapter 36

IN VITRO PROPAGATION OF *COLEUS FORSKOHLII* – A THREATENED MEDICINAL PLANT

Thalavaipatti Swaminathan Mahendran and P. Sampath
Department of Botany, Presidency College, Chennai – 600 005

Plant tissue culture technique for ornamental as well as herbacious plants, have been well established. *In vitro* propagation technique is a powerfull tool for plant germplasm conservation. The ability to generate plants directly for explant is fundamental to clonal multiplication of elite germplasm via micropropagation. A recent advances in plant tissue culture techniques has produced so many varities of plant products which are found to be very useful to man kind. Sharma *et al.* (1991) have reported *in vitro* propagation of *Coleus forskohlii*, a threatened medicinal plant, in which they proved clonal multiplication can be achieved on MS medium using nodal segment, rooting are also achieved on transfering. In present investigation a maiden attempt is done *in vitro* clonal multiplication on MS and N6 medium using leaf nodal region and shoot tip as explants. The shoot multiplied at the rate of twelve folds for every six weeks, then rhizogenesis was achieved upon transfer of shoots on to MS and N6 media using IAA and NAA. Increased concentration of IAA and NAA had given good response on rhizogenesis. Under field condition micropropagated plants were successfully established. Further it is identified that plant produces labdane diterpenoid forskohlin has therapeutic valve which are used in treating glaucoma, congestive cardio myopathy, asthma and has antiinflamatory properties. found in all parts of plants, especially more in root it is also seen in shoot and tissues (Mersinger *et al.*, 1988; Krobholz *et al.*, 1982; Sen *et al.*, 1992). In view of result obtained in present investigation, micropropagation procedure proved to be a useful tool for conservation as well as production of important endangered plant like *Coleus forskohlii.*

INTRODUCTION

Tissue culture is combined as an unique technique in the field of plant biotechnology, which is mainly used for mass multiplication of herbs, shrubs and trees, ornamental, as well as useful medicinal plants. These *in vitro* propagation technique have revealed unique capacity of living plant cell to give rise to whole plants known as cellular totipotency. Plant tissue culture techniques for ornamental as well as herbaceous plants have been well established in three key areas in which plant cell and tissue culture has direct application in ornamental horticulture where large scale propagation of elite clones from hybrid as specific parent lines through micro propagation and somatic embryogenesis of disease free propagules and development of plant varieties through cellular and molecular techniques in conjunction with whole plant breeding (Prakash *et al.*, 1993). *In vitro* propagation technique serves as a powerful tool for plant germplasm conservation. The ability to generate plants, directly from explant is a fundamental technique for clonal multiplication of elite germplasm via., micro propagation. Apart from an increase in production of primary, secondary metabolites, other crop improvement parameters include introduction of organisms, increased resistance, environmental tolerance and greater yield coupled with improvement in grains, oils and secondary metabolites. Moreover, recent advances in plant tissue culture have produced so many important varieties of plant products which are found to be very useful for mankind. The production of secondary metabolites had led to industrial level production of useful compounds. Further more, the important advantage of *in vitro* culture is the production of non pathogenic products. Medicinal plants are sources of various alkaloids and chemical substances which are quite essential to mankind. These compounds are mostly organ and tissue specific. Plant cell culture is a potential source for deriving variety of useful chemical compounds. It is now widely being used as a screening system to investigate the production of specific secondary products as they offer experimental

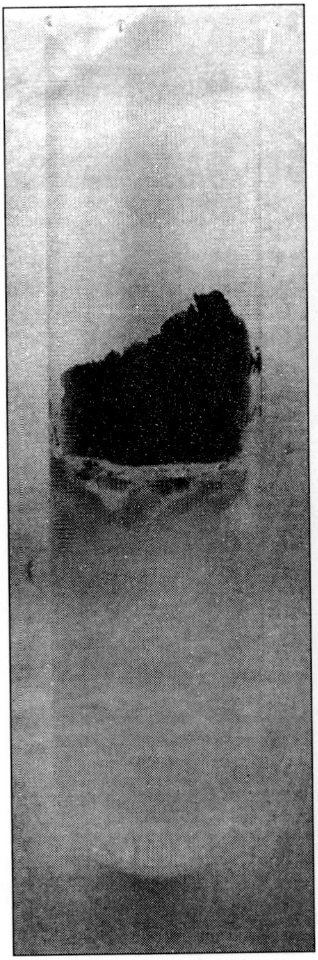

Fig. 1 : Pale yellow friable callus

Fig. 2 : Green compact callus

Fig. 3 : Multiple shoots from nodal segment

Fig. 4 : Multiple shoots from shoot tip

Fig. 5 : Callus in the nodal region

Fig. 6 : Callus and Rhizogenesis in the nodal region

advantages for basic, applied research and to the developmental models with scaled up potentials.

The present study was undertaken to standardize a protocol for the regeneration through *in vitro* cultures of ***Coleus forskohlii,*** a threatened plant due to indiscriminate collection and lack of cultivation practices. Thus it necessitates methods to be developed for the conservation of this threatened species. An attempt was undertaken to induce hairy root in *in vitro* root cultures of ***Coleus forskohlii***. Its medicinal properties were identified as early as 1973. This plant was very little known until recently. It has now gained a status of international importance. Since it is the only source of

"forskohlin", which is an unique Adenylate Cyclase activating drug. ***Coleus forskohlii*** Brig is an aromatic herb ,which belongs to the family **Lamiaceae**, it is an Indian herb, which produces a labdane deterpenoid forskohlin in its tuberous roots and found almost in all parts of the body. Roots of this plant are the main source (Shan *et al.*, 1980) of **forskohlin**, a drug used in treatment of glaucoma congestive cardiomyopathy and asthma. **Forskohlin** has been shown to be a hypotensive agent with spasmolytic, cardiotive and platelet aggregation inhibitory activity.

METHODOLOGY

Different types of culture vessels were used for preparation of media and culture of plants. Glasswares were cleaned with solution containing a mixture of water and potassium dichromate dissolved in concentrated sulphuric acid and they were sterilized. Plants of ***Coleus forskohlii*** were collected from hilly region (Salem district) and maintained in *exsitu,* with the help of sterile scissor nodule segment, shoot tips and young leaves were excised from the plant. All the excised parts were kept in different conical flasks. The explants were kept under running tap water with Teepol solution, for 2 hours. Surface sterilization was done with help of 1% Mercuric chloride and again it was washed thrice with sterile water under aseptic conditions. The nodal region and shoot tip were implanted vertically on to the nutrient media. The media used are Murashige and Skoog (1962) and N6 media with 3% sucrose, 9% agar was used. [MS and N6 Media supplemented with Cytokinin (BAP, Kin) and Auxin (IAA, NAA and IBA) in various combinations]. The pH of the medium was adjusted between 5.6 and 5.8 before dispensing into culture vessels. The medium was autoclaved at 121 °C for 15 minutes. About 15 cultures were maintained for each experiment and all the cultures were sub-cultured every 6 weeks. Cultures were maintained at 25° ± 2°C with the photo-period 3000 lux and 16/8 hours light/dark conditions.

RESULTS AND DISCUSSION

The callus cultures were established from leaf explants using N6 and MS Media. Callus induction and induction of morpho-genetic changes were achieved using various concentrations of plant growth hormones like Kin, BAP, IAA, NAA and 2,4-D either alone or in combination (Table 1) within 20 days of placement of the explants in N6 and MS Media. Blackening, swelling and callus initiation was evident (Fig. 1). Callus initiation was observed along the cut ends and on the entire leaf surface. Within a month appreciable callus was observed. The morphology of callus differed in BAP and NAA and in one supplemented with 2,4-D in terms of color and texture (Fig. 2). Two types of calli viz., dark green compact callus and pale yellowish green friable callus. The dark green compact callus was observed in the

Table - 1 : Different medium used for *in vitro* propagation of *Coleus forskohli*.

Explants Used	Medium Tried (mg/lit)	Response
Leaf Explants	MS + 2,4D (2.0)	++
	MS + BAP (0.5) + NAA (1.0)	+
	MS + BAP (1.0) + NAA (1.0)	++
	MS + BAP (2.0) + NAA (1.0)	++
	N6 + BAP (0.5) + NAA (1.0)	++
	N6 + BAP (1.0) + NAA (1.0)	++
	N6 + BAP (1.0) + IAA (1.0)	++
	N6 + BAP (2.0) + NAA (1.0)	++
	N6 + 2,4D (2.0)	++
Shoot Tip Explants	MS + Kn (3.0) + IAA (1.0)	++
	MS + Kn (2.0) + IAA (1.0)	++
	MS + Kn (1.0) + IAA (0.5)	-
	N6 + BAP (2.0) + NAA (1.0)	++
	N6 + BAP (1.0) + NAA (1.0)	+
	N6 + BAP (1.0) + NAA (0.5)	++
Nodal Explants	MS + Kn (2.0) + IAA (1.0)	++
	MS + Kn (1.0) + IAA (1.0)	++
	MS + Kn (1.0) + IAA (0.5)	++
	N6 + BAP (2.0) + NAA (1.0)	++
	N6 + BAP (1.0) + IAA (2.0)	++
	N6 + Kn (2.0) + NAA (1.0)	++
Rooting Medium	N6 + IAA (0.5)	-
	N6 + IAA (1.0)	+
	N6 + IAA (2.0)	++
	MS + NAA (0.5)	+
	MS + NAA (1.0)	++
	MS + NAA (2.0)	++
	MS + IBA (2.0)	++
Subculture Medium	BAP (1.0) + IAA (2.0) + GA3 (2.0)	Shoot Proliferation
	BAP (1.0) + IAA (2.0) + PVP (50)	Shoot Proliferation
	BAP (0.5) + IAA (2.0) + GA3 (1.0)	Shoot Proliferation

"++" Moderate, "+" Poor, "-" No response

medium supplemented with BAP(1mg/lit) and NAA (0.5 mg/lit). Pale yellowish green color callus was observed in 2,4-D (0.5 mg/lit) (Fig. 1).

According to Haris and Hart (1964) higher concentration of cytokinin were believed to inhibit rhizogenesis in ***Peperonca sandersii***. The present experiment, however, reveals that higher concentrations of Kinetin (1 mg/lit and 2 mg/lit) were effective for rhizogenesis. Rhizogenesis was well established from the callus at higher concentrations of Kinetin (0.5 mg/lit) and IAA (1 mg/lit). Cytokinin has also been reported to be the stimulatory agent for rhizogenesis in potato, Hussey and Starey (1981). In ***Kampheria galanga*** multiple shoots produced roots in the medium containing BAP and Kinetin (Vincent *et al.*, 1992) without being transferred to rooting media. Similarly in ***Coleus forskohlii***, shoots develop roots in N6 medium augmented with BAP (2 mg/lit), NAA (1 mg/lit) without transferring to rhizogenesis medium.

The result demonstrates that in ***Coleus forskohlii*** callus can be successfully induced from the leaf explants in N6 and MS medium containing different concentrations of BAP and NAA and 2,4-D in combination with cytokinin and give better response than 2,4-D alone.

Shoot Induction Medium

Shoot development in ***Coleus forskohlli*** was established from shoot tip as well as from nodal region. Shoot tip inoculated in N6 medium fortified with different concentration of auxins and cytokinins. These explants were inoculated in N6 medium. Out of all combinations tried shoot emerged from shoot tip. In N6 supplemented with different concentrations of BAP and IAA, initiation and elongation was observed within 21 days. Single shoot formation from shoot tip was observed in combination with BAP and NAA. Similar result was obtained by decreasing the concentration of BAP. In medium supplemented with BAP and NAA, dark green compact callus of the base of shoot tip was noticed, whereas increase in concentration of BAP and NAA resulted in rhizogenisis without any shoot development.

The nodal explants when subjected to combination of auxin and cytokinin gave best response in BAP and NAA. Multiple shoots were observed. A single shoot proliferated in BAP, Kinetin and IAA combinations. Multiple shoots were observed with MS fortified with BAP and NAA. Similar results were observed in N6 medium augmented with BAP and IAA. Compact green as well as pale yellow callus from nodal segments on N6 medium fortified with BAP and NAA were also observed. The number of shoots produced per culture was maximal in the media fortified with BAP and NAA. Single node explants cultured in this medium showed 50% but break, upon transfer of these sporulated buds to fresh medium resulted in the initiation of shoots from the node. The nodal explants of ***Coleus forskohlii*** had the potential to form multiple shoot from callus as well as from roots in N6 media containing auxins and cytokinins.

Rooting Medium

Individual shoot developed from shoot tip and nodal explants were transferred to rooting medium supplemented with IAA and IBA. Among the two auxins used, IAA gave the best response. The healthy shoots were removed from shooting media with different concentrations of BAP and NAA and transferred to N6 media with different concentrations of IBA or IAA. The developed shoots were sub-cultured into media containing BAP, IAA and GA3; BAP and IAA; BAP, NAA and GA3. Generally for rhizogenesis, higher concentrations of auxin than cytokinin is needed to get better response. In ***Coleus forskohlii*** IAA (1 mg/lit) gave better response than IBA (1 mg/lit) within 20 days of inoculation. In this experiment both shoots and somatic embryos developed in the presence of BAP alone according to the theory of hormonal regulation of morphogenesis of all types of growth and differentiation response is mediated by relative amounts of growth regulators mainly phyto-hormones.

Acclimatization

The rooted plantlets were transferred to portrays containing **coca peat** and kept under high humid conditions for 12-15 days. Later the plants were transferred to a mix of coca peat and soil (Ratio = 1:1) and maintained under natural conditions.

REFERENCE

Inamdar, P.K., Kanitkar, P.V., Reden, J. and Souza, N.J. (1984). *Planta Med.,* **50** : 30-35.

Mersinger, R., Dornauer, H. and Reinhard (1988). *Planta Med.,* **54** : 200-204.

Murashige, T. and Skoog, F. (1962). *Physiol Plant,* **15** : 473-497.

Shah, V., Bhat, S.V., Bajwa, B.S. and Dornauer de Souza, N.J. (1980). *Planta Med.,* **39** : 183-185.

Inamdar, P.K., Kanitkar, P.V., Reden, J. and Desouza N.J. (1984). *Planta Medica,* **1** : 30-34.

Lundergan, C.A. and Janic, J. (1980). *Hort Research,* **20** : 19-24.

Mersinger, R., Dornauer, H. and Reinhard, E. (1988). *Planta Medica,* **3** : 200-204.

Murashige, T. and Skoog, F. (1962). *Physiol Plant,* **15** : 473-497.

Rahman, M.A. and Blake, J. (1988). *Plant Cell Tiss Organ Cult,* **13** :179-187.

Krombholz, R., Mersinger, R., Kreis, W. and Reinhard, E. (1992). *Planta Med.,* **58** : 328-333.

Recent Advances in Medicinal Plant Research : Vision 21st Century, 2005
Editors : S.K. Prabhuji, G.P. Rao and S.K. Patil
Satish Publication House, New Delhi

Chapter 37

TISSUE CULTURE OF *STEVIA REBAUDIANA* : A POTENT MEDICINAL PLANT FOR 'DIABETES TEA'

Tapan K. Mondal★ and S. Parathiraj
Biotechnology Laboratory, Research and Development Department, Tata Tea Ltd, Munnar - 685 162, Kerala
★*Present address : Department of Plantation Crops and Processing, Faculty of Horticulture, Uttar Banga Krishi Viswavidalaya, P.O. Pundibari, Dist. Cooach Behar - 736 165, West Bengal, India*
e-mail : mondaltk@rediffmail.com

Stevia rebaudiana is an excellent source for low calorie sweet glycoside which is 200-400 fold sweeter than sucrose. It is a cross-pollinated plant and dose not flower under some environmental conditions. Though vegetative propagation is done by stem cutting, however, it is slow. Thus, micropropagation is a better alternative for rapid multiplication of this plant. In the present report a rapid micropropagation and shoot regeneration protocol is developed. With the present protocol more than 15000 plants were produced and hardened at the nursery.

INTRODUCTION

Stevia rebaudiana Bertoni is one of 154 species of the genus *Stevia.* It belongs to family : Asteraceae and as such is related to sunflower, marigolds etc. *Stevia rebaudiana* Bertoni the sweet herb of Paraguay is gaining importance in different parts of the World and is likely to become a major source of high potency sweetener for the fast growing natural food market in the future years. The plant is native of Rio Monday in highlands of Paraguay where it grows near streams. It is a small shrubby perennial plant growing upto 65 to 70 cm tall with sessile, oppositely arranged lanceolate to oblanceolate leaves with serrated margin in the middle. The flowers are small and white with a pale purple throat. However, it does not produce flower at all the places. *Stevia* is cross-pollinated plant and hence the plants are highly heterogeneous. The seeds are small, have very little endosperm, thus, propagation through seeds often encounter low germination percentage.

The sweet compounds found in Stevia leaves are diterpene glycosides (steviol glycosides) and are synthesized, at least in the initial stages, using much the same pathway as gibberellic acid, an important plant hormone. The steviol glycoside are 200-400 times sweeter than sucrose with a very low calorific value. Hence, *Stevia* is widely touted as a remedy for diabetes. Thus it is used as an alternative to sucrose for preparing several food products such as biscuit, chocolates, ice cream, pickles and soft drinks. The precursor compounds are synthesized in the chloroplast, and then from there it is transported to the endoplasmic reticulum, golgi apparatus and then vacuole. Thus, the green leaves are the best source of low calorie sweetner for commercial use.

On the whole plant level, steviol glycosides tend to accumulate in older tissues, so that older lower leaves have more sweetener that younger upper leaves. Since chloroplasts are important in precursor synthesis, those tissues devoid of chlorophyll, like roots and lower stems contain no or trace amounts of glycosides. Steviol glycosides increase in amount when the plants grow under long day condition. The plant cannot tolerate drought conditions and require frequent irrigation. Stevia is harvested just before flowering when steviol glycoside content in the leaves is at its maxium. After the harvest whole plant is dried and the leaves are separated from stems for further processing. Thus, the mature stem which can be used for vegetative propagation of this plant are not only slow growing but also involves the plant's part which contain high amount of steviods for which the plant is valued. Therefore, micropropagation will be a better alternative for meeting the higher requirement of this planting material for large scale cultivation.

Additionally, large number of people suffering from diabetes and high blood sugar, find themselves unable to drink tea with sugar may get the sweetner from *Stevia* as a boon because of its low calorie nature. Therefore, there is a scope of business for tea containing *Stevia* so that people with special need of alternative sweet tea as per their habit may drink the same. Therefore, the present study was undertaken to produce large number of *Stevia* plants through micropropagation for the use of developing "Diabetes tea"

METHODOLOGY

Initiation of Aseptic Cultures

For the present study, twenty five number of *Stevia rebaudiana* plants were purchased from the Institute of Himalayan Bioresource Technology, Palampur, Himachal Pradesh and were grown in a greenhouse of Research and Development Department, Tata Tea Ltd, Munnar, Kerala. Nodal explants (2-3 cm) of green house grown plants were cleaned with Tween-80 and pre-sterilized with 0.001% (w/v) tetracycline for 20 minutes on a gyratory shaker and then surface sterilized in a laminar flow cabinet (Klenzaids, Mumbai) with 0.04% (w/v) mercuric chloride for 7-8 minutes and subsequently washed with distilled water to remove traces of sterilizing agents. These were inoculated on media by Murashige and Skoog (1962) with 30 g/l sucrose, and solidified with 9g/l Difco Bacto agar. The pH was adjusted to 5.8 prior to autoclaving at 121°C for 15 min. Unless stated, all cultures were incubated at cool fluorescent light (52 mmol m^{-2} s^{-1}) at 25±2°C and subcultured at 4 weeks interval. Day-length was 14 h on a 24 h light-dark cycle.

Bud Sprouting and Shoot Proliferation

After initial screening, the aseptic culture (nodal explants) were transferred to MS medium supplemented with various concentrations of BAP (0-4 mg/l) alone or in combination of either 2,4-D (0-5mg/l) or IAA (0-5 mg/l) in agar (0.8%) solidified MS medium fortified with sucrose (2.0%). The data in terms of shoot length and number of shoots formed per explants were recorded after 4 weeks.

Shoot Regeneration

In order to achieve regeneration, *in vitro* leaves were cultured on agar solidified MS supplemented with various concentrations of either 2,4-D (0-5mg/l) or NAA (0-5mg/l). Ten leaves were inoculated per 200 ml flask containing 75 ml of medium. Each treatment consisted of 10 flasks. The data was recorded after 4 weeks.

Ex-vitro Rooting and Transfer to Soil

The *in vitro* microshoots (4-5 cm length) without root were transfered immediately to sand bed in poly-tunnel under a green house for rooting with intermittent watering to maintain a high humidity inside the poly-tunnel. After that the rooted shoots were transferred to polythene sleeves (25 cm length and 15 cm lay flat), filled with virgin soil, sand and well rotten Farm Yard Manure (FYM) (8:1:1) following sieving through a mesh (2 mm size). The data in terms of survival were recorded after each stage.

RESULTS AND DISCUSSION

There are preliminary reports of micropropagation of *Stevia* (Handro *et al.*, 1977; Niztaev and Birnbaum, 1998). However, during the present investigation a compete regeneration system has been developed for *Stevia*. For producing aseptic culture, the explants were cultured on agar solidified MS media for 4 weeks without any growth regulator. It has been observed that contamination, if any, will be appeared with in 2-3 weeks of culturing. Aseptic cultures were, then, used for either micropropagation or regeneration study.

Bud Sprouting and Shoot Proliferation

Among the various media combination, buds of maximum explants (90%) sprouted on MS medium supplemented with 2 mg/l BAP alone within 4 weeks of time. Increasing and decreasing the BAP concentrations leads lower bud sprouting (Data not shown). Longer and sturdy shoots with thicker stems were evident in BAP containing medium. Therefore, in all subsequent experiments, BAP at 2mg/l concentration was employed. Interestingly, it has been observed that further growth and development of the sprouted bud was ceased in BAP containing medium.

Thus, to improve further the shoot multiplication rate, explant with sprouted bud was cultured on 2 mg/l BAP with various combination of 2,4-D or IAA. Among the various combination, a maximum (97%) explant produced (Table-1) 10-15 shoot per explant (Fig. 1A) on BAP (2 mg/l) + IAA (1 mg/l) after four weeks and this continued to increase by three folds with regular subculturing at two weeks interval leads to 40-45 shoots per explant within 6 weeks of time (data not shown). The concentration of BAP showed a marked effect on shoot proliferation. On the other hand, lower concentration produced shoot proliferation but concentration higher than 4 mg/l produced callus formation. In all combination of BAP and 2,4-D leads to callus production in various degree with red color anthocyanine pigmentation. Therefore, we concluded that 2,4-D being a PGR, stimulated production of callus in *Stevia*.

Shoot Regeneration

For regeneration *in vitro* leaves were cultured on MS medium containing either 2,4-D or NAA. However, in all combination no direct regeneration was observed rather a varying degree of callus formation was observed. When different combination of PGR with respect to callus formation were

Table - 1 : Effect of BAP in combination with 2,4-D and IAA on response of axillary buds of *Stevia* nodal segments placed on MS medium.

Concentration (mg/l)		BAP (mg/l)				
		0	1	2	3	4
2,4-D IAA	0	NR★	97.75 + 1.10 CF	97.50 + 1.19 CF	96.75 + 0.05CF	95.00 + 1.22 CF
			17.00 + 1.15 SP	27.50 + 0.86 SP	26.00 + 0.57 SP	22.50 + 1.04 SP
2,4-D IAA	1	29.00 + 2.12 CF	97.25 + 1.18 CF	98.50 + 0.28 CF	99.50 + 0.28 CF	95 + 0.12 CF
		NR★★ + R	89.00 + 2.04 SP	97.50 + 0.28 SP	90.00 + 0.81 SP	58 + 0.45 SP
2,4-D IAA	2	37.75 + 1.10 CF+A	99.00 + .40 CF+A	97.00 + .40 CF+A	98.25 + 0.47 CF+A	91.25 + .10 CF+A
		11.25 + 1.37 SP+R	74.25 + 1.10 SP	78.75 + 0.629 SP	68.25 + 0.75 SP	35.50 + 1.32 SP
2,4-D IAA	3	32.00 + 1.08 CF+A	99.50 + 0.28 CF+A	96.50 + 1.5 CF+A	99.25 + 0.25 CF+A	98.00 0.91 CF+A
		21.00 + 1.29 SP+R	67.50 + 1.19 SP	70.75 + 0.25 SP	66.50 + 0.5 SP	31.50 +1.30 SP
2,4-D IAA	4	32.00 + 1.08 CF+A	99.50 + 0.28 CF+A	96.50 + 1.5 CF+A	99.25 + 0.25 CF+A	31.25 + 1.31 CF+A
		21.00 + 1.29 SP+CF	27.50 + 0.64 SP+CF	24.25 + 1.10 SP+CF	34.25 + 1.70 SP+CF	23.21 + 1.31 SP+CF
2,4-D IAA	5	37.75 + 1.10 CF+A	99.00 + .40 CF+A	97.00 + .40 CF+A	98.25 + 0.47 CF+A	30.25 + 0.85 CF+A
		21.25 + 1.37 CF	24.25 + 1.10 CF	68.75 + 0.629 CF	60.25 + 0.75 CF	16.50 + 0.64 CF

★ = Bud sprouted but did not elongate or multiply, ★★ = sprouted and elongated but did not multiply, SP = Shoot proliferation, R = rooting, A = Anthocyanin production. Data (X + SE) pooled from three independent experiments each with three replicates, CF = Callus formation.

Fig. 1A : Six weeks old *Stevia* shoots growing on BAP (2 mg/l) + IAA (1 mg/l)

Fig. 1B : Two months old *Stevia* plant growing at Nursery.

compared, the best response (i.e. calli formed per week) was observed in 83.3% explants on 2,4-D (2.5 mg/l) supplemented media whereas maximum 10-20% explants were responsive in NAA (2 mg/l). Moreover we observed that callus induced on NAA were hard textured which later failed to even proliferate and regenerate. Among the various concentrations of 2,4-D, the frequency of callus induction and rate of callus growth increased significantly from 5.5% at 1 mg/l to 83.3% at 2.5 mg/l and decreased with increasing the concentration to 43.2% at 5.0 mg/l (data not shown).

In this regard, when calli induced on medium containing 2,4-D were left as such for more than three months, morphogenesis in the form of a large number of roots occurred prior to shoot bud initiation with simultaneous drying of calli. Alternatively, the process of shoot morphogenesis could be hastened to a much earlier date i.e. within four weeks when the calli induced on medium containing 2,4-D were transferred to 2,4-D free medium containing BAP (1.0 mg/l). The critical threshold level of 2,4-D could be achieved by progressive sequencing of 2,4-D with the passage of time either in the form of callus growth or by root formation after being triggered by BAP in the medium. Initiation of shoot buds on BAP (1.0 mg/l) after continued callus growth on 2,4-D depleted medium taken place is further indicative of the role of a critical threshold level of 2,4-D in morphogenesis (data not shown).

The fact that synthetic auxins like NAA and 2,4-D induce callus formation has been well documented (Ratushnyak *et al.*, 1990). In our study also, both NAA and 2,4-D were found to stimulate a callus response. However, regeneration potential was observed only in the callus induced on 2,4-D and this is supported by the fact that responses like somatic embryogenesis (Alizadeh and Mantell, 1991; Kato, 1996) and morphogenesis are evoked by 2,4-D (Mussio *et al.*, 1998). An increase in callus formation with increasing concentrations of 2,4-D in the medium is perhaps due to a proportional increase in the intracellular 2,4-D levels. Variations in the morphogenetic responses due to uptake and accumulation of high concentrations of intracellular 2,4-D from the medium and requirement of a critical intracellular threshold level of 2,4-D are known (Meijer *et al.*, 1999; Mitsuoko *et al.*, 1994).

At higher concentration, i.e. 5 mg/l the intracellular 2,4-D level probably exceeded the threshold value within which a regeneration response could not be induced. Optimal shoot regeneration from callus initiated on 2.5 mg/l 2,4-D on the other hand, is further indicative of the fact that regeneration response requires a critical intracellular threshold level of 2,4-D.

Ex-vitro Rooting and Transfer to Soil

Adventitious roots were initiated at the base of the microshoots after 2 weeks in the sand bed which continued to grow and reached a sizable amount

Table - 2 : A rapid micropropagation protocol of *Stevia rebadiana* using nodal explant.

Response	Medium	Time
Establishment of Explant (Nodal segment)	MSO	4 weeks
↓	↓	
Bud sprouting	MS + BAP (2.0mg/l)	4 weeks
↓	↓	
Shoot multiplication	MS + BAP (2mg/l) + IAA (1mg/l)	4-6 weeks
↓	↓	
Transfer of Shoot of (2-3 cm) to the green house for *ex vitro* rooting	Sand bed inside the poly tunnel with occasional watering	4 weeks
↓	↓	
Rooted shoots are transferred to the nursery	Polythene sleeves (25 cm length and 15 cm lay flat) containing black virgin soil	Kept for 2 months

Table - 3 : Protocol for plant regeneration from leaf tissues of *Stevia rebadiana*

Response	Medium	Time
Callus induction	MS + 2,4-D (2.5 mg/l)	4-6 weeks
↓	↓	
Initiation and growth of shoot buds	MS + BAP (1.0 mg/l)	4-6 weeks
↓	↓	
Transfer of Shoot of (2-3 cm) to the green house for *ex vitro* rooting	Sand bed inside the poly tunnel with occasional watering	4 weeks
↓	↓	
Rooted shoots are transferred to the nursery	Polythene sleeves (25 cm length and 15 cm lay flat) containing black virgin soil	Kept for 2 months

within 4 weeks. About 90% shoots were rooted in the sand bed. The critical factor, what we observed, is the drastic reduction of rooting ability in an open sand bed. It may be due to the fact that close structure leads to high humidity condition — a phenomenon, which is observed in many instances of hardening. After 6 weeks the rooted shoots were transferred in to polythene sleeve in the nursery where an occasional causality has been noticed (Fig. 1B)

In conclusion, this paper reports a rapid micropropagation (Table-2) and shoot regeneration pathway (Table-3) from leaf tissues. This paper also hypothesizes that an intracellular threshold value of 2,4-D and its sequentiality is essential for directing the leaf tissues towards a programme of regeneration through dedifferentiation. Experiments are underway to formulate a black tea product primarily meant for the diabetes patient.

ACKNOWLEDGEMENT

The authors are grateful to Mr P. Haridas, Dr. P. Mohan Kumar, and Mr. P. Ramakrishna for technical suggestion and R. Muragan & Uma Muragan for technical help during the work at Tata Tea Ltd.

REFERENCES

Alizadeh, S. and Mantell, S.H. (1991). Early events during somatic embryogenesis in cotyledon explants of *Solanum aviculare* Forst. *Ann. Bot.*, **67** : 257-263.

Handro, W., Hell, K.G. and Kerbauy, G.B. (1977). Tissue culture of *Stevia rebaudiana*, a sweetening plant. *Planta med Sep.*, **32(2)** : 115-7.

Handro, W. and Ferreira, C.M. (1989). *Stevia rebaudiana* (Bert.) Bertoni: production of natural sweetners. In : Bajaj, Y.P.S. Biotechnology in agriculture and forestry, Medicinal and Aromatic Plants II. Berlin, Springer-Verlag, vol. 7, p 468-487.

Kato, M. (1996). Somatic embryogenesis from immature leaves of in vitro grown tea shoots. *Plant Cell Rep.*, **15** : 920-926.

Meijer, E.A., DeVries, S.C. and Mordhorst, A.P. (1999). Coculture with *Daucus carota* somatic embryos reveals high uptake and release rates of *Arabidopsis thaliana* cultured cells. *Plant Cell Rep.*, **18** : 656-663.

Mitsuoka, K., Honda, H., Xing, X.H. and Unno, H. (1994). Effect of intracellular 2,4-D concentration on plantlet regeneration of rice (*Oryza sativa* L.) callus. *Appl. Microbiol. Biotechnol.*, **42** : 364-366.

Murashige, T. and Skoog, F. (1962). A revised medium for rapid growth and bioassays with tobacco tissue cultures. *Physiol. Plant.*, **15** : 473-497.

Mussio, I., Chaput, M.H., Serraf, I., Ducreux, G. and Sihachakr, D. (1998). Adventitious shoot regeneration from leaf explants of an African clone of cassava

(Mannihot esculenta Crantz) and analysis of the conformity of regenerated plants. *Plant Cell Tiss. Org. Cult.*, **53** : 205-211.

Niztaev, A. and Birnbaum, E.H. (1998). Micropropagation of *Stevia rebaudiana.*, In: A. Altman, M.Ziv and S.Izhar, cds (1999). Plant Biotectnology and *In Vitro* Biology in 21st Century. Kluwer Academic Publishers, Dordrecht, The Netherlands, p 192.

Ratushnyak, Ya. A., Rudas, V.A. and Piven, N.M. (1990). Regeneration of *Lycium barbarum* L. plants from leaf tissue, callus culture and callus protoplasts. *Plant Cell Rep.*, **9** : 84-87.

Soejarto, D.D., Compadre, C., Medon, P.J., Kamath, S.K. and Kinghorn, A.D. (1983). Botanical sweating agents of plant origin. II. Field research for sweet-tasting Stevia species. *Economic Botany,* **37** : 71-79.

Wada, Y., Tamura, T., Kodama, T., Yamaki, T. and Uchida, Y. (1981). Callus cultures and morphogenesis of *Stevia rebaudiana* Bertoni. *Yukagawa,* **36(4)** : 215-219.

Recent Advances in Medicinal Plant Research : Vision 21st Century, 2005
Editors : S.K. Prabhuji, G.P. Rao and S.K. Patil
Satish Publication House, New Delhi

Chapter 38

ALLELOPATHIC EFFECTS OF DIFFERENT MULTIPURPOSE TREE SPECIES LEAF LEACHATES ON SEED GERMINATION OF AROMATIC AND MEDICINAL PLANTS

G.O. Manjunath[1], A. Krishna[2], V. Suryanarayana[3] and S.K. Patil[1]
[1]*Dept. of Forest Utilization, College of Forestry, UASD, Sirsi - 581 401, Karnataka*
[2]*Dept. of Silviculture, College of Forestry, UASD, Sirsi - 581 401, Karnataka*
[3]*Dept. of Forest Protection, College of Forestry, UASD, Sirsi - 581 401, Karnataka*

Laboratory study was undertaken to find out allelopathic effect of leaf leachates of four important, multipurpose tree species *viz.* *Casuarina*, Mango, *Eucalyptus* and *Acacia* on seed germination and growth of aromatic and medicinal plants (Kasturi bendi, Sanka pushpa and Honey plants). The results revealed a varying degree of inhibitory and stimulatory effects on germination percentage and growth of test crops. Inhibitory effect was much pronounced in plumule elongation rather than radical. So, from the study it could be inferred that allelopathic interaction is made possible through leaching of inhibitors due to decay of fallen leaves from the trees. This has several management implications related to agro-forestry system where a large number of intercrops are grown in the space available beneath the tree canopy. Therefore, while identifying suitable trees for agroforestry systems, efforts should be made to select the species with the least allelopathic activity.

INTRODUCTION

Allelopathy is a potential field of research all over the world. Agroforestry systems provide several benefits, but they have some harmful effects also on agricultural crops due to several factors such as competition for space, light, nutrients and organic chemicals released as leaf leachates, which affect the crop plants. Trees are considered to be a rich source of secondary metabolites (Allelo) and these chemicals improve a certain kind of environmental system on other plants growing in their vicinity and such a phenomenon is known as tree allelopathy (Nandal *et al*, 1994). Few researchers consider only the deleterious interactions as allelopathy. However, the latest thinking includes allelopathy to be both harmful and beneficial interactions among the plants. Biological weed control is an area where the allelopathic potential of a species can be successfully exploited for beneficial purposes. Agroforestry system where bund planting of some trees poses a deleterious effect on undestroyed crop grown under it, due to the leachates of fallen leaves, which possess allelochemicals. Since farmers have started the plantation of trees on the bunds, water channels and periphery of their farms to meet the acute shortages of fuel wood, timber and fodder, the study of such interactions would be of paramount importance in agro-ecosystem and in planning agroforestry. As, no information is available on the effect of commonly grown farm trees on aromatic and medicinal crops, the present study was taken up to investigate the influence of allelopathic effects of multipurpose trees like, *Acacia auriculiformis, Casuarina equisetifolia, Eucalyptus hybrid* and *Mangifera indica* on germination behaviour, root and shoot growth of some of the important aromatic and medicinal plants like Kasthuri bendi (*Abelmoschus moschatus*), Sanka pushpa (*Clitoria ternata*) and Honey plant (*Ammi majus*).

HISTORICAL REVIEW

The concept of viewing allelopathy as a beneficial or harmful function was observed by Rizvi *et al.* (1986). In tree-crop interactions of agroforestry sytem, leaf leachates of Albizia, Acasia, teak showed stimulatory effect on germination as observed by Hunsal *et al.* (1998). Similar observations were also made in the studies conducted by Chaturvedi and Jha (1992). The extracts of both green and senescent leaves of neem significantly reduced the germination in pearl millet, sorghum and pigeonpea except in cowpea and black gram where the green leaf extracts increased the germination as observed by Bisla *et al.* (1998). With respect to medicinal plants, effect of treatment of rhizome seed material of *Costus speciosus* with leachates of teak on diosgenin content was studied, but no negative effect was observed in the active principle content (Konar, 1998)

METHODOLOGY

The study was conducted in the College of Forestry, Sirsi situated between 14^0 37' N latitude 74^0 53' E longitude at an altitude of 629 m above sea level. Leaf samples from *Acacia auriculiformis, Casurina equisetifolia, Eucalyptus hybrida* and *Mangifera indica* were collected from a fully matured tree. Leaf leachates of these tree species were prepared by soaking leaves in distilled water for 24 hours at 1:10 proportion (w/v). Preparing the leaf leachates in this proportion has been found to produce solution of low osmotic potential which may not affect the germination. The leaf leachates were filtered and freshly prepared leachates were used for the study using the seeds of Kasthuri bendi, Sanka pushpa and Honey plant. Twenty-five seeds of each species were sown in trays filled with sand. Each tray was added with 200ml of leaf leachates (in case of treatment) and 200ml of water (in case of control). Treatment trays were added with 100 ml of water. All the trays were kept under partial shade. The data on daily emergence was taken to assess the germination capability, shoot length and root length. Germination was expressed on percent basis.

The inhibition and stimulation in the bioassay was studied using the response index (RI) described by Richardson and Williams (1988), which is as follows :

If $T>C$ then $RI = 1- (C/T)$ where C = control and T = treated.

If $T=C$ then $RI=O$

If $T<C$ then $RI=(T/C)-1$

RESULTS AND DISCUSSION

Kasturi bendi germination percent was not significantly influenced with the leaf leachates of any tree species. However, reduction in shoot length and root length was significant with the leachates of all the tree species. Leaf leachates of *Casuarina* had maximum inhibitory effect on shoot length and root length followed by Mango (Table-1).

Honey plant germination percent was not influenced significantly with the leaf leachates of any tree species as compared to control where as shoot length and root length were significantly reduced with leaf leachates of all

Table - 1 : Allelopathic effect of different agroforestry tree species on different aromatic and medicinal plants.

Name of the tree species	Kasthuri bendi			Honey plant			Sanka pushpa		
	Germination (%)	Shoot length (cm)	Root length (cm)	Germination (%)	Shoot length (cm)	Root length (cm)	Germination (%)	Shoot length (cm)	Root length (cm)
Acacia auriculiformis	67.00	4.30	3.10	47.50	3.40	2.20	61.00	4.30	3.30
Casurina equisetifolia	57.00	1.80	2.20	53.00	2.00	1.90	61.50	2.40	1.90
Eucalyptus hybrida	61.00	2.80	3.20	54.00	2.20	3.20	59.50	2.30	3.00
Mangifera indica	56.50	2.60	2.30	52.00	1.90	1.80	66.50	2.20	1.80
Control	64.50	4.40	4.00	57.00	4.00	3.80	70.00	6.00	3.90
SEM±	2.05	0.10	0.15	3.78	0.19	0.12	2.33	0.30	0.23
C.D at 5%	NS	NS	0.45	NS	0.57	0.36	NS	0.60	0.69

Table - 2 : Response index values of different aromatic and medicinal crop seeds exposed to leaf leachates of different tree species.

Name of the tree species	Kasthuri bendi			Honey plant			Sanka pushpa		
	Germi-nation (%)	Shoot length (cm)	Root length (cm)	Germi-nation (%)	Shoot length (cm)	Root length (cm)	Germi-nation (%)	Shoot length (cm)	Root length (cm)
Acacia auriculiformis	0.04	-0.02	-0.22	-0.16	-0.15	-0.42	-0.13	-0.28	-0.15
Casurina equisetifolia	-0.11	-0.59	-0.42	-0.07	-0.51	-0.50	-0.12	-0.60	-0.51
Eucalyptus hybrida	-0.05	-0.36	-0.20	-0.05	-0.45	-0.15	-0.15	-0.61	-0.23
Mangifera indica	-0.12	-0.40	-0.45	-0.08	-0.52	-052	-0.05	-0.63	-0.53

tree species. Leaf leachates of Mango had maximum toxic effect on Honey plant followed by *Casuarina* (Table-1). In Sanka pushpa no differences in germination percent was observed with leachates of tree species. However, leachates of Mango followed by *Casurina* and *Eucalyptus* significantly reduced the root length and shoot length thus indicating the presence of some growth inhibitors in the leaves. These tree species, which may adversely affect the root length and shoot length of Sanka pushpa but have comparatively less effect on its germination.

It could be noted from table-2, that leaf leachates of *Acacia* have different effect on germination of Kasturi bendi. Germination was stimulated (4%) as indicated by response index. However, in other tree species inhibition of germination was exhibited (5-12%). As compared to germination the inhibition of shoot growth (12-59%) and root growth (20-45%) was more pronounced. The suppression in shoot and root growth was more caused by *Casuarina* followed by Mango.

In Honey plant the inhibition as indicated by response index was found to be 5-16%. As compared to germination the inhibition of shoot and root growth (15-52%) was more pronounced by tree species. Mango caused maximum inhibition of root and shoot growth followed by *Casuarina* (Table-2).

The response index range was 5-13% in the inhibition of germination of Sanka pushpa as compared to the root growth inhibition (15-53%). The inhibition of shoot growth was more pronounced (28-63%). The maximum suppression of shoot and root growth was caused by mango followed *Casuarina.*

In general, adverse effect of all the four commonly grown, multipurpose tree species differed with crop plants. The maximum adverse effect on aromatic and medicinal plants was observed with leachates of mango followed by *Casuarina*. Water-soluble inhibitors leaching down, through foliar runoff, added to soil remain for long time especially in dry areas and would have inhibitory effect on germination and subsequent growth of crop plants.

Among aromatic and medicinal plants, Sanka pushpa seeds showed maximum sensitivity followed by Honey plant. Melkania (1984) also observed the different response of leachates of woody species on agricultural crops. The general allelopathic effect of *Casuarina* (Joshi and Prakash, 1992) has been recorded, however, the observed inhibition by Mango might have been due to the phenols present in their leaves. In the present study the data indicated the inhibitory effects of different agroforestry tree species on aromatic and medicinal plants. The inhibition might have been due to the presence of Allelochemicals as reported by Chaturvetdi and Jha (1992) and Swaminathan *et al.* (1989) and not because of osmotic inhibition. Use of 10% leachate has been proved to have no any osmotic inhibition (Richardson and Williams, 1988).

REFERENCES

Bisla, S.S., Nandal, D.P.S. and Husain, Z. (1998). Allelopathic potential of neem on germination and seedling growth of summer crops. Presented in, "Third international congress on Allelopathy in ecological agriculture and Forestry held at UAS, Dharwad, Karnataka from 18-21st August, 1998. Abstarct No. 3.7, pp 72.

Chaturvedi, O.P. and Jha, A.N. (1992). Studies on allelopathic potential of an important agroforestry species. *Forest Ecology and Management,* **53** : 91-98.

Hunsal, C.S., Chennal, H.T., Alagwadi, A.R. and Patil, R.H. (1998). Scope of allelopathy research in agroforestry sytems of South India. Presented in, "Third international congress on Allelopathy in ecological agriculture and Forestry held at UAS, Dharwad, Karnataka from 18-21st August, 1998. Abstarct No. 3.5, pp 70.

Joshi, P.C. and Prakash, O. (1992). Allelopathic effect of litter extract of some tree species on germination and seedling growth of agricultural crops. *In Proc. First National Symposium on Allelopathy in Agro Eco systems.* Hissar, India. Pp 127-128.

Konar, J. (1998). Effect of Tectona grandis leachates on growth and diosgenin content of *Costus speciosus.* Presented in, "Third international congress on Allelopathy in ecological agriculture and Forestry held at UAS, Dharwad, Karnataka from 18-21st August, 1998. Abstarct No. 3.5, pp 70.

Melkania, N.P. (1984). Influence of leaf leachates of certain woody species on agricultural crops. *Indian Journal of Ecology,* **11** : 82-86.

Nandal, D.P.S., Birla, S.S., Narwal, S.S. and Koushik, J.C. (1994). Allelopathic interactions in agroforestry systems. *In: Allelopathy in Agriculture and Forestry* (*eds.*) Jodhapur, pp 93-130.

Richardson and Williams (1988). Allelopathic effects of the sand pine scrub pine and grasses of the sand hill. *Forest Science,* **34(3)** : 592-605.

Rizvi, S.J.H. and Rizvi, V. (1986). Allelopathy: Some new terminological considerations. *Curr. Sci.,* **85** : 191-192.

Swaminathan, K., Sivagnanam and Srimathi (1993). Allelopathic proclivities of multi purpose trees. *My Forest,* **29(3)** : 147-149.

Recent Advances in Medicinal Plant Research : Vision 21st Century, 2005
Editors : S.K. Prabhuji, G.P. Rao and S.K. Patil
Satish Publication House, New Delhi

Chapter 39

BIOCOMPOSTING OF ORGANIC WASTE AND ITS EFFECT ON SHANKHPUSHPI AND PALMAROSA

Varsha Bhargava and Seema Bhadauria
Botany Department, R.B.S. College, Agra - 282 002 (U.P.)

India is considered as one of the richest biodiversity zone, which is going to be lost due to unscientific exploitation of forest resources. Moreover, extensive use of fertilizer and pesticides deteriorates their quality and quantity both especially in field of medicinal plants. To over come this situation, cultivation of medicinal plants has to be done with the help of organic farming. The present study was carried out to evaluate the Vermicomposting of organic waste. An experiment was conducted to prepare Vermicompost using partially degraded organic waste collected from different areas of Agra city, by employing Earthworm species *Eudrilus evgine*ae. The studies on compost quality and growth tests were conducted. The compost was applied to *Evolvulus alsinoides* (Shankhpushpi) and *Cymbopogon martini* (Palmarosa) along with normal soil as a control. The results showed positive impacts in the vermicompost compared to the control sample.

INTRODUCTION

The plants, which are used for medicinal purposes, should not have any residues of chemical fertilizer or pesticides; otherwise they become more of poison than a medicine. Moreover, the medicinal plants have several active ingredients (biochemical) and growing them with chemical fertilizer, pesticides and polluted irrigation may alter or deteriorated their quality. Medicine is always given to a sick person having impaired metabolism and giving a medicine having pesticide residue may further deteriorate his condition. To over come this situation,cultivation of medicinal plants has to be done in near to nature conditions popularily known as organic farming. The concept of organic farming has basically emerged from the biological processes available in nature. With the human efforts these are being utilized at large scale. Decomposition of organic matter is a natural process but with the human methodology it is being efficiently utilized in making compost i.e. vermicompost. So, there is no dearth of nature's biotechnology, the only requirment is to develop methodology for harnessing this, several feasible methods have been developed for utilizing these natural biotechnologies like production and use of biofertilizer, vermicomposting and vermiwash (Pesticides), production of predator and Parasites in artificial environment and this may help a lot for successful organic farming and ultimately sustainability.

Vermicomposting is the application of earthworm in producing vermifertilizer, which helps in maintenance of better environment and results in sustainable agriculture (Senapati, 1996). Vermicompost can be prepared by different organic materials like, vegetable residues, coir, pith, pressmud, weeds, cattle dung etc. The vermicompost is rich in nutrients that can enrich the soil. Present studies revealed positive impact of vermicompost on *Evolvulus alsinoides* (shankhpushpi) and *Cymbopogon martini* (Palmarosa). *Evolvulus alsinoides* is a perennial herb, found during rainy season in gardens, lawns, agricultural fields, along road side etc. (Kasera, 1988; Kasera and Chawan, 2000; Sen, 1981) It is a common weed but also cultivated for medicinal uses. The whole herb is used medicinally and has got anti-dysenteric and antiseptic properties. Shankhpushpi is also used as brain tonic and in increasing memory, nervous debility etc. (Chermexcil, 1992). *Cymbopogon martini,* a hardy, perrenial grass, has antiseptic properties and it has wound healing effects, used in skin diseases and muscular pain. Palmarosa grass yields the commercially important Palmarosa oil, also known as Rusa oil or East Indian Geranium oil. Being important medicinal plants, the present studies were aimed to grow these plants without chemical fertilizers and pesticides.

METHODOLOGY

About 3.5 kg of kitchen waste was collected from different areas of Agra and was allowed for partial degradation (10-15 days) in a plastic tub with regular water sprinkling. It was mixed with 7 kg of cattle dung accurately and transferred to the plastic tubs of 60 × 50 × 24 cm. The earthworms were obtained from Sikandarpur Village, 20 kms from Agra and maintained in the laboratory conditions for about a week for acclimatization. Later, half kg of earthworms of species *Eudrilus eugineae* were released into the plastic tub. Earthwarms fed actively on waste and assimilated it, partially and the rest was excreted out as a loose worm cast. The accumulated Vermicast in the tub was seprated and collected in a seprate glass bottle. The remaining waste was kept in the tub and maintained under optimum condition i.e. 20-25 °C temperature, and 40% moisture by sprinkling of water at regular intervals. After 50 days the fine-grained black coloured compost was separated. The prepared vermicompost was used for experimental test.

For the study of vermicompost on *Evolvulus alsinoides*, seedlings of *E. alsinoides* were uprooted from naturally growing habitats and transplanted in pre-prepared experimental plots filled with compost and soil (1:1) and labelled I No. II contained only control sample. Plots were irrigated as and when required. Regular observations for growth parameters (plant height and biomass (above and below ground) were recorded. Other experiment was conducted to study the effect of vermicompost on the growth of Palmarosa. In a pot experiment the performance of Palmarosa in vermicompost was studied. Seedling of Palmarosa was transplanted in the experimental pot, already containing vermicompost and soil, (1:1). This pot was labelled as No. I. No. II contained only soil sample with normal garden soil and the plant of Palmarosa as control. Regular observations for growth were recorded.

The compost was analyzed for Various parameters like pH, macro (N, P, K) and micro (Mn, Fe, Zn and Cu) nutrients. N was determined by microkjeldhal method, P by olsen and K by flamephotometric method, Micronutrients were measured by atomic absorption spectrophotometer.

RESULTS AND DISCUSSION

The data obtained from the study indicated the adaptability of earthworm species and organic waste as a source of compost. The Earthworm casts contain concentrated phosphorus, exchangeable magnesium, potassium, calcium, all essential plant growth factors and plant nutrients such as N, P, K, Ca and Mg (Lunt and Jacobson, 1944; Graff, 1970). The present study also indicated that the vermicompost was rich in nutrient value (Table 1).

Table - 1 : Nutrient composition of Vermicompost

pH	7.0
organic carbon	13.8%
Available P	3.2 g/kg
Available K	20.0 g/kg
Zinc	1.98 ppm
Copper	2.84 ppm
Manganese	32.10 ppm
Iron	47.47 ppm

Table - 2 : Growth rate of *E. alsinoides*

Date	Plant height (cm.)	
	Soil + Vermicompost (1:1)	Control (soil)
10.8.03	14.12	15
25.8.03	39.00	32
10.9.03	76.18	68
25.9.03	102.12	84
10.10.03	121.17	98

Table - 3 : Effect of Vermicompost on the above and belowground biomass (g dry wt. plant-[1])

Treatments	Above ground		Below ground	
	A	B	A	B
Control	0.7859	4.2100	0.0420	0.1892
Soil + Vermicompost	5.894	10.1842	2.4640	4.4097

A : *E. alsinoides*; B : *C. martini*

Table - 4 : Growth rate of *C. martini*

Date	Plant height (cm.)	
	Soil + Vermicompost (1:1)	Control (soil)
10.8.2003	38.96	36.12
10.9.2003	50.64	44.19
10.10.2003	72.17	56.92
10.11.2003	84.28	68.67

The compost and soil ratio (1:1) was prepared for the present study. Similarly, pure soil sample was used as control. Growth and Biomass of Shankhpushpi was estimated in both the compost and the control soil. The comparitive data showed that the growth rate was significantly higher in vermicompost than the control (Tables 2 and 3). Similarly, the other medicinal plant *Cymbopogon martini* also showed positive impact of vermicompost (Tables 3 and 4).

CONCLUSION

Nainawat (1997) found that the casts produced by decompostion of kitchen waste, garden waste and cattle waste were rich in nutrient values, and compost produced through vermicomposting helps in maintaining the pH of soil and soil properties, and augument the amount of mobile phosphorus and potassium thus increasing the yield (Atlavinyte, 1975; Graft, 1972). Similarly, present work has also shown positive impact of vermicomost on one hand and it saved original medicinal values of important medicinal plant on the other due to lack of harmful chemicals and pesticides etc.

REFERENCES

Atlavinyte, O. (1975). Ekologija dozdevych car vej iich vlijanije na plodorodie pocvyv Litovskoj SSR. Mokslas Vilnius.

Behura, S. and Pradhan, N.K. (2001). Effect of the lime amendment of chromite overburden soil on growth and oil yield of Cymbopogon martini wats. *J. Plant Physiology,* **6** : 295-299.

Chermexcil (1992). Selected Medicinal Plants of India: A Monograph of (Identity, Safety and Clinical Usage, Pharmaceuticals & Cosmetics Export Promotion Cklouncil Bombay, 151-153.

Graff, O. (1972). Stickstoff, phosphor and calcium in der Regen auf der Wiesenversuchs flache des sollingrrojektes. *Ann. Zool. Ecom. Anim. Num. H. Ser.,* **71(7)** : 93-101.

Kasera, P.K. (1988). Arid agro-ecosystem: weed biology and weed management. Ph.D. Thesis, Univ. of Jodhpur, Jodhpur, India.

Kasera, P.K. and Chawan, D.D. (2000). Cultivation and evaluation of medicinal plants under desert conditions of north-west India. Fourth Technical Progress Report, ISMH Project, J.N. Vyas University Jodhpur, pp. 62.

Lunt, H.A. and Jacobson, F.G.M. (1944). The chemical composition of earthworm casts. *Soil Sci.,* **58** : 367-375.

Nainawat, R. (1997). Vermitechnological studies on jorganic solid waste management. Ph.D. thesis., Rajasthan University, Jaipur, India.

Priyasaharan, Jitendra, Shukla, K., Kasera, Pawan and Chawan, D.D. (2001). Effect of different nutritional treatments on Growth and Biomass of shankhpushpi. *J. Indian Botanical Society,* **80** : 127-131.

Sharma, A.K. (2003). Grow medicine not a poison: the organic way. *Agrobios,* **1** : 47-50.

Sen, D.N. (1981). Ecological Approaches to Indian Weeds. Geobios International, Jodhpur.

Senapati, B.K. (1991). In. Proceedings and recommendation national workshop on organic farming for sustainable agricultue, 187-189.

Subject Index

~A~

~B~

~C~

~D~

~P~

~R~

~S~

~T~

~U~

~V~

~W~

~Z~